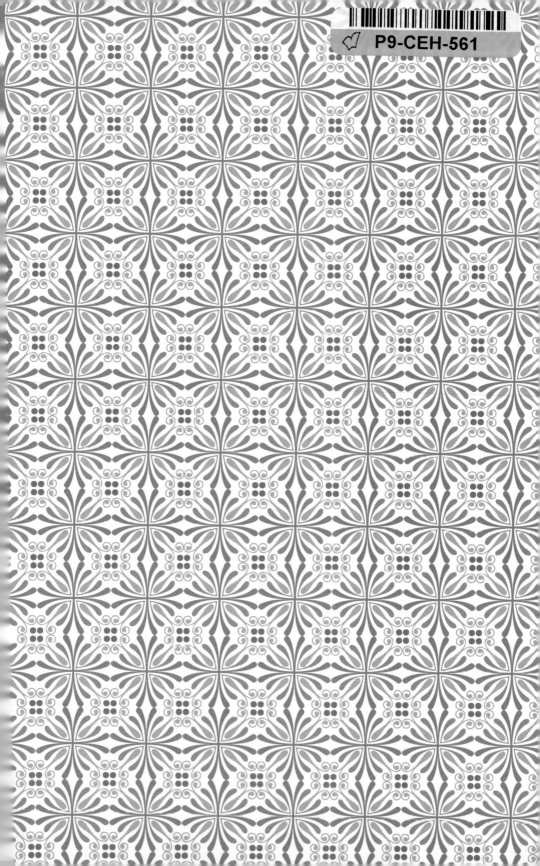

The Book of

Unusual Knowledge

Publications International, Ltd.

Cover Illustration: Images.com/Corbis

Interior Illustrations: Hye Lim An, Art Explosion, Linda Howard Bittner, Erin Burke, Daisy De Puthod, Dan Grant, iStockphoto, Jupiterimages, Nicole H. Lee, Robert Schoolcraft, Shutterstock.com, Shavan R. Spears, Elizabeth Traynor, John Zielinski

Louis Weber, CEO
Publications International, Ltd.
8140 Lehigh Avenue
Morton Grove, Illinois 60053

ISBN-13: 978-1-4508-4580-9
ISBN-10: 1-4508-4580-0

Manufactured in U.S.A.

8 7 6 5 4 3 2 1

Library of Congress Control Number: 2011943204

Contents

✳ ✳ ✳ ✳

You Can Never Get Enough Unusual Knowledge!

✳   ✳   ✳   ✳

FROM THE DAY we're born, we know to eat, sleep, and seek comfort whenever possible. In school, we're taught the basics about how the world works. But sometimes, we just need to dig a little deeper. That's where *The Book of Unusual Knowledge* comes in. The stories within this volume shed some light on the extraordinary things that the human mind can conceive. You'll discover amazing information about animals and intriguing tidbits about some of today's most revered institutions. What could cause a man to bury ten Cadillacs nose down in a pasture? Why did the king of Prussia assemble an "elite" infantry unit composed entirely of unusually tall men? Find the answers to these questions in the pages that follow.

You'll also read about:

✳ Hollywood's urban legends, from Humphrey Bogart as the Gerber Baby to the odd connections between *The Wizard of Oz* and Pink Floyd's *The Dark Side of the Moon*

✳ Moe Berg, the former big-league catcher who became a vital U.S. spy during World War II

✳ A variety of odd hotels, including one that's made entirely out of ice and one that's completely underwater

✳ The odd lives of aviation innovator Howard Hughes, American king James Strang, and jazz musician Billy Tipton

✳ And much more!

So sit back, relax, and expand your perception of what's feasible. *The Book of Unusual Knowledge* may just help you see the world—and the people in it—in ways that you never have before.

Chapter 1

Movies and Television

Exiles in Hollywood

European cinematic talents found a new home in California and, thanks to Warner Brothers, made one of the best films in movie history.

✳   ✳   ✳   ✳

FANS OF CLASSIC American cinema might be surprised to learn that many of their favorite films from the 1930s and '40s benefited from a wave of emigration sparked by the rise of Nazis in Europe. Leading lights from the period—including Peter Lorre, Marlene Dietrich, Billy Wilder, Otto Preminger, and Fritz Lang—came to Hollywood not only to ply their craft but also to escape Nazi Europe. Many had been lured to California in the '20s by the wealth and technical sophistication of the Hollywood system. After the rise of the Nazis, however, some of those who had previously worked in the European film industry sought safety as well as employment in Hollywood.

Searching for a Safe Haven

The 1938 Anschluss, in which Germany annexed Austria, convinced many Europeans that Nazism would not be averted. Then in March, 70,000 Austrians, including many prominent members of Vienna's cinematic community, were arrested because of their political beliefs or ethnicity. Following the 1938 Kristallnacht, or "Night of Broken Glass," during which Nazis destroyed Jewish property and sent as many as 30,000 to prison camps, the fervor to find a safe haven abroad increased. Sadly, the world had not yet recognized the true threat of Nazi Germany, and no country, including the United States, was ready to welcome the flood of Jewish emigrants. Hollywood was no different, and most members of Tinseltown's artistic community saw the refugees as unwelcome competition.

Proving that money trumped morality, nearly all of the Hollywood studios continued dealings with the Nazi regime until the start of World War II. In many cases, studios altered films that might offend the Nazis rather than risk losing the valuable German and Austrian markets. The one exception, Warner Brothers, was headed by the firmly anti-Nazi brothers Harry and Jack Warner, whose family had left Germany at the turn of the century. In 1939, the studio released the controversial film *Confessions of a Nazi Spy*, which starred Edward G. Robinson and was allegedly based on the accounts of former FBI agent Leon G. Turrou, who had investigated Nazi spies in the United States. As a result, studio head Jack Warner, producer Robert Lord, and many members of the cast and crew received death threats from Fascist organizations operating in the United States.

A Cinematic Classic Is Born

Although scores of intellectuals and artists who fled Nazi Europe abandoned their artistic talents during their exile in the United States, others prospered and, through their efforts, changed the course of American cinema. Nowhere is this more apparent than in the classic Warner Brothers film *Casablanca* (1942). Though much of the film's story line and setting are fictional (for example, there weren't any Nazis in Casablanca during World War II), the romanticized story of Europeans seeking refuge in America featured a cast and crew largely comprised of refugees from Hitler's Europe. In fact, 11 of the 14 names that appear in the film's opening credits are European.

The film's director, Michael Curtiz, had left his native Hungary before the rise of Hitler, and his sensibilities were still deeply rooted in the Viennese cinema where he first made his mark. Technical advisor Robert Aisner had fled France using the same route outlined in the film's opening narration. A few of the film's stars—Conrad Veidt, Peter Lorre, and Paul Henreid—as well as many of the bit players had also fled the Nazis. Veidt, who played the villainous Nazi Major Strasser, was a German refugee who had already garnered critical attention by playing menacing Nazis in other films. Indeed, the vehemently anti-Nazi Veidt reportedly agreed to play Nazi characters only if they were thoroughly detestable. The Hungarian Lorre, who played the short-lived Ugarte in the film, was a refugee from Austria and had starred in fellow refugee Fritz Lang's *M*, which had been filmed while both were still in Germany. After narrowly escaping occupied France, Lorre moved to Hollywood where he and Humphrey Bogart became friends and worked together on *The Maltese Falcon* (1941) before being cast in *Casablanca*. Austrian-born Paul Henreid, who played the French resistance leader Victor Laszlo, was a real-life refugee from the Nazis and was extremely critical of the film's numerous plot flaws (for example, why would a man who wanted to remain unnoticed strut around in an all-white suit?).

Other refugees from Hitler's Europe who acted in *Casablanca* include Curtis Bois, who played a pickpocket; Marcel Dalio, cast as a croupier; Helmut Dantine, who played a desperate husband trying to win enough money to purchase freedom for himself and his refugee wife; and S. Z. "Cuddles" Sakall, who played the lovably loose-jowled headwaiter, Carl. The contributions of these actors and numerous extras, technicians, and crew of European extraction lent *Casablanca* a credibility that has made it one of the most beloved films in movie history.

✳ **In 1933, German director Fritz Lang was summoned to the office of Nazi Propaganda Minister Joseph Goebbels, who said that Hitler wanted him to be Germany's "official" film director. Lang agreed to Goebbels's proposal. As soon as he left, Lang prepared to hastily leave the country.**

Behind the Films of Our Times

* Editors left a handful of scenes from *King Kong* (1933) on the cutting room floor. One features the gorilla taking off actress Fay Wray's clothes. Another involves Kong shaking a group of sailors off a bridge and into a spider-infested valley. The scene was so disturbing to a preview audience that it was removed.

* Liza Minelli, the daughter of Judy Garland (Dorothy in *The Wizard of Oz*), ended up marrying Jack Haley Jr., whose father played the Tin Man in the film.

* Actress Charlize Theron first got noticed while throwing a fit at a Los Angeles bank. Legend has it that the A-list actress freaked out when a teller wouldn't accept her check. A talent manager handed her his card as she was being escorted out.

* The cult classic *Monty Python's Life of Brian* (1979) was never shown in some countries. Ireland and Norway were among several nations that banned the movie upon its release.

* The Aston Martin car driven by James Bond in *Goldfinger* was later sold at auction for a cool $275,000, making it one of the most expensive props ever to appear in a film.

* The script of *Raiders of the Lost Ark* (1981) called for Harrison Ford's character Indiana Jones to engage in a sword fight with an enemy—a scene that would have required hours of filming. But Ford, who was weak and in pain with a nasty case of dysentery, asked director Steven Spielberg, "Can't I just shoot the guy instead?" Spielberg complied.

Hollywood: A Teetotaling Town?

Few people realize that although Hollywood became the epicenter of the motion-picture industry, the city's original planner had something far different in mind. Carousing actors and painted ladies may have been de rigueur in certain towns, but H. H. Wilcox's haven was never intended to be one of them.

✳ ✳ ✳ ✳

Straight and Narrow

APART FROM ITS abundant sunshine and favorable climate, Hollywood seemed perhaps the least likely spot to base an industry that's now well known for its seedier side. H. H. Wilcox founded the southern California city not as a progressive hamlet with its eye on an ever-evolving future, but rather as its polar opposite. The city was intended as a preservationist society where modern excesses would not be tolerated. That it turned into something else is a story in itself.

This Is the Place

Wilcox and his wife, Daeida, arrived in Los Angeles in the mid-1880s. A real-estate developer of considerable means, Wilcox purchased large tracts of land in southern California for the purpose of resale. And like any good businessman, he was always on the lookout for more. But in 1887, tragedy struck when the couple's 19-month-old son died of typhoid fever. To ease their despair, the couple took carriage rides into the country. On one of these therapeutic jaunts, Wilcox spied a particularly charming patch of land featuring orange groves and apple orchards. Bordered by the picturesque Santa Monica Mountains, this particular chunk of the Cahuenga Valley struck the developer as the perfect place to start a town.

A Town Develops

The formidable challenge of establishing a town would be daunting to most, but Wilcox was well equipped for the task. Crippled by the same disease that had claimed his infant son,

Wilcox had become accustomed to overcoming hardships on the way to achieving his goals. Certain of the elements that he desired for his settlement—and those that he did not—Wilcox, a strong advocate of the temperance movement (an organized effort to abolish alcohol backed by people who saw drinking as immoral), got busy. Purchasing a 120-acre expanse for his version of utopia, Wilcox offered free lots to anyone who would construct Protestant churches on the land. His goal was a society free from saloons, bordellos, and red-light districts—excesses that he deemed counterproductive to a clean Christian lifestyle. His wife Daeida came up with the name "Hollywood," although it's believed that she copied it from another woman's estate. Either way, the name stuck and the town began its ascendance to infamy.

Slow Growth

Unfortunately, Wilcox's hopes for an idyllic Protestant community never gained much ground, though efforts to keep the original conservatism lasted until about 1911. Population growth was agonizingly slow—there were never more than a few hundred residents in Wilcox's lifetime—and the return on his investment was disappointingly poor. When he died in 1892, Wilcox was up to his ears in land holdings but was surprisingly cash-poor. Nevertheless, Hollywood was poised to become one of the most recognizable places on Earth. Unfortunately for the devoted teetotaler, the town's notoriety would come not for its virtues but for something far, far different.

A Giant Arrives

Like a condor taking nest amongst hummingbirds, the motion-picture industry swooped down upon Hollywood in 1911. Like Florida before it, Hollywood was chosen for its frost-free climate. But the western locale also offered plenty of wide-open spaces at low prices so that studios could own their own property. Plus, California was known for its varying terrain and landscapes—deserts, mountains, beaches, small towns, urban centers—which allowed for a variety of genres to be filmed within a radius of a few miles. The Nestor Film Company was the first studio to set up shop there. Located in a former tavern at the corner of Sunset Boulevard and Gower Street, the modest building would soon be joined by larger studios, which triggered a veritable explosion of commerce. Banks, clubs, restaurants, hotels, movie palaces, and other supporting industries stood at the periphery, ready, willing, and able to feed at the studios' trough of prosperity. Not surprisingly, these new intruders met with considerable resistance and hostility from the town's original residents. Despite such protests, acres of agricultural land gave way to new housing, and existing homes were displaced in favor of commercial buildings. The economic infrastructure and the local government of Los Angeles were in favor of the burgeoning industry, and they offered incentives to the studios that the political leaders and banking institutions in Florida had not. The message was abundantly clear: The motion-picture industry was there to stay. Those opposed would simply have to get out of the way.

Tinseltown

Over time, the once anonymous hamlet in the Cahuenga Valley grew into a movie-making metropolis, with excesses of every sort playing out on screen and off. The thought of people openly enjoying alcohol or walking the streets barely clothed would cause Wilcox to turn over in his grave. But today, the founder's name lives on in a bronze plaque situated at Hollywood and Vine. Curiously, the sign mentions nothing of the town's original goal of purity. Within eyesight of the marker stands an assortment of bars, tattoo parlors, and places that "decent" people like H. H. Wilcox don't talk about. Such, it seems, is the price of progress.

Behind the TV Shows of Our Time

＊ The globe that once appeared in the background of *NBC Nightly News* actually spun the wrong way for years before someone caught on.

＊ The first and last lines ever spoken on *Seinfeld* were exactly the same. Jerry tells George: "See, now to me, that button is in the worst possible spot. The second button literally makes or breaks the shirt. Look at it. It's too high, it's in no-man's land."

＊ Before landing the role of Jerry Seinfeld's quirky neighbor Kramer, Michael Richards auditioned to play Al Bundy on *Married with Children*.

＊ The role of Monica on *Friends* was offered to Janeane Garofalo, who turned it down. Leah Remini, later cast in *The King of Queens*, also tried out for the part.

＊ The empty frame on Monica's door on *Friends* was supposed to have a mirror on it. After a crewmember accidentally broke it, producers decided that they liked the way it looked and left it.

＊ Paul Reiser plays the piano in the opening theme song of *Mad About You.*

＊ A sticker on Dwight Schrute's desk on the American version of *The Office* says "Froggy 101"—which is the name of an actual country music station in Scranton, Pennsylvania, where the show takes place.

＊ *The Office*'s John Krasinski and B. J. Novak went to high school together and were in the same grade in Newton, Massachusetts.

＊ Michael Keaton began his career as a stagehand on *Mister Rogers' Neighborhood*; he helped to puppeteer and operate the trolley.

Queen of the Soaps

When two Chicago radio stations first put soap operas on the air in the 1930s, they revolutionized the broadcasting industry.

✳ ✳ ✳ ✳

STEAMY ROMANTIC TRYSTS, strained family relationships, mysterious medical conditions—these plot twists set soap operas apart from all other programming. While daytime serials have both their dedicated fans and their contemptuous detractors, there's no question that they make up one of the most enduring genres of mass media.

Enter Irna

In the 1920s, the young radio industry was growing in popularity, but the majority of its regular programming went out over the airwaves at night. In both national and local markets, no one believed daytime shows could draw enough listeners to make a profit. That all changed when Irna Phillips, who was earning a living as a radio actress at Chicago's WGN, proposed a daytime program called *Painted Dreams*. Developed with a female audience in mind, the saga followed Mother Moynihan and her troubled family. The station managers took a gamble on the 15-minute serial, which debuted in October 1930, and found an eager audience of housewives who embraced the melodrama as a welcome distraction from their daily routines. Since then, the soap opera has been a staple of daytime programming.

Phillips both wrote and acted in the show for the next two years, until an ownership dispute drove her to WMAQ, where she put on a similar show called *Today's Children*. She soon struck deals with the national networks and created several more daytime dramas, including *The Guiding Light*. Thanks almost entirely to the genre Phillips invented in Chicago, the national networks saw their revenue from daytime programming triple during the 1930s.

Temptresses, Cliff-Hangers, and Organ Music

As her shows evolved, Phillips created many of the familiar conventions we all expect from a soap opera. She structured each episode as a series of brief vignettes that moved back and forth between story lines. Her shows relied on organ music to stress important actions or facial expressions, and cliff-hangers kept audiences coming back day after day. She was also the first to introduce a temptress character—a conniving woman who thrives on using deception or trickery to steal men away from her competition.

By the early 1940s, Phillips had five of her soap operas in national syndication, and she soon moved them to television. She gave up acting but continued to write all the shows; at one point she estimated that she was churning out two million words of script per year. She developed a peculiar working style in which she would create the scenes by acting them out herself, switching back and forth between characters; her assistants would write up the scripts as they watched her.

Over the years, Phillips created some of the best-known soap operas in American history, including *Days of Our Lives* and *As the World Turns.* Two of her employees eventually went on to create enduring soaps of their own: William Bell and Agnes Nixon, both also Chicagoans, were responsible for *The Young and the Restless, One Life to Live,* and *All My Children.* Phillips continued working on her shows from Chicago until shortly before her death in 1973, but her influence has endured to this day. During the 1990s, eight of the eleven nationally broadcast television soaps could be traced directly to her or to one of her protégés, earning Phillips the undisputed title "Queen of the Soaps."

Hollywood's Urban Legends

If ever there was a breeding ground for urban legends, it would be in the Hollywood Hills. Simply put, people love to hear all sorts of gossip about stars and movies... the weirder the better. Here are some of the strangest urban legends to come out of Hollywood.

✳ ✳ ✳ ✳

Humphrey Bogart was the Gerber Baby

EVERYONE KNOWS THE famous black-and-white drawing of the baby that graces Gerber baby food products. Well, there's an urban legend that the baby is none other than actor Humphrey Bogart. This rumor may stem from the fact that Bogart's mother, Maud, was an illustrator who actually did sell drawings of her son to advertising agencies. In fact, she did allow one of her drawings of Humphrey to be used in a baby food advertisement—for Mellin's Baby Food. Gerber did not start producing baby food until 1928, and by that time, Bogart was 29 years old, making it unlikely—but not impossible—that he was the Gerber baby. We now know that Ann Turner Cook was the model. In 1928, she was drawn by artist Dorothy Hope Smith, who submitted the drawing to Gerber.

Disney on Ice

In life, Walt Disney warmed the hearts of millions. In death, Disney is rumored to have had himself frozen until such a time that scientists could warm him up and bring him back to life. What sounds like something out of a sci-fi movie may have been rooted in the fact that Walt Disney liked to keep his personal life private, so when he died, specifics about his burial were kept under wraps, leading to all sorts of speculation. Rumors were further fueled when Disney was buried in Forest

Lawn Cemetery in Glendale, California, which does not publicly list who is interred there. But Disney's (unfrozen) remains are indeed there, in the Freedom Mausoleum, along with those of several family members.

Three Men and a Baby…and a Ghost!

There's a scene in the movie *Three Men and a Baby* (1987) in which Ted Danson's character, Jack, and his mother are walking through Jack's house while the mother is holding the baby. As they walk in front of a window, the ghostly image of a boy is seen standing in the background. When the characters walk by the window a second time, the boy has been replaced by what appears to be a shotgun. Legend has it that the ghost is that of a boy who accidentally shot himself to death with a shotgun in the house where the movie was filmed.

Of course, the truth is a little less spooky. What many people mistake for the boy's apparition is nothing more than a cardboard cutout of Danson, which was supposed to be part of a subplot involving Jack's appearance in a dog-food commercial. And those scenes weren't filmed in a house, either. They all took place on a studio set in Toronto.

Munchkin Suicide in *The Wizard of Oz*

In *The Wizard of Oz* (1939), shortly after Dorothy and the Scarecrow convince the Tin Man to join their posse, they begin singing and skipping down the Yellow Brick Road. As they round the bend in the road and dance off the screen, a strange, dark shape can be seen moving in a bizarre fashion to the left of the road. It is said that one of the Munchkins, heartbroken over a failed love affair, chose to take his own life as the cameras rolled. It makes for a creepy story, but there's no truth to it. What people are actually seeing is nothing more than an exotic bird flapping its wings. Prior to filming, the director decided that adding strange, exotic birds to the scene would give it more color, so he rented several birds from the Los Angeles Zoo and allowed them to roam freely about the set.

Pop Quiz: Famous Movie Quotes

When a line from a movie is really, really good, it's immediately absorbed into popular vernacular. Think about it: You've probably said, "May the Force be with you" a few times, right? See if you can match these famous quotes with their respective movies.

✳   ✳   ✳   ✳

1. "I'm not bad. I'm just drawn that way."

2. "He-e-e-re's Johnnie!"

3. "Here's lookin' at you, kid."

4. "I can't believe I gave my panties to a geek."

5. "I'll never let anyone put me in a cage!"

6. "I have always depended on the kindness of strangers."

7. "I'll have what she's having."

8. "I love the smell of napalm in the morning . . . smells like victory."

9. "Mrs. Robinson, you're trying to seduce me."

10. "Nobody puts Baby in the corner."

11. "Roads? Where we're going we don't need roads."

12. "Shaken, not stirred."

13. "What we've got here is . . . failure to communicate."

14. "Look, you shoot off a guy's head with his pants down, believe me, Texas is not the place you wanna get caught."

15. "You can't handle the truth!"

16. "You had me at 'Hello.'"

17. "You're tearing me apart! . . . You . . . you say one thing, he says another, and everyone changes back again!"

Answer Choices:

A. *A Few Good Men* (1992)

B. *Apocalypse Now* (1979)

C. *A Streetcar Named Desire* (1951)

D. *Back to the Future* (1985)

E. *Breakfast at Tiffany's* (1961)

F. *Casablanca* (1942)

G. *Cool Hand Luke* (1967)

H. *Dirty Dancing* (1987)

I. *Goldfinger* (1964)

J. *The Graduate* (1967)

K. *Jerry Maguire* (1996)

L. *Rebel Without a Cause* (1955)

M. *The Shining* (1980)

N. *Sixteen Candles* (1984)

O. *Thelma & Louise* (1991)

P. *When Harry Met Sally* (1989)

Q. *Who Framed Roger Rabbit* (1988)

Answer Key: 1. Q; **2.** M; **3.** F; **4.** N; **5.** E; **6.** C; **7.** P; **8.** B; **9.** J; **10.** H; **11.** D; **12.** I; **13.** G; **14.** O; **15.** A; **16.** K; **17.** L.

Movies and Television ✳ **21**

Sunny Days Around the World

*Sesame Street isn't just an American address. The iconic kids'
show now reaches more than 100 other countries.*

✳ ✳ ✳ ✳

SUNNY DAYS ARE sweeping clouds away in neighborhoods
far from our familiar Muppet-filled street. The makers of
Sesame Street have brought their award-winning show to countries all across the globe, and each has its own unique twist.

Sesame Street now airs in more than 120 nations. Some just
show overdubbed versions of the American program, but others
have their own original characters and concepts. So grab
Grover and get ready—we're taking a ride around the world.

First Stop: South Africa

South Africa's *Takalani Sesame* tackles some tough issues. With
millions of HIV-positive children in the South African audience, the folks at Sesame Workshop felt AIDS was something
they could not ignore. Thus came the introduction of Kami,
an HIV-positive Muppet. The creature (who is supposed to
be a five-year-old) is said to be shy but friendly and serves as a
strong influence for children living with the deadly virus.

Other local characters include Moshe, a four-year-old Muppet
who likes to dance but who also has heavy concerns for the
environment. Moshe allows the show's creators to address
difficult ideas related to government policy.

Next Up: China

Bet you didn't know that Big Bird has an identical Chinese
cousin. Meet Da Niao, the Asian counterpart to America's
giant yellow-feathered friend. Da Niao shares more than just
his looks with our Big Bird: He, too, is friendly and chipper but
sometimes a little slow to catch on. His friends include Hu Hu
Zhu, a fuzzy blue pig descended from famous philosophers,
and Little Plum, a messy-haired girl with a loud snore.

Moving on to Bangladesh

Five-year-old Tuktuki plays an important role in Bangladesh's *Sisimpur*. Her story is just one of many serious issues covered in the local take on *Sesame Street*, which cleverly includes political and social issues. The purple Muppet with braided pigtails and a bright green dress represents the ideal of gender equality. Her story lines show young viewers that girls can be given the same kinds of opportunities and treatment as boys.

Fourth Stop: Kosovo

Calling it a challenge to create an uplifting children's show in the war-torn nation of Kosovo would be an understatement. But the makers of *Sesame Street* shine through with two locally produced programs, *Rruga Sesam* and *Ulica Sezam*.

Here, the familiar faces of Bert, Ernie, and Elmo offer messages of unity, and segments showcase similarities among children of different ethnic backgrounds. Story lines include various groups of kids playing the same sorts of games or singing similar songs.

Turning to the Netherlands

In the Dutch rendition, known as *Sesamstraat*, Big Bird's avatar is the blue, silly-looking Pino. As a cousin of our yellow avian pal, Pino shares Big Bird's orange beak and endless curiosity. His friends include Tommie, a teddy bear-dog hybrid; Purk, a messy pig; and Ieniemienie, a precocious, giant-eared mouse.

Last Stop: Germany

In Germany, with *Sesamstrasse*, Oscar's relative Rumpel revels in all things dirty, smelly, and grumpy. Rumpel hangs out with caterpillar Gustav, snail Finchen, and brown bear Samson.

That's just a small sample of the multicultural versions of *Sesame* that can be found around the world. From Mexico's giant green parrot, Abelardo, to Palestine's Internet-savvy rooster, Kareem, there's a Muppet for nearly every place and purpose. All you need to know is how to get there—and odds are, you can find a child who can tell you.

Hooray for Hollywood!

Today, Hollywood is almost synonymous with the film industry.
But that wasn't always the case.

✳   ✳   ✳   ✳

THE WORD HOLLYWOOD not only implies the hub of the American film industry but also the arbiter of popular tastes and the heart of the nation's collective dreams and fantasies. Hollywood is so enmeshed with the production and romance of "the movies" that it is difficult to believe that it was not always the center of the industry.

In the early 20th century, the great film centers were New York City and Chicago. The problem with both New York and Chicago was the winter weather, especially for those who did not have studio spaces. Inclement conditions and short daylight hours hampered shooting schedules, which reduced output.

Hooray for... Jacksonville?

Around 1908, several film companies began sending members of their crews to other parts of the country to shoot movies. Called stock companies, these groups were looking for warmer climates, picturesque locations, and more daylight hours.

In 1908, the Kalem Company sent a troupe to Jacksonville, Florida, to start production during the winter months. Other companies quickly followed Kalem to Jacksonville, earning the city the nickname "the World's Winter Film Capital." For a while, Jacksonville rivaled Chicago and New York as a major film center, but in 1917, the city's movie-promoting mayor lost his bid for reelection, and a decline of support by local financial institutions triggered an exodus of production companies to the West Coast.

Among the other locations that attracted production companies were San Antonio, New Orleans, Cuba, Bermuda,

Mexico, and Southern California. In 1908, Francis Boggs of Selig Studios shot the exteriors for *The Count of Monte Cristo* in Los Angeles. This may have been the earliest film shot by a major company in the Los Angeles area. Boggs returned the following year to set up permanent facilities on Olive Street. In 1910, Selig established a studio in Edendale, while American Mutoscope and Biograph—one of the major production companies in the Patents group—arrived in Los Angeles to shoot a few films. By 1911, almost 20 production companies had studios in Los Angeles and its suburbs, including Hollywood, where the Nestor Company had bought property at Gower and Sunset.

Why Hollywood?

As the decade progressed, the Los Angeles area became a permanent film center, primarily because of the stability of the warm climate and sunshine. Another important factor was the variety of landscape—from desert to beaches to mountains—which gave moviemakers a range of scenic choices for picturesque settings. The production companies that relocated to the West Coast found real estate to be inexpensive, which encouraged them to buy large tracts of land to establish studios with huge staging areas, labs, offices, and backlots. Realizing the benefits of this fast-growing industry, the financial and political infrastructures of Los Angeles and its suburbs were supportive of the studios. Another advantage of Southern California was that it was far from the Patents Company, which frequently stirred up trouble for independents. However, as more and more production companies settled in the West, the Patents Company lost its bite and its relevance.

By 1920, weather, landscape, cheap land prices, and cooperative city fathers combined to make Hollywood and the surrounding area the undisputed film capital of America—a title the "City of Dreams" still holds to this day.

Reality TV: Watching Others Look Foolish Since 1948

It became the TV craze of the early 21st century, but reality TV dates back to the medium's infancy. Of course, in those days, networks didn't have to blur and bleep out as many body parts and profanities.

✳ ✳ ✳ ✳

Early Days

I N NEW YORK in 1950, NBC first aired *Truth or Consequences*, a television program that challenged ordinary people to answer trivia questions; participants were required to perform screwball stunts when they blew the answers. Little did the producers realize the size of the boulder they had set in motion.

American Idol isn't a new concept. From 1948 to 1958, *Arthur Godfrey's Talent Scouts* was based on a similar premise: People would perform (some notable participants included Patsy Cline, Pat Boone, and Lenny Bruce), and audience applause would determine the victor. Several reality TV shows evolved from reality radio shows; for example, *Candid Camera* was originally *Candid Microphone*.

"Smile! You're Pranked on National TV!"

That was the theme of Allen Funt's groundbreaking *Candid Camera*, which debuted in 1948. Sometimes the victim was an average Joe or Jane; sometimes it was a celebrity such as Buster Keaton or Ann Jillian. The show would contrive zany situations and a hidden camera would film people's reactions to them. Sample stunts included:

✳ Collecting tolls on hiking trails

✳ Offering to vaccinate workers against a computer virus

✳ Chaining restaurant silverware to the table

For the most part, people took it in good fun—and still do. *Candid Camera* has run intermittently since its debut.

The First Golden Age

In the post–WWII era, reality TV looked like the wave of the future. Game shows mingled the reality concept with prize competitions: *This is Your Life, You Asked For It, Beat the Clock,* and *What's My Line?* all contained reality elements, albeit with some contrivances behind the scenes. Which leads us to reality TV's eternal, dirty secret: There is always some level of stage-managing—often far more than producers want us to know.

Game shows thrived throughout the 1960s and were technically a form of reality TV, but they ultimately comprised their own genre, and reality TV faded into a niche for many years.

1988 and Beyond

Professionals are expensive and may "withhold their professional services" (i.e., strike), but amateurs are cheap and will line up for the opportunity to be on television. So went the networks' thinking during the 1988 Writer's Guild of America strike, which messed up the entire fall schedule. Fox debuted a modern TV version of a 1950s reality radio show called *Cops: Night Watch,* in which ride-along cameras filmed real police work. In 1992, MTV launched *The Real World,* in which seven strangers lived together for months at a time. It was an early example of modern mainstream reality TV in the United States.

Outwit! Outplay! Outlast!

Survivor, The Amazing Race, Fear Factor, Big Brother USA, and *American Idol: The Search for a Superstar* all debuted in the early 2000s. With these shows and dozens of imitators, reality TV became as great a sensation as it ever was in the 1950s.

Evidently, we love to watch ordinary people's suffering and triumph. And networks like reality shows because they're cheaper to produce than scripted TV. It's a win-win situation all around!

Dorothy Meets the Dark Side

"Somewhere Over the Rainbow" isn't the only memorable song tied to the 1939 classic The Wizard of Oz. *Many believe this beloved film has an unusual secondary soundtrack—Pink Floyd's classic 1973 album* The Dark Side of the Moon.

✳   ✳   ✳   ✳

The Dark Side of the Rainbow

JUDY GARLAND AND Roger Waters may not seem like a match made in heaven, but thanks to one of pop culture's stranger urban legends, the leading lady from *The Wizard of Oz* may be forever linked to the lead songwriter from Pink Floyd.

The reason: Plenty of people insist that Pink Floyd's *The Dark Side of the Moon* was crafted with Garland and her Oz-destined gang in mind. Pop in the movie and play the album at the same time, and you'll see a slew of curious connections.

In the phenomenon known as "The Dark Side of the Rainbow," fans cite more than 100 things that seem to sync up when the film and album are played simultaneously. To experience it for yourself, queue 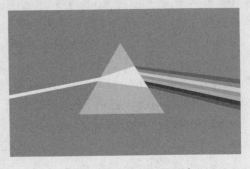 up the movie and the album, and as soon as the MGM lion roars for the third time at the beginning of the film, hit "play" to start *The Dark Side of the Moon* and witness the magic.

Some of the more noteworthy match-ups are as follows:

✳ Dorothy balances on a fence as the lyrics "balanced on the biggest wave" are heard during the song "Breathe."

✳ Dorothy falls off the fence and is rescued as the chaotic intro to "On the Run" begins.

* Dorothy runs away from home as the line "No one told you when to run" from the song "Time" is heard.

* The tornado appears in the sky as the song "The Great Gig in the Sky" plays. Some say the song's slide guitar swells with the tornado and that the operatic vocals vary in intensity to match Dorothy's experiences.

* The song "Us and Them" plays as Dorothy meets the Wicked Witch of the West. The word *black* echoes as the witch first appears. The phrase *black and blue* is sung as the two talk; in the scene, Dorothy wears a blue dress; the witch wears black.

* The lyrics "who knows which is which"—from "Us and Them"—play as the Wicked Witch of the West sees the remains of her sister, the Wicked Witch of the East.

* The song "Brain Damage" plays as the Scarecrow sings "If I Only Had a Brain."

It's worth noting that *The Dark Side of the Moon* is only 43 minutes long, while the run time of *The Wizard of Oz* is more than 100 minutes. Some say the experience ends when the album ends; others suggest you should play the album again to witness more synchronized moments.

Origins and Validity

"The Dark Side of the Rainbow" phenomenon first gained attention in the mid-1990s, though it's not clear who originally discovered it. A radio disc jockey is credited with bringing the concept to the mainstream by discussing it on the air in 1997. MTV News soon followed up with a story of its own, cementing the odd phenomenon's place in movie and music history.

The members of Pink Floyd have long denied any deliberate synchronization. It has also been pointed out that VHS technology didn't even exist when the album was recorded in the early 1970s, so the band would have had a tough time watching the movie during their recording sessions.

From Soap Suds to the Silver Screen

Though much maligned by elitist critics, the soap opera genre represents a unique narrative form. Soaps typically shoot around 100 pages of dialogue per day, and it's a challenge to keep a character fresh for months or years at a time; these are just a couple of reasons why soap operas are a training ground for young actors. Many popular movie stars got their starts on the soaps. Here are some of the most notable.

✳ ✳ ✳ ✳

Kevin Bacon: This prolific actor had a small role in *Animal House* (1978) before his 1979 TV debut on *Search for Tomorrow*. Then, from 1980 to 1981, he played a troubled teen on *Guiding Light*. Since then, Bacon has been involved in numerous films, including *Footloose* (1984) and *A Few Good Men* (1992).

Alec Baldwin: Film and television actor Alec Baldwin (*Malice*, 1993; *Pearl Harbor*, 2001; *It's Complicated*, 2009; *30 Rock*) made an impression on viewers as the strikingly handsome young Billy Allison Aldrich on *The Doctors* from 1980 to 1982.

Taye Diggs: Taye Diggs, star of movies (*How Stella Got Her Groove Back*, 1998), stage (*Rent*), and television (*Private Practice*), lit up the small screen as a talent scout named Adrian "Sugar" Hill on *Guiding Light* in 1997.

Tommy Lee Jones: From 1971 to 1975, this future Oscar winner played a bad seed on *One Life to Live*. As Dr. Mark Toland, Jones portrayed a moody man married to a frigid wife. The combination played out as a recipe for disaster, until he was murdered by a woman while he was running from the law.

Demi Moore: In 1982, long before she starred in *Ghost* (1990), *A Few Good Men* (1992), and *G.I. Jane* (1997), Demi Moore beat out hundreds of contenders for the role of Jackie Templeton on *General Hospital.* She was an instant sensation as a sassy reporter who went to great lengths to get a scoop.

Julianne Moore: A frequent gimmick on soap operas is for an actor to play twins—typically with opposing personalities. In the mid-1980s, Julianne Moore (*Magnolia*, 1999; *The Hours*, 2002) starred as Frannie and Sabrina on *As the World Turns.* Sabrina had been raised in England by a wealthy couple, while Frannie was the daughter of an all-American middle-class family.

Ryan Phillippe: Before tackling serious dramas such as *Flags of Our Fathers* (2006) and *Stop-Loss* (2008), Ryan Phillippe handled the controversial role of Billy Douglas, a gay teenager, on *One Life to Live* from 1992 to 1993. Soaps often feature plots with social messages, and Billy's story line made a strong statement against hate crimes.

Meg Ryan: Before she was the queen of romantic comedies, Meg Ryan was tangled up in a love triangle on *As the World Turns.* From 1982 to 1984, the spunky actress portrayed good girl Betsy Stewart, who was in love with blue-collar Steve Andropoulos. Her stepfather didn't approve, so Betsy married unscrupulous Craig Montgomery instead. But true love prevailed when Betsy left Craig and married Steve in May 1984. Ryan's movie career took off after she left the show, and her memorable role in *When Harry Met Sally* (1989) solidified her place as a leading lady.

Amber Tamblyn: From 1995 to 2001, Amber Tamblyn played Emily Quartermaine, the kindhearted teenage daughter of Port Charles's wealthiest family on *General Hospital.* She later starred in the hit prime-time series *Joan of Arcadia* and the popular films based on Ann Brashares's best-selling *The Sisterhood of the Traveling Pants* young-adult novels.

Kathleen Turner: Kathleen Turner sizzled in *Body Heat* (1981), but fans of *The Doctors* already knew her as trampy Nola Dancy. From 1978 to 1979, Turner played the girl from the wrong side of the tracks who married someone from the right side.

Bugsy Siegel's "Screen Test"

When mobster Bugsy Siegel acted out a scene at the behest of pal George Raft, the results proved eye-opening. Much to the surprise of all, the gangster could really act. Unfortunately, Siegel never pursued acting, choosing instead to remain on his murderous course. This begs the rather obvious question: "What if?"

* * * *

IN THE ANNALS of the underworld, there was perhaps no one more dapper, or more ruthless, than Benjamin "Bugsy" Siegel (1906–1947). Nearly six feet tall, with piercing blue eyes that melted the heart of many a woman, Siegel had movie-star looks and charm that disguised a temperament that could easily be described as "hair-trigger." During his life, Siegel was implicated by the FBI in more than 30 murders.

Born Benjamin Hymen Siegelbaum, the up-and-coming mobster picked up the nickname "Bugsy" (the slang term *bugs* means "crazy") for his high level of viciousness. Siegel hated the tag, considering it a low-class connection to his hardscrabble youth, and threatened to kill anyone who used it in his presence. Still, the mobster was said to be a natural-born charmer who never seemed at a loss for companionship, female or otherwise.

One of Siegel's closest friends was Hollywood actor George Raft, who was known for such memorable films as *Scarface* (1932)and *They Drive by Night* (1940). The two had both grown up on the gritty streets of New York City's Lower East Side. Throughout their lives, the pair would engage in a form of mutual admiration. For example, Raft's movie career featured many mob-related roles. So, when he needed the proper tough-guy "inspiration," the actor would mimic mannerisms and inflections that he picked up from his real-life mobster pals. Siegel, on the other hand, made no secret of the fact that he was starstruck by Hollywood and sometimes wished that he too had become an actor. Hoping to get closer to the

Hollywood action, while at the same time expanding his "operations," Siegel moved to California in 1937.

A Natural-Born . . . Actor?

In no time, Siegel was hobnobbing with major celebrities, even as his deadly business dealings escalated. In 1941, Raft was shooting *Manpower* with the legendary Marlene Dietrich when Siegel showed up on the set to observe. After watching Raft go through a few takes before heading off to his dressing room, Siegel told his buddy that he could do the scene better. An amused Raft told his friend to go ahead and give it a shot; the smirk quickly left Raft's face.

Siegel reenacted the scene perfectly. Not only did he memorize the dialogue line for line, he also nailed Raft's nuanced gestures. This was no small feat, given the fact that Siegel had absolutely no training as an actor. A stunned Raft told Siegel that he just might have what it takes to be a star.

A Dream Unfulfilled

But such Tinseltown dreams were not to be. History shows that Siegel played it fast and loose from that point forward, putting most of his energies into creating the Flamingo Hotel and, along with it, the gaming capital of the world—Las Vegas. Siegel's mob associates from the East Coast put him in charge of construction of the opulent hotel. But when costs soared to $6 million—four times the original budget—Siegel's associates became concerned.

On June 20, 1947, Siegel's dreams of a life on the silver screen came to an abrupt end when a number of well-placed rounds from an M-1 Carbine sent the mobster into the afterworld at age 41. It is believed that Siegel was killed by his own associates, who were convinced that he was pilfering money from the organization. Siegel's life and grisly end are grand pieces of mob drama that got their due on the silver screen in the 1991 flick *Bugsy*, which starred Warren Beatty as the doomed gangster.

Drama, Canadian Style

Most people know that William Shatner, Michael J. Fox, and Mike Myers are Canadian. In fact, Canada has long been a great source of actors and entertainers, as this list shows.

✳   ✳   ✳   ✳

Dan Aykroyd: born in Ottawa, Ontario. Known for: *Saturday Night Live, The Blues Brothers* (1980), and *Ghostbusters* (1984). A member of the Order of Canada, he's a cop buff and the grandson of a Mountie, and he loves riding his own Ontario Provincial Police motorcycle.

Pamela Anderson: born in Ladysmith, British Columbia. Known for: *Baywatch* and five *Playboy* covers (a record). Trivia: Anderson got her big break at a British Columbia Lions football game, where cameras broadcast her in a Labatt's T-shirt. The beer company quickly signed her to help with promotion.

Raymond Burr: born in New Westminster, British Columbia. Known for: *Ironside* and *Perry Mason*. He helped romanticize the profession of lawyer more than any other actor of his era. Trivia: He was badly wounded at Okinawa during World War II.

John Candy: born in Toronto, Ontario. Known for: *Planes, Trains and Automobiles* (1987) and *Uncle Buck* (1989). Loved everywhere as a genuinely nice, principled, and good-spirited man, Candy died of a heart attack at age 43. Trivia: From 1974 to 1991, he made at least one film a year.

Jim Carrey: born in Newmarket, Ontario. Known for: *The Mask* (1994), *The Cable Guy* (1996), and *Bruce Almighty* (2003). Carrey is a two-time Golden Globe winner and has won more MTV movie awards than any other actor. Trivia: Rodney Dangerfield discovered him doing stand-up comedy.

Tommy Chong: born in Edmonton, Alberta. Known for: being half of the comedy duo Cheech and Chong and for the

1978 counterculture comedy hit *Up in Smoke.* Trivia: Before he hooked up with Cheech Marin, Chong was an accomplished R&B guitarist.

Glenn Ford: born in Portneuf, Quebec. Known for: *Gilda* (1946) and *The Blackboard Jungle* (1955). Ford was best at portraying an average Joe in extraordinary situations. Trivia: His real name wasn't Glenn—it was Gwyllyn.

Jill Hennessy: born in Edmonton, Alberta. Known for: *Crossing Jordan* and *Law & Order.* This busy Albertan has an identical twin. Her sister Jacqueline is a TV host and writer in Canada.

Norman Jewison: born in Toronto, Ontario. Known for: directing *Fiddler on the Roof* (1971), *Moonstruck* (1987), and *The Hurricane* (1999). Jewison has directed three actors to Oscar wins and nine others to Oscar nominations. Trivia: His three children also work in show business: associate producer Michael, cameraman Kevin, and actress Jennifer.

Rich Little: born in Ottawa, Ontario. Known for: his dead-on impersonations, particularly of U.S. presidents. He got his start in grade school, where he often answered his teachers' questions in their own voices. Trivia: He also draws celebrity portraits.

Sandra Oh: born in Nepean, Ontario. Known for: *Grey's Anatomy, Under the Tuscan Sun* (2003), and *Sideways* (2004). This Korean Canadian actor turned down a four-year college journalism scholarship to study drama. Trivia: In high school, she started an environmental group dedicated to banning Styrofoam cups.

Shannon Tweed: born in St. Johns, Newfoundland. Known for: her many roles in mainstream erotica and her reign as 1982's Playboy Playmate of the Year. Tweed has two children with Gene Simmons, of the rock band KISS, whom she finally wed in October 2011 after 28 years together. Trivia: She was raised on a Newfoundland mink ranch.

Prehistoric Hollywood

Hollywood was once home to a vast array of Pleistocene ice-age creatures. The La Brea Tar Pits, located near the Miracle Mile district of Los Angeles, is the world's largest repository of fossils from the last ice age, including those of plants, insects, and mammals.

✳   ✳   ✳   ✳

D ESPITE THE NAME, the La Brea Tar Pits are actually a series of asphalt deposits that bubble up from the ground. Over the centuries, oil has oozed to the surface to form sticky bogs that have trapped all manner of animals, condemning them to premature deaths but preserving their skeletons. Since paleontologists began excavating the pits in 1908, remains that date back as far as 40,000 years have been discovered. These include saber-toothed cats, short-faced bears, dire wolves, and even a lion. In early 2009, as workers began excavating an underground parking garage next to the tar pits, they stumbled upon a stunning collection of fossils, including a nearly intact mammoth dating back to the last ice age.

Curators from the George C. Page Museum, which houses fossils collected from the site, estimate that the mammoth—whom they named Zed—stood ten feet tall at the hip and was between 47 and 49 years old, which was young for a mammoth. Zed's three broken ribs indicate that he'd likely been injured fighting with other mammoths—perhaps a precursor to the backbiting that is so common in Hollywood these days.

The La Brea Tar Pits are no stranger to the limelight, having been featured in a number of movies, including *Last Action Hero* (1993), Steven Spielberg's *1941* (1979), and the disaster movie *Volcano* (1997), in which a volcanic eruption originates from the largest tar pit and spews hot lava along the streets of Hollywood.

Behind the TV Shows of Our Time

* Art Carney is best known for his role as dopey Ed Norton on *The Honeymooners*, but he went on to win an Academy Award for his role in the film *Harry and Tonto* (1974).

* In the early 1990s, *Ferris Bueller*, a short-lived sitcom based on the movie *Ferris Bueller's Day Off* (1986), costarred a young Jennifer Aniston.

* In 1998, when NBC aired the series finale of *Seinfeld*, the cable channel TV Land didn't even try to compete. Instead, they broadcast a sign that read, "WE'RE TV FANS; WE'RE WATCHING SEINFELD."

* When Mary Tyler Moore named her production company MTM, she decided to spoof MGM's iconic roaring lion by using a tiny kitten meowing at the tail end of the credits on her shows.

* Homer Simpson's e-mail address is ChunkyLover53@aol.com.

* *Grey's Anatomy* hunk Patrick Dempsey tried out for the part of Dr. Chase on *House M.D.* before accepting his role as "Dr. McDreamy."

* Oprah Winfrey's name was supposed to be "Orpah"—based on a biblical character—but someone made a mistake on her birth certificate. She stuck with the switched-around version.

* The role of Roz on *Frasier* was originally given to Lisa Kudrow, who was later cast as Phoebe on *Friends*. Producers decided to replace her with Peri Gilpin before filming began.

* The outside of the building shown as Dunder-Mifflin on *The Office* is actually located in Scranton, Pennsylvania, coincidentally enough, across from an old bar that was named "The Office."

Protecting the Past

The National Film Preservation Board works to save classic movies.

✳ ✳ ✳ ✳

MOTION PICTURES AREN'T forever. In fact, fewer than 20 percent of the feature films made in the 1920s survive in complete form, and the percentage drops to just 10 percent for movies made in the 1910s.

That's a lot of lost movies, and the number increases every year. In 1988, in an effort to save cinema's cherished heritage, Congress established the National Film Preservation Board, and in 1996, it created the National Film Preservation Foundation to find, restore, and preserve motion pictures of all types.

The Foundation's Role

The National Film Preservation Foundation awards grants and raises private funds to help American film archives preserve movies. The Foundation places special emphasis on what it calls "orphan films," which don't have owners, such as studios, to pay for their restoration and preservation. The most at-risk films are silent films, documentaries, and important amateur footage, such as the Zapruder film of John F. Kennedy's assassination.

Film preservation is a race against time. Many of the earliest motion pictures were lost because they were made with volatile nitrate stock, which disintegrates with age. As a result, many had become gelatinous masses when their film canisters were opened decades later. Sadly, the so-called "safety film" onto which many older movies were transferred had problems of its own, including an irreversible film decay called "vinegar syndrome." In addition, many Technicolor movies made during the 1950s and '60s are fading fast.

Many of today's movies are projected digitally and released on DVD within months of their theatrical release, so their preservation is not an issue; as they are transferred from one new medium to the next, they should continue to look as crisp and vibrant as the day they were "born." But storing digital data is costly, and technology changes so rapidly that equipment used to store and view films today may be obsolete tomorrow.

But saving older movies—even those made as recently as the 1950s and '60s—is a much more immediate concern and an expensive and time-consuming endeavor. In many cases, movies must be pieced together from prints found around the world. Sometimes, they must be meticulously restored frame by frame.

A National Treasure

Every year, the Library of Congress selects 25 movies for addition to the National Film Registry. More than 500 films are now included in the collection, ranging from *Blacksmith Scene*, a film from 1893, to the Coen Brothers' dark comedy *Fargo* (1996). The collection is far-ranging in type and subject matter and includes many well-known cinematic classics as well as a large number of films that hold little public interest but have great historical significance, such as *San Francisco Earthquake and Fire, April 18, 1906* (1906).

The Registry discourages copyright owners of classic movies from altering the films by cutting, colorization, or other means. If a film on the Registry is altered by the copyright holder, the DVD or video packaging must feature a disclaimer that informs consumers that it has been altered. Much of the manipulation of older films has stopped, partly due to the efforts of the Registry.

Movies are sometimes taken for granted as a mere diversion, but their decay is inevitable unless they are properly preserved. Thanks to the efforts of the National Film Preservation Board and the Library of Congress, more movies that might have been lost will now be around for the world to enjoy forever.

The Odd Genius of Ed Wood

For years, he was considered the worst director in the history of cinema, but in recent decades, his work has gained new respect. Despite their shaking sets, loopy lighting, awkward editing, questionable casting, abhorrent acting, incoherent dialogue, and overall cinematic calamity, the films of Ed Wood are now regarded as entertaining glimpses into the mind of Hollywood's greatest B-movie oddball.

✴ ✴ ✴ ✴

THE STRANGE SAGA of Edward Davis Wood Jr. began with his birth on October 10, 1924, in Poughkeepsie, New York. Allegedly, the junior Wood's mother desperately wanted a daughter, so she dressed her son in girls' clothing until he reached puberty. Aside from acquiring a fetish for angora sweaters and a tendency to wear female finery, Wood had a fairly normal childhood. He was not a homosexual cross-dresser but a heterosexual who just happened to prefer frilly undergarments to cotton briefs. Wood loved movies, worked as an usher in the local cinema as a teen, and carried a Kodak "Cine Special" camera wherever he went.

In 1942, Wood signed up for the Marine Corps and fought in World War II, where he took a slew of slugs in the leg that left him permanently maimed. He also lost a pile of pearly whites when he was on the receiving end of an enemy's rifle butt. After the war, he returned stateside, joined a traveling freak show in which he played a bearded lady and, in 1947, eventually made his way to Hollywood.

The Knock on Wood

In Tinseltown, Wood wrote scripts for a couple of low-budget TV Westerns that were quickly forgotten. Despite these failures, he was commissioned to direct a film based on the life of transsexual Christine Jorgenson, which was released under the title *Glen or Glenda* (1953) and proved to be Wood's big break.

Wood's girlfriends and wives often helped out with his movies. His first wife, Norma McCarty, costarred in *Plan 9*. Second wife Kathy O'Hara served as art director on *Night of the Ghouls*. And girlfriend Dolores Fuller costarred in *Glen or Glenda*, *Jail Bait*, and *Bride of the Monster* before becoming a songwriter; she composed "Rock-a-Hula Baby" for Elvis Presley, among others.

Wood went on to create more than a dozen unique pieces of cinematic history. These "epics" include *Plan 9 from Outer Space* (1959), which is often described as the worst movie ever made; *Jail Bait* (1954); *Bride of the Monster* (1955); *Night of the Ghouls* (1959); and *The Sinister Urge* (1960). The common denominators in all of Wood's films are technical mishaps and continuity errors, crude special effects, erratic and outrageous dialogue, eccentric casts, bizarre plot elements, and, of course, their failure to make any kind of impact at the box office.

The Steep Decline

By the 1960s, Wood's career had taken a serious nosedive, and he was forced to write sex novels and dabble in pornography in order to survive. A heavy drinker—his pen name was Akdov Telmig (vodka gimlet spelled backward)—he died penniless and forgotten in 1978 at age 54.

In the 1980s, the advent of film festivals revolving around bad movies revived the work of Ed Wood, catapulting the eccentric director to a level of fame that he never attained during his lifetime. In 1994, Tim Burton and Johnny Depp teamed up to make the film *Ed Wood*, a loving tribute to one of Hollywood's kookiest writers and directors.

Chapter 2

American Landmarks

Unusual Tourist Attractions

Jaunting around America can be a visual adventure. But to truly experience the kitschy, sometimes you need to meander the back roads. That's where you'll find giant roadside statues, fascinating collections, and these unusual attractions.

✳ ✳ ✳ ✳

1. **World's Largest Ball of Twine:** Determining the world's largest ball of twine can be difficult. But the hands-down winner in the solo winder category has to be the nearly 9-ton 11-foot-tall hunk of string on display in Darwin, Minnesota. From 1950 to 1979, Francis Johnson spent four hours a day rolling the ball. He used a crane to hoist the ever-expanding ball as it grew, to ensure uniform wrapping.

   Also in the running is the 1,300-mile-plus length of string originally rolled by Frank Stoeber of Cawker City, Kansas. From 1953 until his death in 1974, Stoeber diligently wound this twine ball. Every August, Cawker City hosts a festival during which anyone can add a bit of twine to the ball. It now outweighs the one in Darwin, but it has had more than one person working on it.

2. **Paul Bunyan Statues:** There are enough Paul Bunyan statues around the continent to delight any teller of tall tales. Representations of the big fella—known for his ability to lay down more trees in a single swing of his ax than any

contemporary logging firm—can be found wherever there have been logging camps. One of the most memorable is in Bangor, Maine—the lumberjack's alleged birthplace—where a 31-foot-tall, 37,000-pound Paul shows off his ax and scythe. Other statues, such as those in Klamath, California, and Bemidji, Minnesota, show Bunyan accompanied by his faithful companion, Babe the blue ox.

3. **Corn Palace:** The city of Mitchell, South Dakota, proudly calls itself the "Corn Capital of the World," and it even has a palace in which to celebrate. The Mitchell Corn Palace, originally constructed in 1892, is now an auditorium with minarets and murals that local artists create each year out of corn and other local grains. After the annual fall harvest, pigeons and squirrels are allowed to devour the palace's murals until the next year, when the process begins anew.

4. **Coral Castle:** The Coral Castle was the brainchild of Edward Leedskalnin, who was jilted by his fiancée the day before their wedding. Crushed, Leedskalnin left his home in Latvia and set out to build a monument to his lost love. The result is the Coral Castle in Homestead, Florida. Without any outside help or heavy machinery, the distraught man sculpted more than 1,100 tons of coral into marvelous shapes. The entry gate alone is made of a single coral block weighing nine tons. The fact that Leedskalnin was barely five feet tall and weighed only 100 pounds adds to the feat.

5. **Crazy Horse Memorial:** The Crazy Horse Memorial in South Dakota, is a labor of love that sculptor Korczak Ziolkowski began in 1948 to honor the great Native American leader. Ziolkowski's life's work (until his passing in 1982), the sculpture is likely the most ambitious roadside project ever undertaken. Ziolkowski's family continues the project, but the statue is still a work in progress. The carving depicts the legendary warrior on horseback and will measure 641 feet long by 563 feet high when completed.

6. **Lucy the Elephant:** Looming 65 feet over the beach at Margate, New Jersey, Lucy is one of the only examples of "zoomorphic architecture" left in the United States. With staircases in her legs leading to rooms inside, the wide-eyed elephant was built in 1881 as a real-estate promotion. Over the years, Lucy has served as a summer home, a tavern, a hotel, and a tourist attraction. Lucy was spared from demolition in 1970, and she received a loving restoration in 2000.

7. **Albert, the World's Largest Bull:** Located in Audubon, Iowa, Albert stands 30 feet tall and weighs in at 45 tons of concrete. Named after local banker Albert Kruse, the monster Hereford statue was built in the 1960s for Operation T-Bone Days, an event held each September to honor the past, when local cattle would board trains to the Chicago stockyards. As an interesting side note, Albert's internal steel frame is made from dismantled Iowa windmills.

8. **Superman Statue:** Metropolis, in far southern Illinois, has nothing to fear these days because Superman lives there. In 1972, the town decided to capitalize on its famous name and subsequently adopted the moniker, "Hometown of Superman." A seven-foot-tall statue was erected in 1986, and replaced in 1993 by a more impressive 15-foot bronze monument. In 2010, a statue of Lois Lane was unveiled next to her hunky beau in Superman Square.

9. **House on the Rock:** Resting atop a 60-foot stone formation in Spring Green, Wisconsin, the House on the Rock is one of the best-known architectural oddities in the United States. Built by eccentric artist Alex Jordan in the 1940s, the House has 14 lavishly decorated rooms—including the Infinity Room with 3,264 windows—and a surrounding complex that features a miniature circus and the world's largest carousel. The House on the Rock is at once wacky, tacky, innovative, and elegant.

I'm Going to Disney World!

With 6 parks, 23 hotels, 5 golf courses, and a 120-acre shopping complex, Walt Disney World is a city unto itself.

* * * *

* Disney World is located on 40 square miles of Disney-owned land. That's enough room to fit San Francisco or two Manhattans. Less than 35 percent of that land is developed.

* Disney includes 4,000 acres of maintained landscapes and gardens. To care for 2,000 acres of turf, landscapers log 450,000 mowing miles per year. That's the equivalent of 18 trips around the world.

* The resort's monorail trains have traveled the equivalent of 30 round-trips to the moon since 1971.

* Spaceship Earth (Epcot's giant sphere) weighs 16 million pounds. The surface comprises 11,324 individual triangles made from an aluminum and plastic alloy.

* Walt Disney Imagineering, the creative team behind the parks, holds more than 100 patents.

* Every year, park guests consume about 75 million Cokes, 10 million hamburgers, 6 million hot dogs, 1.6 million turkey drumsticks, and 9 million pounds of french fries.

* With around 62,000 "Cast Members," Walt Disney World is the largest private single-site employer in the United States. The total payroll is more than $1.3 billion per year.

* The complete wardrobe for all cast members includes 2,500 different designs and about 1.8 million separate pieces. Mickey alone has more than 290 outfits in his closet.

* Every year, the lost-and-found team collects around 3,500 digital cameras, 6,000 cell phones, 18,000 hats, 7,500 autograph books, and 76,000 pairs of sunglasses.

Get Your Kicks

*Route 66 calls forth that part of the American spirit that is restless,
adventurous, and longs to hear the song of the open road.*

✻　✻　✻　✻

ROUTE 66 IS ARGUABLY the most famous highway in the
United States. Celebrated in history, novels, and songs, it
carries a mystique that no modern interstate can surpass.

During the 1920s, the American Association of State Highway
Officials recognized that the country's road system was not
advancing at the same rate as automobile ownership. Cyrus
Avery, a member of the Association, thought the existing
system of named roads (such as the Lincoln Highway and the
National Road) was antiquated and should be replaced by an
integrated network of numbered interstate routes. He also
pushed for an east-to-west route that would stretch more than
2,000 miles from Chicago to California. This roadway was
formally proposed in 1925. By the following year, it had been
approved, allocated the number "66," and opened.

Much of Route 66 was still unpaved in 1926, but it connected
small towns to larger cities, making it easier for rural residents to
escape failed farms for urban centers. As dirt roads gave way to
paved ones, a network of motels, diners, gas stations, and oddball
attractions sprang up along Route 66, making long-distance
travel not only possible but also comfortable and entertaining.

The Mother Road in American Pop Culture

Writer John Steinbeck was the first to immortalize Route 66 in
American culture when he nicknamed it "the Mother Road" in
his 1939 novel *The Grapes of Wrath*. Thousands of farmers had
left the southern Great Plains region because of the devastation
and financial ruin caused by dust storms and the Depression.
Steinbeck chronicled their situation through the fictional Joad
family, who, like their real-life counterparts, packed up all of

their belongings and hit the road seeking better opportunities in California. After World War II, Route 66 remained the road to golden opportunities when returning soldiers and their families traveled West to make new lives for themselves.

In 1946, bandleader Bobby Troupe celebrated the Mother Road in his song "Route 66," which invited listeners to get their "kicks" by taking to the highway and heading for California. A quarter-century later, the rock band the Eagles paid tribute to Winslow, Arizona—a major stop on Route 66—while extolling the virtues of the open road in their song "Take It Easy."

The roadway's mythic status reached a zenith in 1960, when the television drama *Route 66* debuted. The series captured the footloose spirit of American youth, following two friends who leave behind the drudgery of the nine-to-five world to tool around the country in a 1960 Corvette, reveling in their freedom while finding adventure and romance on the road.

Pull Up and Have a Rest at Wigwam Village

Another reason for the legendary status of Route 66 was the number of unique, privately owned restaurants, motor camps (motels), and attractions along the way. Unlike the chain restaurants and hotels of today, these establishments were all one-of-a-kind and afforded travelers of the postwar era memories that have steeped Route 66 in nostalgia. From the Launching Pad Drive-In restaurant in Wilmington, Illinois, with its giant fiberglass spaceman in the parking lot, to the Wigwam Village motor court in Holbrook, Arizona, where guests stay in concrete teepees, Route 66 was lined with colorful and charming examples of Americana.

Beginning in the 1950s, interstates began to replace Route 66, and by 1984, the last part of the original highway was finally bypassed. Sections of the old highway are still maintained as "Historic Route 66," and the spirit of the fabled roadway is embedded in every American who revels in the freedom of the open road.

6 Peculiar Museums in the United States

1. **Circus World Museum, Baraboo, Wisconsin:** This national historic landmark is located on the banks of the Baraboo River, where the Ringling Bros. Circus spent the winter months from 1884 to 1918. Circus World Museum is a not-for-profit educational facility that includes a museum, a library, and a research center that showcase the historic role of the circus in American life. Other attractions include a miniature circus, a clown exhibit, and the world's largest collection of antique circus wagons. Live circus performances take place from May through September.

2. **Lizzie Borden Museum, Fall River, Massachusetts:** The Fall River Historical Society has a collection of items related to Lizzie's alleged slaying of her parents—gruesome crime scene photos, bloodstained linens and clothing, and a hatchet purported to be the murder weapon itself. If that's not enough, tourists can spend the night at the scene of the crime when they stay at the Lizzie Borden Bed & Breakfast, which has been faithfully restored to appear just as it did at the time of the murders.

3. **The World's Largest Collection of the World's Smallest Versions of the World's Largest Things, Various Locations:** Artist Erika Nelson is the owner of this mobile attraction. She drives a van around the country visiting the world's largest roadside attractions—ball of twine, kachina doll, Paul Bunyan and Babe—adding data to her archive of information. Then she crafts and displays miniature renderings of the world's largest things.

4. **Sing Sing Prison Museum, Ossining, New York:** Around 1,700 inmates call Sing Sing Prison home. They may not think it's worth celebrating, but a museum down the street does just that. Sing Sing Prison Museum houses a variety of

artifacts from the town of Ossining and Sing Sing itself. A re-creation of two cell blocks and a replica electric chair are among the highlights, along with a display of confiscated prison weapons. Plans are also under way to turn the original cell block (built in 1825) into a museum.

5. **Liberace Museum and Foundation, Las Vegas, Nevada:** The Liberace Museum houses the entertainer's world-famous collection of 18 rare and antique pianos, including a rhinestone-covered Baldwin grand and a mirror-encrusted concert grand. Also on display are the showman's bejeweled, sequined, and rhinestone-encrusted costumes; jewelry; and cars, including a rhinestone-laden roadster and a mirror-tiled Rolls-Royce. In addition, Liberace's lavish bedroom from his Palms Springs estate is re-created in all its glittering splendor. The Liberace Foundation, which is located in the museum, offers scholarships to talented students pursuing careers in the performing and creative arts.

6. **National Museum of Health and Medicine, Washington, D.C.:** In a city of stellar museums, this one often gets over-looked. But for those interested in the effects of injuries and disease on the human body, this is one you won't want to miss. The National Museum of Health and Medicine was established in 1862 to research and document the effects of war wounds and disease on the human body. Exhibits include more than 5,000 skeletons, 10,000 preserved organs, and 12,000 historical objects, such as the bullet that killed Abraham Lincoln and bone fragments and hair from his skull. Visitors can compare a smoker's lung to a coal miner's lung, touch the inside of a stomach, and view kidney stones and a brain that's still attached to its spinal cord.

Frank Lloyd Wright's Chicago

Frank Lloyd Wright has long been considered one of America's greatest architects. Many of his most important works were created in Chicago and its suburbs.

✳   ✳   ✳   ✳

STAND ON THE corner of Woodlawn Avenue and East 58th Street on any given day in the "Windy City" and you'll likely see a gathering of architecture fans from around the world. That's because 5757 South Woodlawn is the site of the Robie House, a crowning achievement of Frank Lloyd Wright.

More than 50 years after his death, Wright remains one of the world's best known architects. He is most closely associated with Chicago, and for good reason: He got his start there, lived in the nearby suburb of Oak Park, and designed and built some of his greatest works in and around the city.

Few public figures lived a life as full of artistry and achievement as Wright, and simply cataloging his buildings can be dizzying: He designed more than 1,000 structures (of which 400 were built), making him one of history's most prolific architects. While his personal life was often in turmoil—and he ran his business in a haphazard fashion—as an architect, he is considered unmatched.

Taking a Cue from the Surroundings

The list of Wright works in and around Chicago is long, but any appreciation of Wright in the Windy City has an easy starting point: Robie House. Long and low, it hugs the ground and employs a projecting cantilevered roof, continuous bands of stained-glass windows, and thin bricks that emphasize the building's horizontal design. Seen by many as the best example of Prairie-style architecture, the Robie House brings to life Wright's belief in the need for a design philosophy that emphasizes the relationship between architecture and nature.

For other examples of Wright's Prairie style, check out the 1915 Emil Bach House, a private residence located at 7415 North Sheridan Road; the J. J. Walser, Jr. House at 42 North Central Avenue; the Raymond W. Evans House at 9914 South Longwood Drive; and the James Charnley House at 1365 North Astor Street. All in all, there are nearly 20 surviving examples of Wright's designs within the city's boundaries.

Oak Park Treasures

A treasure trove of Wright's work lies beyond Chicago's western border in Oak Park. The village has the largest collection of Wright-designed residential properties in the world. The best place to start is the Frank Lloyd Wright Home and Studio, which served as the prolific architect's abode and workplace for the first 20 years of his career. In addition, there are more than 25 Wright-designed residences within the town, as well as Unity Temple, a revolutionary Cubist structure. Wright took special interest in this project because he was a member of the congregation. This unique place of worship was made from poured concrete. Its high windows allow for plentiful natural light but also ensure privacy and encourage contemplation.

Unity Temple is currently endangered. It was constructed without expansion joints, and the walls are beginning to crack. The congregation is working to raise funds to renovate the worship space, but it will not be cheap or easy.

A World of Wright

In the Chicago metropolitan area as a whole, more than 100 Wright historical sites can be found, not to mention countless examples of his influence and legacy when it comes to architectural design. Since his death in 1959, Wright has taken on an iconic image in the world of architecture, one who left behind a truly American design style.

The Sixth Floor Museum

The Sixth Floor Museum at Dealey Plaza carries a certain amount of baggage in some circles.

✳ ✳ ✳ ✳

THE ASSASSINATION OF John F. Kennedy was one of the most significant events of the 20th century, and its site is one of the most visited spots in North Texas. More than six million people have come to downtown Dallas to learn more about the events of that unforgettable day.

Soon after Kennedy was assassinated, evidence was found that Lee Harvey Oswald had fired shots from the sixth floor of the Texas School Book Depository building. As time passed, it only seemed natural to memorialize the location with a museum dedicated to the life, death, and legacy of JFK.

Planning the Museum

In 1972, the idea to house a museum in the School Book Depository was met with mixed reactions. Many felt that the memories were too painful and that the building should be torn down altogether. Others were concerned that the museum would be viewed as a memorial to an assassin rather than JFK. Conover Hunt, the institution's founding director, continued to lobby for the museum despite public outcry. But just when it looked as if the city council was going to give its approval, John Hinckley Jr. attempted to assassinate President Ronald Reagan. Hinckley had grown up in Dallas, and the local ties begged comparison to Oswald. The museum was delayed again.

The Dream Is Realized

Finally, on Presidents' Day 1989, the Sixth Floor Museum was opened as a direct response to the number of visitors coming to learn more about the assassination and see the sites where the events took place. Exhibits were designed to highlight the impact of Kennedy's death on the nation and the world.

On Presidents' Day 2002, the museum expanded to include a gallery for temporary exhibits, public programs, and special events. The current collection holds 35,000 items, including manuscripts, documents, photographs, films, audio recordings, newspapers and magazines, and oral histories. These are a few of the most popular exhibits:

* The Abraham Zapruder Collection, which contains the world-famous 8mm film of the Kennedy motorcade just before, during, and after the shooting.

* The Orville Nix Collection, which contains his film of the motorcade shot from Dealey Plaza, the opposite angle of the Zapruder film.

* The Jay Skaggs Collection, which consists of 20 slides taken by an amateur photographer in Dealey Plaza just before and immediately after the shooting.

* The Phil Willis Collection, which represents the most extensive record of the day's event in Dealey Plaza with 30 color slides, including a single photo of JFK as he was shot that has been studied by investigators and researchers.

* The Parkland Hospital Collection, which includes medical reports, doctors' summaries, and employees' recollections of the events on the day of the assassination.

The museum is nonprofit and receives all its revenue from ticket sales. And despite its macabre exhibits, the museum continues to be the second-most-visited historical site in Texas behind the Alamo, proving that the legacy of JFK and the public's fascination with the circumstances surrounding his death are eternal.

23 Silly City Names

1. Bird-in-Hand, Pennsylvania

2. What Cheer, Iowa

3. Ding Dong, Texas

4. Elbow, Saskatchewan

5. Monkeys Eyebrow, Kentucky

6. Flin Flon, Manitoba

7. Goofy Ridge, Illinois

8. Hell, Michigan

9. Intercourse, Pennsylvania

10. Joe Batt's Arm, Newfoundland

11. Cut and Shoot, Texas

12. Jackass Flats, Nevada

13. Owls Head, Maine

14. Peculiar, Missouri

15. Placentia, Newfoundland

16. Saint-Louis-du-Ha! Ha!, Quebec

17. Suck-egg Hollow, Tennessee

18. Swastika, Ontario

19. Tightwad, Missouri

20. Toad Suck, Arkansas

21. Truth or Consequences, New Mexico

22. Wahoo, Nebraska

23. Paint Lick, Kentucky

Hallelujah!

Divine inspiration led a Mansfield, Ohio, pastor to found a museum that brought the Bible to life.

✳ ✳ ✳ ✳

THOSE WHO ARE eager to visit the Holy Land but find the cost prohibitive might consider a trip to BibleWalk in Mansfield, Ohio, where holy scripture comes to life.

This unique attraction was conceived by Pastor Richard Diamond after he was deeply moved by a depiction of the Ascension of Jesus Christ that he saw in a wax museum. Diamond and his flock worked for years to make the pastor's vision a reality, and the museum finally opened to the public in August 1987.

More than One Way to Spread the Word

There is a lot to see at BibleWalk, so those who visit should plan to spend the day. The museum offers several tours inspired by the Bible, including a New Testament tour, an Old Testament tour, a Christian Martyrs tour, and a Reformation tour. In total, the facility features 70 distinct tableaus containing more than 300 individual figures, all designed to inspire the faithful.

Other popular exhibits include a collection of rare Bibles from around the world (some date back to the 1500s), and a collection of meticulously crafted woodcarvings by artist John Burns that depict important events from the Bible. The artwork, which is stunning in its intricacy, took Burns 16 years to complete.

BibleWalk is also home to an exhibit of Korean word art made by Rhee Kwang Hyuk. Scenes from the Bible are recreated entirely out of words of scripture—a closer look will reveal letters instead of lines—and are a marvel to behold. BibleWalk's special take on traditional dinner theater, called Dinner with Grace, treats diners to live portrayals of gospel events. So if a trip to Jerusalem is out of reach, this town in north-central Ohio offers the next-best thing.

At the Center of It All

A stroll through New York City's Central Park might lead you to believe that it is the one remaining slice of nature amid the towering skyscrapers of steel and glass that flank it. In fact, this urban park was almost entirely manufactured. And even though Manhattan's northern half was laid out in the early 19th century, the park was not part of the Commissioners' Plan of 1811.

✻ ✻ ✻ ✻

BETWEEN 1821 AND 1855, the population of New York City nearly quadrupled. This growth convinced city planners to build a large open-air space. Initial plans mimicked the public grounds of London and Paris, but it was eventually decided that the space should evoke feelings of nature—complete with running water, dense wooded areas, and rolling hills.

The original park layout included the area from 59th to 106th Streets as well as a stretch between 5th and 8th Avenues. The land itself cost about $5 million. This part of Manhattan featured an irregular terrain of swamps and bluffs and included rocky outcrops left from the last Ice Age 10,000 years earlier; it was deemed unsuitable for private development but was ideal for creating the park that leaders envisioned. However, the area was home to about 1,600 poor residents, most of them Irish and German immigrants, as well as a thriving African American community. Ultimately, these groups were resettled, and the park's boundaries were extended to 110th Street.

In the 1850s, the state of New York appointed the Central Park Commission to oversee the development of the area. A landscape design contest was held in 1857, and writer and landscape architect Frederick Law Olmsted and architect Calvert Vaux won with their "Greensward Plan."

Olmsted and Vaux envisioned a park that would include "separate circulation systems" for its assorted users, including horseback riders and pedestrians. To accommodate crosstown traffic while still maintaining the sense of a continuous single park, the roads that traversed Central Park from east to west were sunken and screened with planted shrub belts. Likewise, the Greensward Plan called for three dozen bridges, all designed by Vaux, with no two alike. These included simple granite bridges as well as ornate neo-Gothic conceptions made of cast iron. The southern portion of the park was designed to include the mall walk to Bethesda Terrace and Bethesda Fountain, which provided a view of the lake and woodland to the north.

Central Park was one of the largest public-works projects in New York during the 19th century; some 20,000 workers reshaped the topography of nearly 850 acres. Massive amounts of gunpowder (more, in fact, than was used in the Battle of Gettysburg) were used to blast the rocky ridges, and nearly three million cubic yards of soil were moved. At the same time, some 270,000 trees and shrubs were planted to replicate the feeling of nature.

Despite the massive scale of work involved, the park first opened for public use in 1858; by 1865, it was receiving more than seven million visitors a year. Strict rules on group picnics and certain activities kept some New York residents away, but by the 1880s, the park was as welcoming to the working class as it was to the wealthy.

Over time, the park welcomed a number of additions—including the famous carousel and zoo—and activities such as tennis and bike riding became part of the landscape. Today, Central Park plays host to concerts, Shakespearean plays, swimming, and ice-skating. It also features a welcoming bird sanctuary—for watchers and their feathered friends alike—and is a pleasant urban retreat for millions of New Yorkers.

The National Museum of Funeral History

This museum takes visitors "six feet under" with style and panache.

❋ ❋ ❋ ❋

WHAT HAPPENS TO us after we die is a matter completely open to faith and conjecture. And although people possess only scant knowledge of the afterworld, there's no reason why they shouldn't still plan to leave this life with a modicum of dignity and style. Jumping on the bandwagon of self-expression, funeral directors have long answered this most basic human desire. From no-frills pine boxes to elaborate diamond-inlaid caskets, from simple funeral handcarts to highly stylized hearses, these keepers of the deceased merrily ply their macabre trade and turn a tidy profit for their efforts. Celebrating the art of the final send-off, Houston's National Museum of Funeral History (NMFH) covers nearly all.

Dearly Departed

The NMFH might not exist if survivors didn't feel a pressing need to memorialize their loved ones. When Uncle John passes on, for example, what loving family would commit his earthly remains 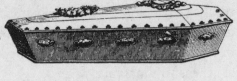 to a potter's field when a full-blown mausoleum with a polished granite vault is within their grasp? The NMFH examines nearly all forms of such loving expression, from the simplest coffin to the most elaborate sarcophagus.

A Deadly Combo

Robert L. Waltrip's museum was to be a place that would "educate the public and preserve the rich heritage of the funeral industry." Some 25 years in the making, his dream became a reality in 1992 at a 20,000-square-foot site in Houston.

Packed to the rafters with caskets, coffins, hearses, and other items associated with the post-death process, the space has grown to some 35,000 square feet. Because of its vast size, the NMFH is billed as the largest educational center on funeral heritage in the United States. As a not-for-profit organization, the museum relies totally on contributions from funeral aficionados and ordinary citizens. The NMFH hosts a pair of golf tournaments each year to further the effort, funneling the proceeds into additional exhibits and upkeep, which breathes new life into a museum firmly committed to death.

A Grave Situation

To think of the NMFH simply as a repository for old caskets and a few old hearses, however, is to sell the operation short. This facility traces the death journey in myriad ways and features world-class exhibits that any mainstream museum would—pardon the pun—kill for.

While moving amongst the museum's exhibits, visitors will find a full-size mock-up of an old casket factory. They will also learn how woods were chosen, the various tools and glues used in the process, and how the craft was passed down from generation to generation. The museum also displays its share of burial oddities: Coffins shaped like fish, automobiles, a chicken, an outboard motor, and even a KLM airliner prove that people can be just as bizarre in death as they are in life.

Life—for the Living and the Dead

The museum's motto is: "Any day above ground is a good one." Though this sounds like a reasonable statement, it does run counter to the museum's mission. After all, the NMFH demonstrates that "six feet under" isn't necessarily the gloomy affair it's been made out to be. Whether or not such glorious departures will signal a safe arrival on "the Other Side," no one can say. But if a glimpse into life's big ending is what a person is after, he or she need look no further than the National Museum of Funeral History. When it comes to the end, their coverage is dead-on.

Cadillac Ranch

Although it's no longer on Route 66, the Cadillac Ranch has become almost as famous an icon as the old highway itself, symbolizing America's continuing love affair with the automobile, the freedom of travel it affords, and whatever excesses the journey might bring.

✳   ✳   ✳   ✳

PARIS HAS THE Eiffel Tower; Rome, the Colosseum; and China, its Great Wall. In West Texas, Amarillo's claim to fame is the Cadillac Ranch, an impressive roadside monument to motoring that's caused passing cars to stop and wonder for nearly 40 years now.

Ironically, the curious car-sculpture-turned-tourist-attraction came to life in 1974 at the height of the Arab oil embargo, when gasoline dried up at service stations nationwide. While car owners fumed in long lines waiting for their rations of fuel, Texas helium mogul and millionaire Stanley Marsh 3 (he thought Roman numerals were pretentious) commissioned a San Francisco artists group (known as the Ant Farm) to create the ostentatious sculpture for his Amarillo ranch. The medium? Cadillacs—actual cars. The canvas was a dusty wheat field.

The Ant Farm Goes to Work

Artists Doug Michels, Chip Lord, and Hudson Marquez jointly labored on the project, collecting both running and derelict Caddies from around the Panhandle area. Models representing the "Golden Age" of American automobiles from 1949 through 1963 were chosen. It's no coincidence that this span highlighted the birth and death of the tail fin, the Cadillac's most defining feature. After all, the tail fin represented America, the space race, and the nation's emergence to prosperity during the 1950s. It was the perfect symbol for a sculpture constructed with full-scale automobiles.

To anchor the cars, massive eight-foot holes were dug into the Texas prairie. The vehicles were put to rest nose down and were positioned so that they would face west, at the same angle as the Great Pyramid of Cheops in Egypt.

Location, Location, Location

Because Marsh's homage to the Cadillac was located directly along America's famed "Main Street"—U.S. Route 66—there was no shortage of gawkers speeding past. The Stonehenge-like assemblage quickly gained a loyal following. Photographs were taken of it, articles written about it, songs sung about it, and movies made that featured it. If you planned to motor west on Route 66, making a pilgrimage to see the fins was a must.

Relocation, Relocation, Relocation

In 1997, Marsh became worried that Amarillo's urban sprawl would endanger his beloved Caddies, so he had the pop art homage to Detroit steel quietly moved two miles further west, along Interstate 40 (which replaced the original Mother Road in Texas). The only evidence left behind at the original site was ten huge holes. It was an eerie reminder of the death of Route 66 itself, which was decommissioned a dozen years earlier.

Today, Cadillac Ranch continues to draw the curious. Although located on private property, it's easy to access: Just drive along the frontage road and enter the pasture through an unlocked gate. Part of the ritual of visiting Cadillac Ranch is to leave a personal touch on the decaying cars with spray paint, a practice that Marsh doesn't seem to mind. Be sure to bring along a can.

As the seasons change and the travelers come and go, the Cadillacs mutate through a range of colors and messages left by modern-day explorers who are out to discover the off-the-wall attractions that bring life to the roadside carnival.

Where Madness Meets Medicine

Welcome to the Glore Psychiatric Museum: where people can view a Lunatic Box, a Bath of Surprises, and a human-size gerbil wheel.

❋ ❋ ❋ ❋

INTERESTED IN THE history of mental health care? Curious about what goes on in the minds of the clinically insane? A visit to the Glore Psychiatric Museum in St. Joseph, Missouri, provides a shocking look at the profession's past. In many cases, what today appears to be a medieval torture device was—until recently—employed in the "treatment" of the mentally ill.

Foreign Objects

As far as medical museums go, a curator is fairly certain to strike gold when his or her featured attraction is a case of rusty nails retrieved from a woman's stomach. Nearly 1,500 objects— including safety pins and saltshaker tops—comprise the former intestinal contents of a mental patient who suffered from a form of allotriophagy, a bizarre desire to eat foreign objects.

This is just a taste of what you'll find at the Glore Psychiatric Museum, which is housed in what was known as State Lunatic Asylum No. 2 when it opened in 1874. George Glore, a long-time state mental health employee, founded the museum in 1968; today it features replicas of a multitude of 16th-, 17th-, and 18th-century treatment devices. The exhibit proved so perversely successful that it became a permanent fixture.

This Will Only Hurt a Bit

Some items at the museum date back to a time when the mentally ill were treated more like prisoners than patients. Devices that originated with the hospital's late 19th-century beginnings represent a dark age of psychiatric medicine, a time when doctors held the belief that mental illness could be driven out of a person purely by physical torment.

One example is the Fever Cabinet, which is basically a lightbulb-filled box designed to enclose a patient's body and drive up his or her body temperature. Conversely, there was the Dousing Tub, in which inmates were bound and blasted with water, and the "Bath of Surprise," a gallows device that dropped patients suddenly into a pool of shockingly cold water "to break the chain of delusional ideas and ... create conditions favoring sane thinking."

Other contraptions include the Lunatic Box, a narrow enclosure that forced patients to remain confined indefinitely in a standing position; the Tranquilizer Chair, a seat with a built-in toilet, leg irons, and a box to cover the head; and O'Halloran's Swing, which spun its victims at up to 100 revolutions per minute. Patients who needed some exercise could also walk in the giant wooden enclosed treadmill—a human-size hamster wheel.

While some visitors find the museum disturbing, Glore clearly believes the collection serves an important purpose. "We really can't have a good appreciation of the strides we've made if we don't look at the atrocities of the past," he said in a 1995 *Los Angeles Times* interview.

Let the Healing Begin

Visitors are also treated to a peek at the artistic and compulsive sides of the mentally ill. For example, a wall features a selection of more than 500 handwritten notes that had been stuffed into a television set. Although the author's intention is unknown, he had placed the slips of paper one by one through a slot in the back, possibly as a method of transmitting his thoughts to the outside world. Nearby, a cage holds 100,000 cigarette packs amassed by another patient, who was working toward an imagined redemption scheme to win a new wheelchair.

Glore served as the museum's curator for three decades and retired in the mid-1990s. Due to its popularity, the collection eventually moved out of the asylum and found an even bigger home. If you're ever in St. Joseph, Missouri, be sure to stop by. You'd be crazy not to!

Chapter 3

Games People Play

Iraq Embraces America's Pastime

Baseball-starved American troops find a way to play during wartime and, in the process, spread the love of the game.

※　※　※　※

BATS, BALLS, AND gloves were not high on the U.S. military's list of equipment to bring to Iraq. But that didn't stop the Hawaii-based 25th Infantry Division's 2nd Brigade. With a bit of free time on their hands and a desire to get a pick-up game started, the soldiers improvised as well as any stickball-playing youngsters on a Brooklyn street ever could. The setting: a soccer field in Altun Kupri, a small and historically safe town north of Baghdad. The equipment: wadded paper wrapped in duct tape for a ball and an aluminum cot leg for a bat. With Humvees parked at each "outfield pole" and another serving as a center-field wall the troops were ready to "Play Ball!".

After word of the game got back to the sister of Captain Deron Haught in West Virginia, the soldiers became better equipped for games more closely resembling traditional baseball. "She felt bad," Haught noted in 2004. "We were over here serving our country, and we were playing baseball with a tape ball and a cot leg. So she started 'Operation Home Run.'" Soon, balls, bats, and mitts poured in from the States. And as the equipment arrived, some of the locals began taking interest in a game that differed vastly from the traditional sports of choice in Iraq—soccer and volleyball.

Haught went one step further, convincing the Altun Kupri city council to help interested teenagers form teams for an organized game. The rules were not easy to teach (try to explain baseball to youngsters who have never been exposed to the sport), but the kids caught on quickly to the basics of throwing, hitting, and running. Mainly, what the American soldiers-turned-coaches stressed were traits like teamwork and sportsmanship. "I think baseball is a great example of democracy," Haught said.

Haught served as home plate umpire and public-address announcer when Nawruz (Kurdish for New Year's Day) took on Brusik (Team Lightning) in the game that served as the crowning moment of Operation Home Run's efforts. Haught announced: "We'd like to welcome you to the first Iraqi baseball game. This game has been played in America for over a hundred years, and we want to share it with Altun Kupri and with this country."

As the 25th Infantry Division's 2nd Brigade played—and taught—the game, more and more youngsters came out to participate. It was Haught's hope that baseball would continue to spread not only in the Altun Kupri area but also in other parts of the country.

And for the record, Nawruz defeated Brusik 10–7. Said Diller Fakhraddin, the winning pitcher, "I like this game. It's better than soccer."

✳ Fiery yet deaf and mute pitcher Luther "Dummy" Taylor got by with signing and mouthing vile tirades at umpires for years. But in 1902, a lip-reading plate umpire ejected him from one of those contests.

✳ Brooks Robinson hit into triple plays on four occasions, a major-league record. The first was on June 2, 1958.

Curious About Curling?

*If you've ever caught a curling match on TV, you might have
wondered: Who invented that?*

❊   ❊   ❊   ❊

History

CURLING BEGAN IN Scotland in the 1500s and was initially
played with river-worn stones. Over the next century,
enterprising curlers began to fit the stones with handles.
Curling was a perfect fit with Canada's heavily Scottish culture
and northerly climate. The Royal Montreal Curling Club began
in 1807, and in 1927, Canada held its first national curling
championship. Today, curling has millions of enthusiasts
around the world. Canadian curlers routinely beat international
competition, which provides a source of national pride.

How to Curl

The standard curling rink measures 146 feet by 15 feet. At the
end is a 12-foot-wide series of concentric rings called the *house,*
the center of which is the *button.* There are four curlers per
team. Each throws two rocks (shoves them, rather; you don't
really want to go airborne with a 44-pound granite rock) in an
effort to get as close to the button as possible. When all eight
players have thrown two rocks each, the *end* (which is similar
to a baseball inning) is concluded. A game consists of eight
or ten ends.

Curling Strategy

The goal in curling is to try to knock the other team's rocks out
of the house—and thus out of scoring position—while getting
yours to hang around close to the button. After all throws have
been completed, the team with the stone closest to the button
scores a point for each rock that's closer to the button than the
opponent's nearest rock (and inside the house).

Proper Ice-keeping

When you see curlers sweeping the ice with brooms, they aren't trying to keep the area clear of crud. The team captain determines strategy and advises the players using the brooms in the fine art of *sweeping*. Players can guide the stone with surprising precision by skillfully sweeping in front of it with their brooms, but they can't touch (*burn*) that rock or any others in the process.

"Good curling!" "Thanks, you too."

Curling is a game that values good sportsmanship. Curlers even call themselves for burns. When a team is so far behind that it cannot win, it is considered proper sportsmanship to concede by removing gloves and shaking hands.

Curling Jargon

Bonspiel: a curling tournament

Curler: a curling player

Draw: a shot thrown to score

Hack: a foot brace that curlers push off from, like sprinters

Hammer: the last rock of the end (advantageous)

Hog line: a blue line in roughly the same place as a hockey blue line. One must let go of the rock before crossing the near hog line—and the rock must cross the far hog line—or it's hogged (removed from play)

Pebble: water drops sprayed on the ice between ends, making the game more interesting

Takeout: a shot meant to knock a rock out of play

Up! Whoa! Off! Hurry! Hard!: examples of orders the team captain might call to the sweepers

Weight: how hard one slides the rock

Fast Facts from Around the Diamond

* In 1999, Fernando Tatis of the St. Louis Cardinals became the only player to hit two grand slams in the same inning.

* The first major-leaguer to hit a home run under the lights on was Babe Herman of the Cincinnati Reds in July 1935.

* Major-league umpire Cal Hubbard is the only person in both the football and baseball Halls of Fame.

* Pitcher Phil Niekro holds the record for having the most wins without ever appearing in a World Series (318). He played 24 seasons, mostly with the Braves.

* Eddie Mathews is the only person to play for the same franchise in three different cities: He played for the Braves in Boston, Milwaukee, and Atlanta.

* Chuck Finley of the Indians and Angels is the only pitcher to strike out four batters in one inning more than once. He did it three times in 1999 and 2000.

* In 1882, Paul Hines became the first ballplayer to wear sunglasses on the field. They weren't corrective; he just didn't like having the sun in his eyes.

* When Cy Young started pitching, the pitcher was allowed to stand five feet closer to the plate than he can today. He also had to throw underhand and fouls weren't strikes. Young didn't even wear a glove during his first few seasons—that's how "old school" the old school was.

* In a 1920 Reds–Giants game, an argument at the plate lasted so long that Reds center fielder Edd Roush lay down. When the napping Roush didn't answer to the ump's eventual call of "Play ball!," he was given the rest of the day off to nap at will.

Convict Corral

Hardened criminals—cowboy convicts—risk life and limb to complete in prison rodeos.

✳ ✳ ✳ ✳

Spectator's Sport

BETWEEN 1940 AND 2009, hotel rooms were hard to come by on on particular weekend in McAlester, Oklahoma (population 18,000). In late August, or on Labor Day weekend, masses of people visited the town, or more precisely, the maximum-security Oklahoma State Penitentiary, to watch an odd competition. The curious watched and cheered from behind a thick, razor wire-topped chain-link fence within the prison's walls, as inmates from ten state prisons competed in classic rodeo events such as steer wrestling and bull riding.

The top crowd-pleasers, though, were the prison's own signature events. In one competition, inmates attempted to grab a ribbon from between the horns of a 2,000-pound bull for a prize of $100 (about ten times what they made in a month laboring behind bars).

Dressed in borrowed Western wear and often lacking real-world rodeo experience, the inmates competed in teams of ten. Only the well behaved were eligible.

Prison Rodeos

Oklahoma's neighbor Texas put on the nation's first "behind the walls" rodeo at Huntsville in 1931. A lack of funding forced that rodeo to close down in 1986. The McAlester rodeo, too, fell prey to funding gaps, and shut down after the 2009 event.

Angola

Visitors seeking a similar event now need to travel to Louisiana, where prison rodeos at "Angola," the Louisiana State Penitentiary, began in 1965 and still run twice a year, in April and in October (with exceptions for COVID-19).

When Baseball and Beer Don't Mix

An all-American game turned into a major-league disaster when rowdy Cleveland fans managed to give cheap beer a bad name.

✳ ✳ ✳ ✳

THE IDEA SEEMED sound enough: Entice Cleveland Indians fans to a game with ten-cent cups of beer. Nothing could go wrong, right? That's apparently what organizers thought back in 1974 when they conceived the now-infamous "Ten-Cent Beer Night" promotion at Cleveland Municipal Stadium. Needless to say, they couldn't have been more off base.

To be fair, the Indians' beer bash did work in one way: More than 25,000 people—twice the team's typical crowd—packed the stands. But that's where the success ended.

Foul Behavior

The worst of it started in the second inning. First, a presumably intoxicated woman ran onto the field and bared her breasts for the entire stadium to see. Then, in the fourth inning, a man charged the field fully naked and—believe it or not—slid into second base (ouch!). The indecent exposure continued in inning five, when a couple of fans flashed their fannies at the Texas Rangers, the opposing team.

The nudity, however, was nothing compared to what came next: People started chucking everything from tennis balls to batteries onto the field. Someone even tossed lit firecrackers into the Rangers' bullpen. Then all hell broke loose, with people ripping apart seats and using pieces of metal as weapons. By the time the players got involved, an all-out riot had begun.

In the end, nine people were arrested, and the Indians forfeited the game. The crowd stole all the bases (and not in the normal sense of the term). The bags have never resurfaced—nor has the dignity of the various drunkards who showed way, way too much skin that warm June night in 1974.

Beware! Flying Cabbage!

Heads fly—but they don't roll—in Shiocton, Wisconsin, each autumn at the World Championship Cabbage Chuck, a fundraiser for a local church that pays homage to the cabbage-growing community.

✳ ✳ ✳ ✳

THE SHIOCTON AREA is known for its cabbage, albeit in a more refined form: sauerkraut, which is finely shredded, fermented cabbage. Zesty, tangy sauerkraut was a favorite of the German immigrants who originally settled the area. The Great Lakes Kraut Company, located in nearby Bear Creek, dates back to 1900 and is the world's largest sauerkraut plant, processing more than 170,000 tons of raw cabbage each year. That's a lot of kraut!

Cabbage Chuck entrants come in two varieties: those using air cannons to launch the leafy greens into orbit and those who go medieval by using a trebuchet to catapult their heads of cabbage.

From the firing line, entrants launch their cabbages into the air to see who can send the cabbage careening the farthest. Shiocton isn't far from Green Bay, home of the Packers, so team pride runs rampant. An outhouse painted purple and labeled "Vikings Draft Room" (for one of the Pack's main NFL rivals) is a hot target for flying cabbages.

The record cabbage chuck has remained intact since the event began in 2006—unlike the hundreds of heads of cabbage that have been sacrificed during that time. That record is a whopping 1,171 feet, courtesy of a cabbage shot from an air cannon built by Pat Peeters of Bear Creek. That should be good enough for Guinness, as there's no cabbage-throwing record on the books.

Still, there's a challenge to tracking the distance these heads fly. They tend to disintegrate as they move through the air, essentially changing from cabbage to coleslaw in a matter of seconds.

Odd Sporting Events

Many modern sports have become mundane, mainstream activities, so it's refreshing to discover pursuits that fall well outside those confines. Here are a few of those "untamed" sports.

✳ ✳ ✳ ✳

Roach Racing

AT LEAST TWO U.S. venues offer roach racing, a sport that's definitely not meant for the squeamish. New Berlin, Wisconsin, joins in the fun with an annual contest held by Batzner Pest Management, and they do it with style. The contestants at this event are boisterous and ever-so-huge Madagascar hissing cockroaches. At the seventh annual event in 2007, a competitor named Rocky went the distance, thus ensuring that a local charity would become $200 richer. During the Roachingham 500, held each year at Bugfest in Raleigh, North Carolina, cockroaches race in front of huge crowds that come to see this and other bug-related events.

Segway Polo

Why play ordinary polo on the back of an outdated, analog horse when you can play on a Segway, a two-wheeled electric vehicle that is ridden while standing? This unique scooter was first put into polo-playing use during a Minnesota Vikings halftime show in 2003. Since then, this funky sport has attracted many people, and organized groups have popped up. Segway polo's greatest advantage over regular polo is that there's nothing to shovel off the field when the game is over.

Underwater Hockey

Aside from the obvious difference in playing surfaces, another major distinction between ice hockey and underwater hockey is the uniforms. Ice hockey players layer up with heavy padding and clothing, but underwater hockey players must don a swimsuit, fins, a snorkel, and a mask, and then submerge themselves in water.

Invented in 1954 by Alan Blake of England, underwater hockey has evolved into an international sport. Dozens of teams from around the world compete at the World Championships, which are held every other year. The sport's rules are similar to those of traditional hockey, but the equipment varies greatly between the two sports, most notably the sticks and pucks. Since underwater hockey requires that its participants push the puck along the bottom of a pool, the disk is suitably hefty, weighing about three pounds. On the other hand, the stick is a wimpy device, approximately one foot long.

Ferret Legging

Ferret legging is a bizarre English sport with roots tracing back to the 1970s. It encourages interplay between humans and ferrets and has many critics up in arms, citing animal cruelty.

The premise is straightforward: Two ferrets are dropped into a competitor's trousers after the bottoms have been tied off. This precludes an easy escape and forces the wily critters to feverishly climb the competitor's legs in search of an exit. Yet, the contestants also tightly cinch their belts, just for added fun. As the ferrets become panicked, they begin to bite, a behavior that the elongated critters indulge in with alarming ferocity. And did we mention that participants must compete sans underwear? The winner is the player who ends up "keepin' 'em down" the longest, to use the sport's jargon. For many years, the ferret-legging record stood at just under one minute, but the current record is more than five hours!

A Woman on the Mound

Pioneers who brought new meaning to "ERA."

✳ ✳ ✳ ✳

T HE YEAR WAS 1931, and the New York Yankees were heading home from spring training, stopping to play exhibition games along the way. Facing the Chattanooga Lookouts, they must have been surprised when they saw that their mound opponent was a 17-year-old girl named Jackie Mitchell. Was she serious? She certainly thought she was, and she bent off several wicked curveballs, striking out Babe Ruth and Lou Gehrig consecutively on only seven pitches. A few days later, Commissioner Kenesaw Mountain Landis voided her contract, insisting that baseball was "too strenuous" for a woman. Disappointed and defeated, Mitchell began barnstorming with the House of David team. But she eventually grew tired of the circuit's sideshow antics and retired to work at her father's optometry office. Although she continued to play with local teams, Mitchell did not play professional baseball again.

It would be 66 years until another woman played in a professional game. Her name was Ila Borders, and she got her chance with the St. Paul Saints of the independent Northern League. Borders was signed by son-of-a-legend Mike Veeck, who claimed that it wasn't a publicity stunt. Unfortunately, Borders didn't pitch well for the Saints and was traded to the Duluth-Superior Dukes. In her third appearance of the 1998 season, she earned a 3–1 win with six scoreless innings, becoming the first woman to win a men's pro game.

Borders was praised for her "curve with a variety of breaks, a changeup that worked like a screwball," her control, and her "pitching smarts." However, her fastball wasn't strong enough for her to compete consistently. Although she had a 1.67 ERA in 15 appearances in 1999 for the Madison Black Wolf, she retired after the 2000 season.

Smooth Operator

Before Frank Joseph Zamboni Jr. invented his self-propelled ice-resurfacing machine, cleaning and clearing a sheet of ice was laborious, time-consuming, and inefficient.

✳   ✳   ✳   ✳

ZAMBONI, AN AMATEUR inventor and the owner/operator of the Iceland Skating Rink in Paramount, California, needed a new method for sweeping the surface of his ice. Artificial ice rinks were still a novelty when Zamboni opened his facility in 1939, and it took a team of three people to repair and resurface the ice after it was gouged by hundreds of skaters, a procedure that took up to 90 minutes. Zamboni came up with an idea for a motorized machine that could do all the necessary work—sweep, scrape, and saturate—and could be operated by one man. He stripped an old Jeep down to its nuts, bolts, and bare underbody chassis and placed a blade on the undercarriage to shave the ice smooth. He devised a device to sweep up the shavings and deposit the icy debris into a tank that melted the scrapings and used the water to rinse the rink. After several attempts and numerous prototypes, he perfected the "mechanical monster" and it became a tourist attraction in its own right.

In 1950, Sonja Henie, a three-time Olympic figure skating champion and one of Hollywood's top box-office attractions, was rehearsing her new Hollywood Ice Revue at the Iceland Skating Rink when she saw the revolutionary resurfacer at work. She commissioned Zamboni to build her a new model for her upcoming performances in Chicago. That endorsement allowed Zamboni to mass-produce the machines that now bear his name.

Zamboni does a slick business. As of 2011, the company has sold more than 9,000 ice-resurfacing machines.

Baseball Pitches

Ever wonder about the mechanics of the various types of pitches?
Here are the tricks (as they apply to right-handed pitching).

Q: How does one throw a curveball, and how does it move?

A: Several throwing grips begin with holding the ball in a peace sign (index and middle fingers making a V, thumb below). In contrast to the fastball, which is held with the fingertips, you cram the ball deep in the hand for a curve. Deliver the ball fully overhand with mostly top-to-bottom motion, snapping the hand as though pounding a nail with a hammer. Let the ball squirt out. If done correctly, the ball will arc sharply down and to the pitcher's left. Good curveball pitchers produce dozens of annoying little curveball variants, which is important because a good hitter can see the curve coming.

Q: How about a slider?

A: Very different. Start with the peace-sign grip, but put the fingers together like a Cub Scout salute with the thumb alongside the ball rather than below it. Cock the wrist right. Instead of extending your arm to throw, pull the ball back behind your ear the way a quarterback throws a football. As you throw, cut your fingers away viciously to the right, creating a fierce sidespin. The ball should curve away left. A great slider, thrown at the batter, breaks across the plate for a strike while the batter jackknifes away from it. Now you see why it's a hated pitch. Surgeons are familiar with the slider—its effect on the elbow can be devastating. Kids should avoid throwing it.

Q: What about the famous splitter?

A: It's also known as the split-fingered fastball. Put the ball in the peace-sign fastball grip, but curve the fingers down the sides of the ball to make a kind of a wilted peace sign. That crams the ball between those fingers, splitting them wide apart. Now throw your normal fastball, flicking your wrist down a bit.

Done right, it looks like a fastball and drops slightly just before the dish. Good hitters can hit anything they can pick up, and the splitter is hard to see coming.

Q: Many Little League coaches say that the ability to change speeds is the most important pitching tool. How is a changeup thrown?

A: The changeup can be thrown many ways, but here's a mainstream approach. Add your ring finger to the peace-sign grip, thumb under the ball, but don't curl your fingers to enclose it. Throw it with all three fingers extended in a Boy Scout salute, and aim at the catcher's head (some pitchers have to aim at the umpire's head). Throw it like a fastball but with more downward motion. It should look like a fastball, but it should arrive late and tend to fall off at the end (the reason you aim high). To upset a hitter's timing, good pitchers use different speeds.

Q: How do you chuck the knuck?

A: The knuckleball is difficult, but it's so easy on the arm that its masters can pitch professional ball into their late 40s. Experiment with grips; no two are alike. Start with the Boy Scout salute changeup grip previously described, but curl the three fingers back on themselves so the tips dig into the ball (align them behind a lace). Stiffen your wrist and throw the ball, flicking your fingers forward. The idea is to flick just hard enough to defeat the ball's natural backspin so that it hardly spins at all, letting wind currents cause the ball to float around. Catchers like watching the batter's head bob up and down trying to follow it. You haven't lived until you've watched someone try to bunt the knuck, moving his bat up and down.

* **Fastballs:** four-seam, two-seam, cutter, splitter, forkball, sinker
* **Breaking balls:** curveball, knuckle curve, slider, screwball
* **Changeups:** changeup, palmball, circle changeup
* **Others:** knuckleball, Eephus pitch, spitball, gyroball, Shuuto

Old-school Olympics

Imagine attending a sporting event where blood and broken limbs are the norm. It's hot out but water is scarce; the food is overpriced and lousy. Motels are few, pricey, and crummy; almost everyone has to camp out. Your bleacher seat feels like freshly heated limestone. Forty thousand drunken, screaming savages surround you.

✳ ✳ ✳ ✳

N O, YOU AREN'T at a modern college football game—you're at the ancient Greek Olympic Games! Millennia later, nations will suspend the Games in wartime; for these Olympics, Greek nations would (for the most part) suspend wartime.

What, When, Where, Why

Olympia was a scenic religious sanctuary in the boonies of western Greece. The nearest town was tiny Elis, 40 miles away. According to chroniclers, the ancient Olympic Games started in 776 B.C. That's about a century after Elijah and Jezebel's biblical difference of opinion. Rome wasn't yet built, and the Assyrian Empire ruled the Near East. Greece's fractious city-states waged constant political and military battles. In 776, with disease and strife even worse than usual in Greece, King Iphitos of Elis consulted the Delphic Oracle. She said, roughly translated: "Greece is cursed. Hold athletics at Olympia to lift the curse."

"Done deal," said the king. Greek legend spoke of games of old held at Olympia in honor of Zeus, occurring perhaps every four or five years, so Iphitos cleared some land at Olympia and held a footrace. The plague soon petered out.

Play It Again, Iphitos

It's unclear why they decided to repeat the Olympics every four years; that was probably the most prevalent version of the ancient tradition. Likely, they lacked the resources to do it more often. Whatever the reason, the Games became Elis's reason for being.

Its people spent the intervening years preparing for the next Olympiad. Given the amount of feasting and drinking that occurred at the Games, they probably needed that long to recover.

Over time, the event was extended to five days. Any male Greek athlete could compete. Winners became stars, with fringe benefits to match. There were no silver or bronze medals; losers slunk away in shame.

Olympia soon became the site of a building boom, as a great arena, shrines, and training facilities were constructed. Historians believe that 40,000-plus people would converge on Olympia to see events ranging from chariot racing to track-and-field to hand-to-hand combat.

Except for the boxers, wrestlers, charioteers, and *pankratists* (freestyle fighters)—who were frequently maimed or killed—the athletes had it easy compared to the attendees. The climate was hot and sticky, without even a permanent water source for most of the ancient Olympic era (until someone finally built an aqueduct); deaths from sunstroke were common. Most people had no way to bathe, so everyone stank, and disease ran rampant.

Married women couldn't attend, except for female charioteers. The only other exception was a priestess of Demeter, who had her own special seat. Athletes competed nude, so there was no chance that a woman could infiltrate as a competitor. Unmarried girls and women, especially prostitutes, were welcome.

An Endurance Event for All

For five days every four years, Olympia combined the features of a carnival, a track meet, martial arts, a banquet, racing, and a brothel into a scene of organized bedlam. The Olympics weren't merely to be experienced—they were to be survived.

In A.D. 393, Roman Emperor Theodosius I banned all pagan ceremonies. Since the Games' central ritual was a big sacrifice to Zeus, this huge heathen debauch clearly had to go. The party was over. It wouldn't start again until 1859.

The Origins of the Game

The Abner Doubleday Fan Club isn't going to like this.

❊   ❊   ❊   ❊

IT WAS LONG believed that Abner Doubleday invented baseball in 1839. While we now know this isn't true, we still don't know exactly how baseball came about. Games involving sticks and balls go back thousands of years. They've been traced to the Mayans in the Western Hemisphere and to Egypt at the time of the Pharaohs. There are historical references to Greeks, Chinese, and Vikings "playing ball." And a woodcut from 14th-century France shows what seem to be a batter, pitcher, and fielders.

Starting with Stoolball

By the 18th century, references to "baseball" were appearing in British publications. In an 1801 book titled *The Sports and Pastimes of the People of England,* Joseph Strutt claimed that baseball-like games could be traced back to the 14th century and that baseball was a descendant of a British game called "stoolball." The earliest known reference to stoolball is in a 1330 poem by William Pagula, who recommended to priests that the game be forbidden within churchyards.

In stoolball (which is still played in England, mostly by women), a batter stands before a target—perhaps an upturned stool— while another player pitches a ball to him or her. If the batter hits the ball (with a bat or his or her hand) and it is caught by a fielder, the batter is out. Ditto if the pitched ball hits a stool leg.

The Game Evolves

It seems that stoolball eventually split into two different styles. One became English "base-ball," which turned into "rounders" in England but evolved into "town ball" when it reached the United States; the other side of stoolball turned into cricket. From town ball came the two styles that dominated baseball's development: the Massachusetts Game and the New York

Game. The former had no foul or fair territory; runners were put out by being hit with a thrown ball when off the base ("soaking"), and as soon as one out was made, the teams switched sides. The latter established foul lines, and each team was given three "outs" per inning. Perhaps more significantly, soaking was eliminated in favor of the tag. The two versions coexisted in the first three decades of the 19th century, but when Manhattanites codified their rules in 1845, it became easier for more and more groups to play the New York style.

A book printed in France in 1810 laid out the rules for a bat/base/running game called "poison ball," in which there were two teams of eight to ten players, four bases (one called "home"), a pitcher, a batter, and fly-ball outs. Variations such as "Tip-cat" and "trap ball" were notable for how important the bat had become. It was no longer used merely to avoid hurting one's hand; it had become a real cudgel, used to swat the ball a long way.

The Knickerbocker Club

In the 1840s, Alexander Cartwright, a New York City engineer, was one of a group that met regularly to play baseball, and he may have been the mastermind behind formalizing the rules of the game. The group called themselves The Knickerbocker Club, and their constitution—which was enacted on September 23, 1845—led the way for the game we know today.

The Myth Begins

Although baseball's origins are murky, there's one thing we know for sure: Abner Doubleday did not invent it. Albert Spalding organized the Mills Commission in 1905 to search for a definitive American source for baseball. They "found" it in an ambiguous letter spun by a Cooperstown resident (who turned out to be insane). But Doubleday wasn't even in Cooperstown when he supposedly invented the game. Also, *The Boy's Own Book* presented the rules for a baseball-like game ten years before Doubleday's alleged "invention." Chances are, we'll never know for sure how baseball came to be the game it is today.

Wacky Sports Injuries

✳ **Ryan Klesko:** In 2004, this San Diego Padre was in the middle of pregame stretches when he jumped up for the singing of the national anthem and pulled an oblique/rib-cage muscle, which sidelined him for more than a week.

✳ **Freddie Fitzsimmons:** In 1927, New York Giants pitcher "Fat Freddie" Fitzsimmons was napping in a rocking chair when his pitching hand got caught under the chair and was crushed by his substantial girth. Surprisingly, he only missed three weeks of the season.

✳ **Clarence "Climax" Blethen:** Blethen wore false teeth, but he believed he looked more intimidating without them. During a 1923 game, the Red Sox pitcher had the teeth in his back pocket when he slid into second base. The chompers bit his backside and he had to be taken out of the game.

✳ **Chris Hanson:** During a publicity stunt for the Jacksonville Jaguars in 2003, a tree stump and ax were placed in the locker room to remind players to "keep chopping wood," or give it their all. Punter Chris Hanson took a swing and missed the stump, sinking the ax into his non-kicking foot. He missed the remainder of the season.

✳ **Lionel Simmons:** As a rookie for the Sacramento Kings, Simmons devoted hours to playing his Nintendo Game Boy. In fact, he spent so much time playing it that he missed a series during the 1991 season due to tendonitis in his wrist.

✳ **Sammy Sosa:** In May 2004, Sosa sneezed so hard that he injured his back, sidelining the Chicago Cubs' outfielder and precipitating one of the worst hitting slumps of his career.

✳ **Gus Frerotte:** In 1997, Washington Redskins quarterback Frerotte had to be treated for a concussion after he spiked the football and slammed his head into a foam-covered concrete wall while celebrating a touchdown.

✳ **Jaromir Jagr:** During a 2006 playoff game, New York Ranger Jagr threw a punch at an opposing player. Jagr missed, his fist slicing through the air so hard that he dislocated his shoulder. After the Rangers were eliminated from the playoffs, Jagr underwent surgery and continued his rehab during the next season.

✳ **Paulo Diogo:** After assisting on a goal in a 2004 match, newlywed soccer player Diogo celebrated by jumping on a perimeter fence. He accidentally caught his wedding ring on the wire, and when he jumped down, he tore off his finger. To make matters worse, the referee issued him a violation for excessive celebration.

✳ **Clint Barmes:** Rookie shortstop Barmes was sidelined from the Colorado Rockies lineup for nearly three months in 2005 after he broke his collarbone when he fell while carrying a slab of deer meat.

✳ **Darren Barnard:** In the late 1990s, British professional soccer player Barnard was sidelined for five months with knee ligament damage after he slipped in a puddle of his puppy's pee on the kitchen floor. The incident earned him the unfortunate nickname the "Whiz Kid."

✳ **Marty Cordova:** A fan of the bronzed look, Baltimore Orioles outfielder Cordova was a frequent user of tanning beds. However, he once fell asleep while catching some rays, resulting in major burns to his face and body that forced him to miss several games with the team.

✳ **Jamie Ainscough:** A rough-and-ready rugby player from Australia, Ainscough's arm became infected in 2002, and doctors feared they might need to amputate. But after closer inspection, physicians found the source of the infection: The tooth of an opponent had become lodged under his skin, unbeknownst to Ainscough, who had continued to play for weeks after the injury.

The Tennis Star Who Wasn't

No one can accuse Sports Illustrated *of not having a sense of humor. For laughs, it invented an attractive, camera-ready tennis star to rival Anna Kournikova. Her name was Simonya Popova.*

✳ ✳ ✳ ✳

Sports Satire

A SEPTEMBER 2002 ISSUE OF *Sports Illustrated* told of a 17-year-old tennis force named Simonya Popova, who hailed from Uzbekistan and was a media dream: 6′1″, brilliant at the game, busty, and blonde. But she wouldn't be competing in the U.S. Open—her father forbade it until she turned 18.

The magazine rhapsodized as it compared her to other tennis beauties. Editors claimed that, unlike Popova, all of those women were public-relations disappointments to both the Women's Tennis Association (WTA) and sports marketing firms because they avoided media attention to concentrate on playing a good game. As a result, U.S. tennis boiled down to Venus and Serena Williams, trailed by a pack of hopefuls and wannabes. The article concluded with this line: "If only she existed."

Just Kidding!

Popova *was* too good to be true. The article was fiction, and her confident gaze simply showcased someone's digital artistry. Some people got it, but many didn't, including the media. They bombarded the WTA with calls, wanting to know more about this phenom. The article emphasized what many thought—the WTA was desperate for the next young tennis beauty. WTA spokesperson Chris DeMaria called the story "misleading" and "disrespectful to the great players we have." He added, "We're a hot sport right now and we've never had to rely on good looks."

Sports Illustrated claimed that it was all in grand fun. It hardly needed to add that it was indulging in puckish social commentary on the sexualization of women's tennis.

The Catcher Was a Spy

When it comes to character assessments, you gotta listen to Casey Stengel, who claimed Moe Berg was "the strangest man ever to put on a baseball uniform." But Berg wasn't just strange in a baseball uniform, he was strange and mysterious in many ways—some of them deliberate.

✳ ✳ ✳ ✳

MOE BERG LIVED a life shrouded in mystery and marked by contradictions. He played alongside Babe Ruth, Lefty Grove, Jimmie Foxx, and Ted Williams; he moved in the company of Norman Rockefeller, Albert Einstein, and international diplomats; and yet he was often described as a loner. He was well-liked by teammates but preferred to travel by himself. He never married, and he made few close friends.

"The Brainiest Guy in Baseball"

Moe was a bright kid from the beginning, with a special fondness for baseball. As the starting shortstop for Princeton University, where he majored in modern languages, Moe was a star. He was fond of communicating with his second baseman in Latin, leaving opposing baserunners scratching their heads.

He broke into the majors in 1923 as a shortstop with the Brooklyn Robins (later the Dodgers). He converted to catcher and spent time with the White Sox, Senators, Indians, and Red Sox during his career. A slow runner and a poor fielder, Berg nevertheless eked out a 15-season big-league career. Pitchers loved him behind the plate: They praised his intelligence and loved his strong, accurate arm. And while he once went 117 games without an error, he rarely nudged his batting average much past .250. His weak bat often kept him on the bench and led sports writers to note, "Moe Berg can speak 12 languages flawlessly and can hit in none." He was, however, a favorite of sportswriters, many of whom considered him "the brainiest guy in baseball."

He earned his law degree from Columbia University by attending classes in the off-seasons and even during spring training and partial seasons with the White Sox. When Berg was signed by the Washington Senators in 1932, his life underwent a sudden change. In Washington, Berg became a society darling, delighting the glitterati with his knowledge and wit. Certainly it was during his Washington years that he made the contacts that would serve him in his espionage career.

Time in Tokyo and on TV

Berg first raised eyebrows in the intelligence community at the start of World War II, when he shared home movies of Tokyo's shipyards, factories, and military sites that he had secretly filmed while on a baseball trip in 1934. While barnstorming through Japan along with Ruth, Foxx, and Lou Gehrig, Berg delighted Japanese audiences with his fluency in their language and familiarity with their culture. He even addressed the Japanese parliament. But one day, he skipped the team's game and went to visit a Tokyo hospital, the highest building in the city. He sneaked up to the roof and filmed Tokyo Harbor. Some say those photos were used by the U.S. military as it planned an attack on Tokyo eight years later. Berg maintained that he had not been sent to Tokyo on a formal assignment, and that he had acted on his own initiative to take the film and offer it to the U.S. government upon his return. Whether or not that was the case, Berg's undercover career had begun.

On February 21, 1939, Berg made the first of several appearances on the radio quiz show *Information, Please!* He was an immense hit, correctly answering nearly every question he was asked. Commissioner Kenesaw Mountain Landis was so proud of how intelligent and well-read the second-string catcher was that he told him, "Berg, in just 30 minutes you did more for baseball than I've done the entire time I've been commissioner." But Berg's baseball time was winding down, as 1939 was his last season.

Secret Agent Man

Berg's intellect and elusive lifestyle were ideal for a post-baseball career as a spy. He was recruited by the Office of Strategic Services (the predecessor of the CIA) in 1943 and served in several capacities. He toured 20 countries in Latin America early in World War II, allegedly on a mission to bolster the morale of soldiers there. But he was really trying to determine how much the Latin countries could help the U.S. war effort.

His most important mission for the OSS was to gather information on Germany's progress in developing an atomic bomb. He worked undercover in Italy and Switzerland and reported information to the States. One of his more daring assignments was a visit to Zurich, Switzerland, in December 1944, where he attended a lecture by German nuclear physicist Werner Heisenberg. If Heisenberg indicated that the Germans were close to developing nukes, Berg had been directed to assassinate the scientist. Luckily for Heisenberg, Berg determined that German nuclear capability was not yet within the danger range.

Life After the War

On October 10, 1945, Berg was awarded the Medal of Freedom (now the Presidential Medal of Freedom) but turned it down without explanation. (After his death, his sister accepted it on his behalf.)

After the war, he was recruited by the CIA; it is said that his is the only baseball card to be found in CIA headquarters. After his CIA career ended, Berg never worked again. He was often approached to write his memoir. When he agreed, in 1960 or so, the publisher hired a writer to provide assistance. Berg quit the project in a fury when the writer indicated that he thought Berg was Moe Howard, founder of the The Three Stooges. But his unusual career turns were later immortalized in the Nicholas Dawidoff book *The Catcher Was a Spy*. At age 70, Berg fell and injured himself; he later died in the hospital. His last words were to ask a nurse, "What did the Mets do today?"

Unlikely Super Bowl Heroes

Most of the gridiron gladiators who grasped glory in the Super Bowl brought a storied and successful pedigree into the decisive game, with numerous accolades and accomplishments on their résumés before they stepped into the sport's brightest spotlight. Most, but not all.

✳   ✳   ✳   ✳

From the Supermarket to the Super Bowl—Kurt Warner: Before he marched the St. Louis Rams to victory in Super Bowl XXXIV, Kurt Warner was barely a name in his own household. A backup throughout much of his collegiate career at Northern Iowa, Warner was signed by the Green Bay Packers in 1994 but was handed a pink slip before the season even began. Out of prospects and cash, he took a job at a local grocery store in Cedar Falls, Iowa, for $5.15 an hour. Undaunted by his early failures, Warner began to slowly climb the football ladder. He rode the buses in the Arena Football League and played before an average of 15,000 fans with the Amsterdam Admirals of NFL Europe.

After returning to the United States, Warner landed a job as the backup quarterback of the St. Louis Rams in 1998, which meant that he stood on the sideline and held a clipboard on game days and took most of the hits during practices. An injury to starter Trent Green early in the 1999 campaign put the ball in Warner's hands, and he ran, threw, and excelled with it. *Sports Illustrated* immediately realized that he was unrecognizable and put him on the cover of their October 18 edition with the caption, "Who Is this Guy?" With Warner calling the signals, the Rams became known as "The Greatest Show on Turf" and rode Warner's arm and calm grace under pressure all the way to the Super Bowl, where they dispatched the Tennessee Titans by a 23–16 score. Warner set a litany of Super Bowl records—including most passing yards and most pass attempts without an interception—and was named the game's MVP.

Hung On, Hungover—Max McGee: In January 1967, McGee—who'd caught only four passes all season—wasn't expected to play for the Green Bay Packers in the inaugural Super Bowl, so he spent the night (and most of the morning) throwing back more than a few cold ones. As fate would have it, Boyd Dowler, the Packers' top receiver, was injured on the third play of the game, forcing McGee into the fray. After borrowing a helmet—he hadn't bothered to bring his own—McGee caught nine passes, including two touchdown tosses, helping the Packers defeat the Kansas City Chiefs, 35–10.

Rookie to the Rescue—Tim Smith: After rushing for only 126 yards in the regular season, Washington Redskins rookie Tim Smith set a Super Bowl record by gaining 204 yards on the ground, helping Washington down the Denver Broncos, 42–10, in Super Bowl XXII.

Up Right, Right Up—Jim O'Brien: After connecting on only 19 of 34 field goal attempts during the season, the Baltimore Colts' Jim O'Brien was hardly a lock to deliver the championship-winning boot when he lined up for a 32-yard try in the waning seconds of a 13–13 tie in Super Bowl V. An early indication that the young kicker was nervous came when he tried to check the wind direction by attempting to pull up a few strands of grass from the artificial turf. Still, O'Brien popped the pigskin through the middle of the uprights to give Baltimore a 16–13 victory over Dallas.

❋ Since Super Bowl XXXIV in 2000, game footballs have been marked with synthetic DNA to prevent sports-memorabilia fraud. Souvenirs from the 2000 Summer Olympics were marked with human DNA in the ink.

❋ When the Japanese attacked Pearl Harbor, many U.S. politicians and military officers had to be paged over the public-address system at Griffith Field, where the Washington Redskins were playing the Philadelphia Eagles, and at New York's Polo Grounds, where the (football) Giants faced the Dodgers.

Double No-No

With a little luck, Johnny Vander Meer made lightning strike twice.

✳ ✳ ✳ ✳

A T EBBETS FIELD in Brooklyn on June 15, 1938, Johnny
Vander Meer reached the stars with his unprecedented—
and unmatched—second consecutive no-hitter.

It's a record that is highly unlikely to be tied and almost certain
never to be broken: no-hitters in back-to-back performances.
Cincinnati's Johnny Vander Meer was just 23 years old when
he faced the Boston Bees on June 11 and notched a 3–0 victory
without allowing a hit. It was the 48th no-hitter since the
National League had been founded in 1876. Four days later,
he faced the Brooklyn Dodgers at Ebbets Field. The game was
already historic—it was the first night game ever played there.
In fact, some claim that the poor
lighting had as much to do with
Vandy's success as his furious fastball did. Vander Meer was a notoriously wild, hard-throwing lefty;
sometimes the ball went where it
was supposed to, but plenty of other
times, it didn't.

In his first no-hitter, on June 11, Vandy
walked only three batters. Things were different
on the night of June 15. He had already given free passes to
five Dodgers when the ninth inning began. One out came
quickly—and then the torture began. He walked three men
in a row to load the bases. A ground ball forced one man out
at home. The next batter was the veteran Leo Durocher. On a
2–2 count, Durocher tagged a Texas Leaguer into center that
Harry Craft nabbed at the knees. Thanks to his own skill and
the help of his teammates, Vandy had claimed a new spot in the
record books—one that will probably never be broken.

Pickled Eels, Dirty Shirts, and Lucky-Sounding Wood

No one is more superstitious than a player on a winning streak.

✳ ✳ ✳ ✳

WHAT MAKES A baseball team successful? Is it a player who can slug dingers over the fence, a fielder who snags the ball with ease, a pitcher who blazes fastballs over the plate? Or is there something more? Many players are firm believers in luck, and they aren't satisfied to let it come to them. Instead they participate in elaborate rituals designed to keep their luck flowing freely.

On-Field Rituals

Many well-known baseball superstitions are practiced by everyone from Little League players to Hall of Famers. Step on one of the bases as you leave the field between innings, or spit on your hand before picking up your bat for good luck. Comment on a pitcher's performance when it looks like he might throw a no-hitter and you're guaranteed to jinx it. These traditions are as much a part of baseball as chewing tobacco, batboys, and the seventh-inning stretch. Said Ralph Kiner, who never stepped on the foul lines, "It didn't help or hurt me. I just didn't want to take any chances."

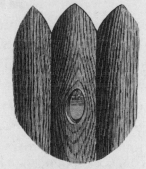

But other players don't court luck as casually as Kiner. Instead they perform intricate rituals. Some may have evolved from those well-known superstitions, but others are, well, out of left field. For example, Hall of Famer Clark Griffith thought shutouts were unlucky and went through his first 127 starts without one (in the dead-ball era). He once ordered rookie Frank Chance to drop a pop-up to allow a score and break up his own shutout in a one-sided game. He finally completed one in 1897 and then went on to tie Cy Young for the most shutouts in two different leagues in 1900 and 1901. Not so unlucky after all.

Eat, Drink, and Be Lucky

Food can be the key to ensuring a good game. Baltimore Orioles pitcher Jim Palmer had to eat pancakes before every start, while Wade Boggs stuck to chicken. Pitcher Turk Wendell would chew licorice (and brush his teeth) between innings. After Lou Gehrig's mom sent a jar of pickled eels to the clubhouse, the Yankees went on a hitting streak—and an eel streak. To ensure that the streak continued, each team member would have at least a few bites of eel before every game.

For Denny McLain of the Tigers, Pepsi-Cola was a magic elixir. McLain would sometimes drink more than 20 bottles a day, even taking a bottle to bed so he could sip it when he woke up during the night. Despite drinking himself to 31 victories during the 1968 season, McLain was never appointed official spokesperson for Pepsi.

Magical Equipment

Bats receive royal treatment because of the luck they hold. Every baseball fan knows you never lend your bat to another player, and if you need an extra shot of luck, you can always try sleeping with your bat. Some players would make an annual trip to Hillerich & Bradsby Company in Louisville, Kentucky, to ensure that their bat was lucky. Both Ted Williams and Al Simmons would roam the warehouse to pick out the perfect piece of wood for their bat. Williams liked a narrow grain while Simmons swore by the widest. Apparently they were both on to something: Their lifetime batting averages were .344 and .334, respectively.

For Hugh Duffy, whose career average was .324, the wood had to sound just right. Duffy would bounce pieces of wood off concrete, listening for the sound of success only he could discern. Orlando Cepeda wasn't that particular about wood or sound, just as long as his bats were "productive"; he would only use bats that had never made an out. Len Dykstra felt the same way about his batting gloves, throwing them away if he failed to get a hit in a single at-bat.

Since Mark "The Bird" Fidrych was a pitcher, it seems logical that he centered his superstitions around the pitcher's mound. In addition to playing with the dirt, Fidrych would talk to the ball before a pitch.

Don't Go Changing...

Routine is the driving force behind many players' superstitions. If you have good luck, you must do everything exactly the same; if you have bad luck, change, change, change! Many of the routines involve clothing. Rafael Palmeiro wore the same T-shirt under his jersey every day if he was hitting well. Charlie Kerfeld of the Houston Astros was another member of the T-shirt club. He wore the same George Jetson shirt the entire 1986 season and finished with 11 wins and 2 losses. Astros second baseman Craig Biggio kept his luck in his cap; he didn't wash it all season.

For some, their superstitions were their undoing. Blue Jays catcher Rick Cerone wore long johns under his uniform for an April game and promptly began a hitting streak. He refused to abandon the thermals responsible for his good luck even as the season wore on and the temperature soared. Although he kept his thermals, he didn't keep his luck: His batting average that season was a career-low .239. Cerone could have learned a thing or two from Milwaukee Brewers pitcher Pete Vukovich, who stopped the final game of the 1982 World Series so he could change a shoe. (Mismatched shoes had brought him luck in the past.) Perhaps he should have changed both shoes: The Brewers lost to the St. Louis Cardinals.

Sometimes the rituals work, and sometimes they don't. But that doesn't stop players from believing in the old ones and creating new ones. After all, you never know...

Hey Batter, Batter...

"Is that the best game you ever pitched?"

—QUESTION POSED BY AN ANONYMOUS REPORTER TO DON LARSEN
FOLLOWING HIS PERFECT GAME IN THE 1956 WORLD SERIES

"It was a cross between a screwball and a changeup. It was a screw-up."

—CUBS RELIEVER BOB PATTERSON ON A LOUSY PITCH HIT OFF HIM BY CINCINNATI'S BARRY LARKIN
FOR A GAME-WINNING HOMER

"Well, you can't win them all."

—CONNIE MACK ON HIS 1916 PHILADELPHIA A'S, WHO WENT 36–117

"I believe the sale of Babe Ruth will ultimately strengthen the team."

—RED SOX OWNER HARRY FRAZEE, WHO IN JANUARY 1920 SOLD THE GREATEST PLAYER IN
BASEBALL HISTORY TO THE NEW YORK YANKEES

"I never questioned the integrity of an umpire. Their eyesight, yes."

—LEO DUROCHER, *NICE GUYS FINISH LAST*

"Allen Sutton Sothoron pitched his initials off today."

—LEAD IN ST. LOUIS NEWSPAPER, 1920S, AS QUOTED IN *THE PITCHER*

"With this batting slump I'm in, I was so happy to hit a double that I did a tap dance on second base. They tagged me between taps."

—FRENCHY BORDAGARAY OF THE BROOKLYN DODGERS

"Age is a question of mind over matter. If you don't mind, it doesn't matter."

—SATCHEL PAIGE, WHO WAS 42 DURING HIS ROOKIE SEASON IN THE MLB

"All I want out of life is that when I walk down the street, people will say, 'There goes the greatest hitter who ever lived.'"

—RED SOX SLUGGER TED WILLIAMS

"A couple of years ago, they told me I was too young to be President and you were too old to be playing baseball. But we fooled them."

—JOHN F. KENNEDY (AGE 45) TO STAN MUSIAL (AGE 41) AT THE 1962 ALL-STAR GAME

"The only man who could have caught it, hit it."

—SPORTSWRITER BOB STEVENS AFTER WILLIE MAYS BELTED A DRIVE
OVER THE CENTER FIELDER'S HEAD

"Throw strikes. The plate don't move."

—SATCHEL PAIGE, *STRIKEOUT: A CELEBRATION OF THE ART OF PITCHING*

"Maybe a pitcher's first strikeout is like your first kiss—they say you never forget it."

—BOB FELLER, *NOW PITCHING, BOB FELLER*

"Well, Mr. Barrow, Lou Gehrig is badly underpaid."

—JOE DIMAGGIO'S RESPONSE WHEN YANKEE GENERAL MANAGER ED BARROW TOLD JOE THAT
HIS 1937 CONTRACT DEMAND OF $45,000 FOLLOWING HIS ROOKIE SEASON WAS HIGHER THAN
WHAT GEHRIG WAS MAKING AFTER 15 YEARS

"So I swing, and would you believe it's a bases-loaded home run? I really sped around those bases to get back to the dugout and those candy bars in a hurry."

—RON SANTO ON HITTING WHILE HAVING A DIABETIC REACTION

"Many baseball fans look at an umpire as a sort of necessary evil to the luxury of baseball, like the odor that follows an automobile."

—CHRISTY MATHEWSON

"To do what he did has got to be the most tremendous thing I've ever seen in sports."

—PEE WEE REESE ON THE ROOKIE SEASON OF DODGER TEAMMATE
JACKIE ROBINSON, *THE BOYS OF SUMMER*

"One of the secrets of the Babe's greatness was that he never lost any of his enthusiasm for playing ball, and especially for hitting home runs. To him a homer was a homer, whether he hit it in a regular game, a World Series game, or an exhibition game."

—SPORTSWRITER FRANK GRAHAM ON BABE RUTH, *THE NEW YORK YANKEES*

"Baseball's unique possession, the real source of our strength, is the fan's memory of the times his daddy took him to the game to see the great players of his youth."

—BILL VEECK, *THE HUSTLER'S HANDBOOK*

Moonshine and NASCAR Are Kin?

White lightning, hooch, mountain dew, hillbilly pop—whatever you called it, it was the lifeblood of the South. It was also illegal, but it was the juice that jump-started NASCAR.

✳ ✳ ✳ ✳

Model "T" for "Tripper"

HENRY FORD WAS an adamant teetotaler, but without him, the South's illicit moonshine business could never have been so successful. The mass production of Ford's Model T and later his V-8 coincided with the dawn of 1920s Prohibition. Southerners demanded bootleg booze, and "whiskey trippers" were equipped to deliver as many as 100 gallons a night at 30 cents a gallon. To crack down was futile: Local cops often had kinfolk in the biz and besides, they appreciated "corn likker" as much as anybody else.

Despite the end of Prohibition in 1933, moonshine was still in demand. Alcohol was legal but only from legitimate, tax-paying distilleries. Selling untaxed liquor, or moonshine, was illegal. Moonshine was also cheaper and stronger than the legal stuff.

Federal agents were sent in to hunt down the shiners, who knew that speed was their best defense. For an extra leg up, shiners sought the help of "whiskey mechanics" who souped up V-8s by sawing cylinders to boost horsepower and tweaked the rear suspensions with heavy-duty springs and steel wedges to keep liquor bottles in place while driving on twisty mountain roads.

Shiners vs. Shiners

Informal races soon cropped up among the shiners to see who had the fastest cars and meanest driving skills. Local entrepreneurs started cashing in by sponsoring races at horse tracks and fairgrounds across the South.

The appeal was clear: Until then, car racing was the domain of the American Automobile Association (AAA) and the wealthy

Northern elite. But stock-car racing's everyman appeal struck driver and race promoter Bill France as a golden opportunity.

NASCAR Is Born

Southern stock-car racing picked up momentum after World War II, when several regional organizations formed. AAA had been flirting with stock-car racing for years, and it decided it wanted to control the action, but France wanted races for regular working folk.

In 1947, representatives from southern stock-car groups met and formed the National Association of Stock Car Auto Racing (NASCAR), named France president, and agreed to uniform rules. To gain respect as a legitimate organization, France felt that NASCAR should be distanced from its moonshine roots. That was easier said than done, since most of his drivers had cut their teeth running whiskey. So he made the races "strictly stock," meaning souped-up cars were not allowed. In February 1948, several drivers showed up at the first race with no car to drive, looking for willing spectators to hand over their keys.

Glenn Dunaway finished first in a borrowed Ford coupe—that is, until a post-race inspection showed signs of "bootlegger souping," and he was disqualified.

Keep Shining

Many of NASCAR's early stars had roots to moonshining. Legendary Junior Johnson was eight years old when he started running whiskey in 1939; he gave it up in the late '50s after a yearlong prison stint interrupted his racing career. Curtis Turner, another great, claimed he started running hooch at age ten.

NASCAR is now a multibillion-dollar industry with a family-friendly image. Yet glimmers of its whiskey-soaked past still "shine" through: In March 2009, retired driver Carl Dean Combs was arrested in North Carolina for making and selling hooch.

Who's Too Old for the Olympics?

Think of the average Olympic athlete and the following images likely come to mind: physical perfection, drive, determination—and youth. Not necessarily. It could be just a matter of time before the AARP holds its own Olympic trials.

✳ ✳ ✳ ✳

Oscar Swahn: Swedish shooter Swahn participated in three Olympic Games. At age 60, he won two gold medals and a bronze at his first Olympics, which took place in London in 1908. Four years later, at the Sweden Games, he won a gold in the single-shot running-deer team, making him the world's oldest gold medalist. Swahn returned to the Olympics in 1920 at age 72 and managed to win a silver medal in the double-shot running-deer competition.

Anders Haugen: Even at the ripe age of 72, Swahn is not the oldest person to have won an Olympic medal. At the first Winter Olympic Games in Chamonix, France (1924), U.S. ski jumper Anders Haugen placed fourth with a score of 17.916 points. The third-place finisher, Norway's Thorlief Haug, received a score of 18.000 points. Fifty years later, a sports historian determined that Haug's score had been miscalculated and that he should have finished behind Haugen. At a special ceremony in Oslo, Haugen was finally awarded the bronze medal when he was 83 years old, making him the "eldest" recipient of an Olympic medal and the only American to ever win a medal in the ski-jump event.

Hilde Pedersen: When Norway's Pedersen took home the bronze in the ten-kilometer cross-country-skiing event at the 2006 Turin Winter Olympics, she became the oldest woman to win a Winter Games Olympic medal. It was an impressive achievement for the 41-year-old, but as she and other "older" competitors have proved in the past, age is no barrier to claiming an Olympic medal.

DL? I Don't Think So

These players overcame physical limitations to leave their marks on the game.

✳   ✳   ✳   ✳

BASEBALL IS A game in which pitchers work to exploit hitters' weaknesses, hitters seek vulnerability in fielders, and physical shortcomings are routinely punished. Here are profiles of some players who entered the game with physical limitations and worked to overcome them in an unforgiving sport.

William Ellsworth "Dummy" Hoy

The day that William Hoy made his major-league debut in 1888, his Washington Nationals teammates arrived to find a handwritten note posted on the clubhouse wall.

"Being totally deaf . . . and some of my teammates being unacquainted with my play, I think it is timely to bring about an understanding between myself, the left fielder, the shortstop, and the right fielder," the note began. Hoy's note explained that, as center fielder, his teammates should listen for him to yell, indicating that he would make the play on a fly ball. Though teammates would describe Hoy's yell as more of a squeak, they understood him perfectly.

Rendered deaf and mute as the result of a childhood bout with meningitis, Hoy enjoyed a 14-year career in which he amassed 2,044 hits, 1,426 runs, 40 home runs, 594 stolen bases, and 726 RBI with six teams. As his teammates quickly learned, he was a magnificent fielder and a fine hitter (with a .287 career batting average)—one of baseball's first stars to overcome a physical handicap.

On May 16, 1902, Hoy came to bat against Luther "Dummy" Taylor of the Giants in the first matchup in baseball history of a deaf pitcher versus a deaf hitter. The opponents exchanged greetings in sign language, and then Hoy singled.

Mordecai "Three-Finger" Brown

Mordecai Brown turned childhood tragedy into professional success. Growing up on an Indiana farm in the late 1800s, Brown lost his right index finger just below the knuckle when he got his hand caught in a corn grinder. With his hand still in a cast, he fell and broke the pinky and middle fingers on the same hand. Those fingers grew bent and misshapen.

Brown's deformed hand was an impediment to gripping a baseball, but through ingenuity and practice he developed a unique grip that produced baffling pitches with incredible movement. Not blessed with great velocity, Brown relied on movement, smarts, and remarkable control.

In an era of low scoring, Brown stood out as a run preventer: He consistently ranked near the top of his league in fewest walks per innings pitched. Over a 14-year Hall of Fame career, he won 239 games and fashioned an ERA of 2.06—sixth-best in the history of baseball—and led the Chicago Cubs to two world championships and four National League pennants.

Giants ace Christy Mathewson was one of Brown's toughest opponents. The Hall of Famers faced one another 25 times, with Brown winning 13 of the games, including a stretch of nine victories in a row. After one such game, exasperated Giants manager John McGraw reportedly examined Brown's hand and remarked, "I'm going to have the first finger on the throwing hands of every one of my damned pitchers cut off tomorrow."

Pete Gray

An accidental fall off a wagon when he was six years old cost Pete Gray nearly his entire right arm. As a result, he developed a fierce determination to overcome this obstacle.

Gray channeled that determination toward a career in baseball. He taught himself to make solid contact swinging a bat with only his left arm, and he used a customized glove that he would deftly tuck beneath his stub when throwing. He worked his way from semipro ball through the minor leagues, and in 1945, with many of the best major-leaguers off to war, Gray was signed by the St. Louis Browns, who saw the addition of a disabled player as both a gate attraction and a message to wounded veterans. Gray understood this and made the most of the opportunity, hitting .218 and striking out just 11 times over 77 games and 253 plate appearances.

Jim Abbott

"I wanted to be like Nolan Ryan. I didn't want to be like Pete Gray."

Such were the boyhood dreams of Jim Abbott, a native of Flint, Michigan, who sought a baseball career despite having a right arm that ended just above the wrist. Abbott didn't want to be accepted into the majors as an oddity, the reason many believe that Gray was signed during the war-depleted 1945 season. Abbott yearned to succeed on his talent alone, and in the end, he would do so—but not without becoming a reluctant hero.

Born in 1967—just as his own hero, Nolan Ryan, was breaking in with the Mets—Abbott taught himself to transfer his glove from his left arm to his right by throwing against a brick wall. He later said that his missing hand "wasn't really an issue when [he] was a kid." By becoming an expert fielder, he kept hitters from taking advantage of him by using the logical tactic of bunting against him. During his college career he racked up a 26–8 record at the University of Michigan, took home the 1987 Sullivan Award as the nation's best amateur athlete, and won a gold medal in the 1988 Olympics.

California's first pick in the 1988 draft, Abbott became one of a handful of players in history to completely bypass the minor leagues, and then overcame a media circus to go 12–12 as a rookie with the 1989 Anaheim Angels. More great moments would follow—including a 1993 no-hitter for the Yankees—and it wasn't long before he had gotten his wish. He was simply Jim Abbott: pitcher.

Chucking Cow Chips

Wisconsin has its chocolate chips and potato chips—and then there are the cow chips.

✳   ✳   ✳   ✳

COW CHIPS ARE the waste products of a cow's herbivorous diet. You know...cow droppings. Wisconsin is "the Dairy State," so it's no surprise that there are a lot of cows around. At the Wisconsin Cow Chip Throw in Prairie du Sac, cow chips are front and center. There are a lot of activities to please the family at this festival, but the chip-throwing contest is the highlight.

Whatever Gave Them that Idea?

For the early settlers of Wisconsin, winters were cold and lonely. When trees were scarce—or too big to cut down—the pioneers looked for alternate heating fuel sources. They discovered that cow chips (chock full of grasses and minerals) could be used for this purpose. After the chips dried, they were odorless, and they burned with intense heat and left no sooty residue. They were so valuable to the settlers that they were even used to trade for food and other necessities.

Wisconsinites don't use cow chips to heat their homes these days, but each year thousands of people gather to celebrate them. There's plenty to do if you're not into chucking chips—just look for the Trojan Cow. It's 20 feet tall and kids can climb inside it. There's also live music, good food, races, and a parade.

But let's get to the main event. Contestants select their chips from the Meadow Muffin Wagon. The best ones are about six inches around and fairly dense. Chips are usually tossed like a flying disk, and each competitor gets two throws. The winner is simply the competitor with the longest throw. There are two things to remember when competing. First, contestants can't wear gloves. Second, if they're brave enough, they can lick their fingers for a better grip.

No Man's Land

Imagine a ballplayer who, in one season, reaches base 215 times, attempts 203 stolen bases, and is caught only twice!

✳ ✳ ✳ ✳

THERE WAS SUCH a ballplayer, and *her* name was Sophie Kurys. She was a star player in the All-American Girls Baseball League (AAGBL) of the 1940s and '50s. Kurys's 201 steals in 1946, a year the basepaths were actually lengthened, were six times more than Brooklyn Dodger Pete Reiser's major-league high of 34 that year—and Pistol Pete didn't have to slide with bare legs. Sliding was excruciating while wearing the uniform's short skirts, especially in Fort Wayne, where the grounds crew burned the infield with gasoline to dry it on rainy days. But the players all slid, and never headfirst. It was a running league, and the constant sliding caused deep cuts—known to ballplayers as "strawberries"—that would painfully stick to clothes after games. "I had strawberries on strawberries," Kurys recalled.

Kurys stole 1,114 bases in just eight seasons—most of them as a Racine Belle—and her 140-steal annual average was higher than major-league record-holder Rickey Henderson's best season. Kurys's most memorable moment, though, was beating the throw home for the only run of a 14-inning game to win the 1946 league championship. Max Carey, president of the AAGBL and the all-time National League stolen-base leader back then (with 738), called it "the best game I've ever seen."

The league was started in 1943. Most able-bodied men were serving in the armed forces, and the major leagues—not to mention the minors—were forced to use many players who would've been laughed out of the ballpark before World War II started. The AAGBL wasn't just a wartime phenomenon, though. The league lasted until 1954, making several attempts to branch out. A 1949 tour of Central America featured exhibition games in four different countries. There were also junior teams—most notably the Junior Belles in Racine—where teenage girls learned fundamentals, participated in a short-season schedule, and occasionally played before the start of AAGBL contests.

The league had some outstanding players. Two-time batting leader Dorothy Kamenshek (.292 lifetime) was sought by a men's minor-league team in Fort Lauderdale, but she turned them down. Joanne Weaver, at 14, followed her sister Betty into the league and later became the only .400 hitter in AAGBL history. Bonnie Baker, a Saskatchewan-born catcher and former model, was the only player ever hired as a manager in the league. Pitcher Jean Faut, whose husband would eventually be her manager, had a baby in March 1948 and then went out and won 16 games on the season while also playing third base; she later pitched two perfect games and led South Bend to consecutive championships. Rose Gacioch once had 31 outfield assists in a season and converted to pitcher, helping the Rockford Peaches win four titles. Dottie Schroeder was the only woman to play every season—and nearly every game—of the league's existence, winning her only title in 12 seasons by driving in the winning run for the Kalamazoo Lassies in the last game in league history.

The managers were better known: ex-Cub Woody English, former Pirate Carson Bigbee, Bill Wambsganss (who was noted for turning an unassisted triple play in the 1920 World Series), and future Hall of Famers Carey, Dave "Beauty" Bancroft, and Jimmie Foxx. Heavy-drinking Foxx, who retired from the majors in 1945 at second place on the all-time home run list, managed the Fort Wayne Daisies and was reportedly the inspiration for the character Jimmy Dugan, played by Tom Hanks in the 1992 film *A League of Their Own*.

Hollywood likes to stretch the truth, but *A League of Their Own* wasn't too far off base. If anything, the movie condensed the 12 seasons of the league's existence into one year and pieced many real-life stories together to create characters with fictional names. The most memorable aspects of the film are mostly true: the Midwestern locale, the mass tryout in Chicago, the compulsory charm school, the chaperones, the bus trips, the camaraderie, a player's child in uniform on the bench, and the candy magnate who started the whole thing (the fictional Walter Harvey, proprietor of Harvey Field, as opposed to Chicago Cubs owner Philip Wrigley). A brief 1993 television series with the same name as the film could have shed more light on other facets of the league, but its six-episode run was shorter than some homestands.

Contrary to what was depicted in the movie, though, the league began with a 12-inch softball thrown underhand from 40 feet away; the ball got smaller, the mound got farther away, and the distance between the bases grew as the league wore on. By 1954, the ball was $9\frac{1}{2}$ inches, or the same size as a regulation major-league ball; the mound was 60 feet from home plate; and the bases were 85 feet apart. But all these changes weren't enough to keep the league from folding.

Mismanagement by independent owners, too much expansion, and dwindling attendance eventually killed off the AAGBL. In its last year, the players drove their own cars to games when teams could no longer afford a bus, and they even played without pay toward the end. The ride had been a good one, though. The $45–75 average weekly salary was far more than most women would have earned in occupations they enjoyed far less. What's more, the 600 women who made the grade earned the right to play baseball professionally. That's something only a handful of women have been able to claim in the half-century since then.

When Bicycles Ruled the (Sporting) World

What sport lasted a day longer than the ancient Olympics, broke the race barrier before baseball, and caused more injuries than modern football? Turn-of-the-century bicycle racing, of course.

✳ ✳ ✳ ✳

Blood, Guts, and Determination

I N 1900, THE most popular sport in North America was a grueling six-day bicycle race. Usually held at indoor velodromes with wooden tracks, teams of two riders would compete for 144 hours, taking turns accruing laps and competing in sprinting events. This was not for the  faint of heart. As many as 70,000 fans watched these powerful riders push themselves to the limits of endurance, often sustaining serious—even fatal—injuries in the process. Here are some of the sport's major players.

Reggie McNamara (1887–1970)

Dubbed the "Iron Man" of cycling, Australian Reggie McNamara had a seemingly inhuman capacity for the punishment and exertion that defined the six-day event. On the fourth day of a competition in Melbourne, McNamara underwent an emergency trackside operation without anesthesia to remove a large abscess "from his side." Though he lost a considerable amount of blood, he rose from the dust and, ignoring the advice of his doctor, resumed the race. In fact, injuries put him in the hospital so often that he wound up marrying an American nurse after a 1913 race in New York. He set several world records and beat the French so badly that they refused to ride against him.

Bobby Walthour (1878–1949)

During his career, bicycling champion Bobby Walthour of Atlanta, Georgia, suffered nearly 50 collarbone fractures and was twice assumed to be dead on the track—only to rise and continue riding. By the time he was 18, he was the undisputed champion of the South; soon after, he was international champion—a title he kept for several years. In addition to making cycling familiar to people all over the world, Walthour brought a great deal of prominence to his native Atlanta. Invigorated by his accomplishments, Atlanta built the Coliseum, one of the world's preeminent velodromes at the time.

Marshall "Major" Taylor (1879–1932)

African American cyclist Major Taylor proved that endurance cycling was a sport in which individual talent could not be denied. In an era of overt racism, he rose through the ranks to become one of the highest-paid athletes of his time. After moving from Indiana to Massachusetts, Taylor began to rack up victories in the six-day and sprinting competitions. Taylor toured the world, defeated Europe's best riders, and set several world records during his professional career.

Enter the Machines

Like modern stock-car racing, six-day cycling events used pacing vehicles. Originally, these were bicycles powered by two to five riders. But in 1895, English races began using primitive motorcycles. These new pace vehicles allowed the cyclists to travel faster, owing to the aerodynamic draft produced by the machines. Crowds thrilled to the speed and noise of these mechanical monsters, which weighed about 300 pounds each. It took two men to operate the motorcycles—one to steer and one to control the engine. They were also quite dangerous: A tandem pacer forced off the track in Waltham, Massachusetts, on May 30, 1900, killed both riders and injured several fans. The advent of motorcycles increased the popularity of the six-day races for a time, but it waned with the arrival of a new vehicle that spectators preferred over bicycles: the automobile.

The Weird World of Sports

Professional athletes are often viewed as heroes with super powers and epic strength. But as this list shows, no matter how much money or prowess they have, they're still only human.

Dubious Distinction

During the 1998 home-run battle between Sammy Sosa and Mark McGwire, pitcher Rafael Roque had the distinction of giving up round-tripper number 64 to both hitters. McGwire hit his 64th dinger off the Milwaukee Brewers slinger on September 18, and Sosa did the same on September 23.

Davey Decks Goliath

During a 1976 basketball game between the Boston Celtics and the Houston Rockets, 6'9" Sidney Wicks threw an elbow at 5'9" Calvin Murphy. It was a costly mistake. With a ferocity that belied his smaller stature, Murphy "chopped down" the giant with dozens of unanswered blows thrown in flurries and combinations. It took four players to dislodge Murphy from his victim. Wicks, who appeared dumbfounded, needed several stitches to close a gaping wound on his nose. He honestly didn't know what hit him.

Cantankerous Coaches

In 1978, two college football coaches from different teams stepped way out of bounds. Ohio State's legendary coach Woody Hayes tackled Clemson middle linebacker Charlie Bauman as he ran out of bounds during the Gator Bowl. To top it off, the crazed coach then slugged the player.

At another game that year, Fairfield University's football coach, Ed Hall, did Hayes one better. When he saw that Western New England's Jim Brown had eluded every Fairfield defender while on his way to a touchdown, Hall took action. The angered coach bolted from the sideline and tackled Brown at midfield. A stunned Brown asked, "Are you out of your mind, coach?"

Later, Hall explained the incident. "Something just happened to me and the next thing I knew, the referee was standing over me and screaming at me to get out of the game. As I started walking to the rear of the bleachers, with the crowd booing me, I broke down and cried."

Wrong-Way Nicholl

On March 20, 1976, while playing for Britain's Aston Villa soccer team, Chris Nicholl began scoring like crazy. Amazingly, the footballer scored every goal in a 2–2 draw against Leicester City, including two "own goals," or goals for the opposing team.

The Sultan of Swat

On June 23, 1917, while pitching against the Washington Senators at Fenway Park, Red Sox pitcher Babe Ruth got into an argument with umpire Brick Owens after walking the first batter. Owens ejected Ruth, and the furious Sultan of Swat lived up to his nickname by slugging the ump. As luck would have it, Ruth's replacement, Ernest Shore, retired the next 27 Senators—thus pitching a no hitter—and the Red Sox won the game 4–0.

That's the Way the Ball Bounces

During a May 26, 1993, night game between the Cleveland Indians and the Texas Rangers, the Indians' Carlos Martinez hit a long fly ball. Rangers right fielder Jose Canseco tried to make the catch, but he lost the ball in the lights and it struck him on top of the head, took an improbable bounce, and landed over the outfield wall for a home run.

❋ On November 17, 1968, in a game referred to as the Heidi Bowl, the New York Jets were leading the Oakland Raiders 32–29 with 1:05 left on the clock, when NBC cut away to the classic children's movie *Heidi*, preventing millions of viewers from seeing Oakland's 43–32 comeback.

❋ In 1980, Green Bay Packers kicker Chester Marcol ran in his own blocked field-goal attempt for a touchdown to beat the Chicago Bears 12–6.

Chapter 4

Art and Literature

The Night Frankenstein Was Born

If you think the drama and anxiety in Mary Shelley's Frankenstein *is intense, you should get a load of the lives that hovered in and around its creation. What a soap opera!*

✳ ✳ ✳ ✳

BORN IN 1797, Mary Wollstonecraft Godwin was the only child of intellectuals William Godwin and Mary Wollstonecraft, who died only 11 days after giving birth to her. Although Mary didn't have an extensive formal education, she was tutored by her father and given free rein of his library.

Mary and Percy

Percy Bysshe Shelley had been bullied in school and had few friends, so he read—a lot. Through his reading, he came up with ideas that society considered heretical and dissident. When Oxford University caught wind of Percy's pamphlet *The Necessity of Atheism*, he was expelled, which caused a falling-out with his father. Shortly after that, he married a woman named Harriet Westbrook. Percy delved deeper into radical politics, publishing poems about atheism, vegetarianism, and free love. He also gave revolutionary speeches on religion and politics and traveled throughout Europe.

When he made his way to the home of William Godwin, one of his intellectual heroes, he found inspiration—as well as Godwin's brilliant 16-year-old daughter, Mary. Against Godwin's wishes,

Mary and Percy—who were both advocates of free love—began a romantic relationship. Mary may have believed in it more as a theory, because although Percy encouraged her to take other lovers, she had little interest in pursuing anyone but him. Despite her objections, Percy engaged in amorous relations with women both inside and outside of their social circle.

Enter Lord Byron

Lord George Gordon Byron was born with a clubfoot, which was a source of shame his entire life. But that didn't keep him from becoming a well-known poet. Byron famously ripped and roared through life—writing poems, speaking eloquently at the House of Lords, piling up debt, and burning through love affairs with both men and women. In 1816, Byron decided to leave England temporarily until the rumors about his lifestyle died down and some of his debts had been forgotten. But what he expected to be a trip across Europe ended up being a permanent exile from England.

Claire Clairmont, was the stepsister and very close friend of Mary Wollstonecraft Godwin. Despite this, Claire became involved with Percy while Mary was pregnant with his child. At 18, after an on-and-off affair with Percy, Claire moved on to a powerful crush on Byron. Under the pretense of looking for advice on how to become a poet, Claire wrote him letters daily and eventually became his lover. When Byron left England in 1816, Claire followed, bringing Mary and Percy with her for a visit. Unbeknownst to the rest of the group, Claire was pregnant with Byron's child. To say the least, tensions were brewing.

The Year Without a Summer

Meteorological moodiness was on tap for the summer of 1816. The weather was so dank and cool that it was almost as if summer never came at all. Many nights at Byron's Swiss villa on Lake Geneva were spent inside because of the cold and rain. On one particular night, Mary, Percy, Byron, and John Polidori (Byron's physician) were reading ghost stories. Byron challenged the members of the group to each write a spooky tale. Mary delighted in the challenge. Perhaps it was her need to be accepted by the group, or her frustration with fitting into a role in which she couldn't express her considerable intelligence. Or maybe she simply wanted to write, just as she had desired all her life. Or maybe, just maybe, it was the alleged rampant opium usage among the group on this dark and stormy night.

The story that Mary wrote started as a wisp—a few pages about a doctor who creates a monster by snatching body parts from morgues; it vaguely explained the process in which these parts are fused together and animated. While the initial story was short, there was enough there for Percy to encourage her to develop it. Over the next few days, Mary completed a draft, and she finished writing *Frankenstein* the following year.

On January 1, 1818, the first edition of *Frankenstein* was published anonymously, with a foreword by Percy. The second edition, which was released in 1823, was credited to Mary Shelley, and in 1831, the first popular edition of the book was published after the text was heavily revised by Mary. The novel became a classic despite the initial criticism it received. To this day, people around the world love to read about the sad monster that may or may not represent the feelings floating around a lavish Swiss home one stormy night.

The Finale

The next few years remained drama-filled for the four characters in this soap opera. Claire had a daughter by Byron—a daughter she eventually left for him to raise; he put the child in a convent, where she died of typhus at age five. Mary married Percy after his first wife died; they had four children, all but one of whom died in childhood. Percy drowned when his boat capsized during a sudden storm in Italy in 1822, just six years after the summer at Byron's villa on Lake Geneva. And just two years after that, Byron died of a violent fever in Greece while fighting for Greek independence from the Turks.

Overdue

Just how long have you had the library's copy of War and Peace? And when exactly are you planning to pay that hefty fine you've accrued? Read on to see how your library misdeeds stack up.

✳ ✳ ✳ ✳

THE MOST OVERDUE library book in U.S. history was never officially checked out. And the man who took it thought it belonged to another library entirely. That's the story that emerged in February 2009 when book collector Mike Dau of Lake Forest, Illinois, showed up at the Washington & Lee University Library with volume one of W.F.P. Napier's *History of the War in the Peninsula and in the South of France.* According to a handwritten note on one of the pages, the book was taken from the nearby Virginia Military Institute on June 11, 1864, by Union soldier C. S. Gates during a raid on the area.

The book passed to Gates's descendents and eventually ended up with Dau, who noticed that the title page indicated that it had originally belonged to Washington & Lee—at that time known as Washington University. He returned the book to the university's Leyburn Library 52,858 days after Gates had removed it. Librarian Laura Turner graciously waived the fine.

If you think you can ignore library fines, think again. Rabbi Avrohom Sebrow of Queens, New York, found that out the hard way when he was denied a mortgage loan due to an overdue library fine. The Queens Public Library had turned the matter over to a collection agency, which reported Sebrow's fine of $295.40 plus $66 in late fees to credit agencies. The embarrassed rabbi returned the late materials and cleared his record.

Perhaps we should all take a page from the book of 91-year-old Louise Brown of Stranraer, Great Britain. As of July 2009, she had borrowed (and returned) nearly 25,000 books from her local library. And she's never had to pay a single late fee.

The Unlikely Art Career of Grandma Moses

*Late bloomers, take heart: You haven't missed your chance.
That's one of the most significant lessons to be learned from the
incredible life of Anna Mary Robertson Moses (1860–1961), better
known as "Grandma" Moses.*

✳   ✳   ✳   ✳

SHE WAS BORN before Abraham Lincoln took office, and
when she died, President Kennedy offered tribute: "Both
her work and her life helped our nation renew its pioneer
heritage and recall its roots in the countryside and on the
frontier. All Americans mourn her loss." Grandma Moses was
one of the most successful and famous artists in her home
country, and possibly the best-known American artist in
Europe. Featured on radio, on television, and in magazines,
Moses was the first artist to become a media superstar, and
the influence of her unique folk-art style continues to this day.
Indeed, it would not be outlandish to say that Grandma Moses
was, and remains, the most famous female artist of all time.

Unexpected Success

It all started in 1938, when a New York City art collector drove
through Moses's hometown of Hoosick Falls, New York, and
spied her paintings in a drugstore. He bought them all, and
then he found her at home, where he snatched up all of her
remaining work. The following year, she was represented in an
exhibition of contemporary unknown painters at the Museum
of Modern Art in New York. And the year after that, "What a
Farmwife Painted" opened at New York's Galerie St. Etienne.
The exhibition stopped people in their tracks—and not just
because the artist first began to paint seriously when she was
76 years old (due to arthritis, which made her give up her
beloved embroidery). It was her subject matter—the nostalgic
remembrance of things past in what appeared to be a much

simpler time: ordinary farm activities, such as maple-sugaring and making candles, soap, and apple butter. But it was also the charming optimism, luminous colors, and realism that emanated from the canvases, which then cost three to five dollars. These paintings were later reproduced on everything from ceramic plates to greeting cards.

Not All as Easy as It Seemed

But Grandma Moses's life was no rose garden. She was indeed a farmwife, with all that the role entailed, and a "hired girl" before that. At age 12, she left school and home, where she had four sisters and five brothers, to go to work. At 27, she married the hired hand on the farm where she did the housework, and they had ten children, five of whom died as infants. Grandma Moses eventually outlived her husband and all of her children, but she did enjoy her 9 grandchildren and more than 30 great-grandchildren.

Grandma Moses painted more than 1,000 pictures, including 25 after her 100th birthday. The paintings have increased dramatically in value: Today, her smallest works begin at $15,000, whereas a large oil painting can command in excess of $100,000.

In 2001, the Galerie St. Etienne created a traveling exhibition titled "Grandma Moses in the 21st Century." It earned rave reviews from the public and critics, including *The New York Observer*'s notoriously tough Hilton Kramer, who gushed, "Grandma Moses is back, and she's enchanting."

Grandma Moses was a tough old gal, but her basic philosophy was one that remains relevant: "I look back on my life like a good day's work; it was done and I feel satisfied with it. I was happy and contented, I knew nothing better and made the best out of what life offered. And life is what we make it, always has been, always will be."

The Death of Sherlock Holmes

"I am in the middle of the last Holmes story," Sir Arthur Conan Doyle wrote to his mother in April 1893, "after which the gentleman vanishes, never to reappear. I am weary of his name." But Doyle was wrong: It would take several more months for the author to figure out a suitable demise for his world-famous and wildly popular detective. But why kill Holmes at all?

✳ ✳ ✳ ✳

It's a Mystery

AFTER BURSTING ONTO the scene in 1887, Sherlock Holmes became one of the most enduring literary characters of all time. His popularity gave Doyle the financial and artistic freedom to pursue whatever creative avenues he chose. But as the years went by, Doyle began to feel strangled by his own creation.

There are many theories about Doyle's decision to kill off Holmes. Some postulate that publishing deadlines were tight, and the pressure of always having to come up with intricate plots was wearing on Doyle and affecting his overall literary output. "The difficulty of the Holmes work," Doyle wrote, "was that every story really needed as clear-cut and original a plot as a longish book would do. One cannot without effort spin plots at such a rate."

Another theory is that because he was so busy writing Holmes stories, he did not notice the beginnings of an illness in his wife that eventually became tuberculosis. "As a doctor, [Doyle] should have recognized her condition long before it developed advanced symptoms," wrote biographer Martin Booth. It has been suggested that Doyle felt guilty for this and blamed Holmes. Perhaps killing the character was a form of revenge.

Whatever the reason, Doyle had been thinking about killing Holmes for quite some time. Toward the end of 1891, he mentioned his plans to his mother, and she frantically pleaded with him to change his mind. Doyle did so, albeit briefly.

Fearful Falls

Doyle knew that Holmes needed a proper death. "A man like that mustn't die of a pin-prick or influenza," he said. "His end must be violent and intensely dramatic."

In the summer of 1893, Doyle told English cleric and novelist Silas K. Hocking of his plan to kill Holmes. "Why not bring him out to Switzerland and drop him down a crevasse?" Hocking said. "It would save funeral expenses."

Doyle laughed, but the conversation stuck with him. During a trip to Switzerland, he visited the famous Reichenbach Falls, and he decided the foaming waters there would be a suitable end for his detective. Doyle then set about writing *The Final Problem*, in which he created the criminal mastermind Professor Moriarty. At the end of the story, the two adversaries, locked in combat, apparently plunge into Reichenbach Falls.

"Killed Holmes," Doyle noted in his diary. Or so he thought.

A Cry Heard Around the World

The death of Sherlock Holmes unleashed worldwide protest. *The Strand* magazine, publisher of the Holmes stories, lost 20,000 subscriptions, and people in London wore black mourning armbands. Even the British royal family was upset.

For years, Doyle resisted the pressure to bring back Holmes. Finally, in 1901, he released *The Hound of the Baskervilles*, but set it before Holmes's death. However, the public still wasn't satisfied. In 1903, Doyle relented and published *The Adventure of the Empty House*, which brought Holmes back safe and sound (it turns out he never went into the falls at all). Ultimately, Doyle's creation had a life of its own, and he went on to write Holmes stories for another two decades.

Losing Philip K. Dick's Head

Consider the sorts of things an air traveler might accidentally leave in an overhead bin: a novel, a jacket, a laptop computer... or perhaps the $750,000 android head of author Philip K. Dick.

❋ ❋ ❋ ❋

Meet David Hanson

IN THE WORLD of robotics, David Hanson is known as the genius inventor of "frubber"—an authentic-looking synthetic skin. With this breakthrough, Hanson has created robots modeled after popular figures ranging from Albert Einstein to rocker David Byrne. In 2004, his firm, Hanson Robotics, led a team of artists, scientists, and literary scholars to create an android head modeled after science-fiction writer Philip K. Dick, who died in 1982. To build the android's body of synthetic knowledge, a team led by artificial intelligence (AI) expert Andrew Olney scanned 20 of Dick's novels as well as interviews, speeches, and biographical information into its computer brain.

After six months of work, the completed android was an impressive achievement. Using a camera for its eyes, the android could follow movement, make eye contact, and recognize familiar faces in a crowd. The android's cutting-edge AI applications allowed it to respond to queries using its inputted source material (though its replies were often off-topic and bordered on the surreal—not unlike the real-life Dick, according to friends). The team called it "Phil."

Meet Philip K. Dick

Hanson and his team chose an apt subject for their robot. Many of Dick's stories are peopled with synthetic beings who call into question what qualifies as human. Hanson's team was particularly inspired by Dick's novel *We Can Build You*, in which two unsuccessful electric-organ salesmen construct a lifelike robot of Edward M. Stanton, Secretary of War under Abraham Lincoln. The competing needs of the human characters, coupled with the android's increasing humanity, play out in a rich tale that challenges the reader's imagination. Little did Hanson's team realize that their robot would also become part of its own odd legend.

David Hanson Loses His Head

In June 2005, Phil made his debut appearance at Chicago's NextFest technology exhibition. The technically minded marveled at the feat of engineering, the sci-fi fans thrilled at the wonderful irony of interacting with an android replication of their long-dead hero, and the curious were rewarded by the eerie sense of humanity that Phil inspired.

On the heels of this success, Hanson, who was juggling time between his growing firm and his doctoral work, carted Phil around the country. In Pittsburgh, Hanson received the Open Interaction Award for Robotics from the American Association for Artificial Intelligence. At the San Diego Comic-Con, Phil took part in a panel discussing the upcoming release of the film *A Scanner Darkly* (which was based on the Dick novel of the same name). There was talk of a tour to promote the movie, an appearance on the *Late Show with David Letterman*, and a stint in the Smithsonian Institution's traveling collection. It was awfully heady stuff for a head.

In early 2006, Phil was on his way to Mountain View, California. He was stored facedown in Styrofoam, bundled in a gym bag, and stowed in the overhead bin on the plane. Hanson, bleary from sleep, changed flights in Las Vegas. Soon afterward, he realized that Phil was still on the other plane. The airline confirmed that Phil had traveled with the plane to Orange County. The android was supposed to be put on a flight to San Francisco, but it never arrived. Phil has not been seen since.

All Is Not Lost

Though the unique Philip K. Dick head had vanished, Hanson's laptop, containing Phil's brain, was safe. Despite the devastating loss, Hanson eventually built another head for Phil.

What's in a Name?

A rose by any other name may smell as sweet, but a book with a bad title could end up on the sale table instead of the bestseller list. After all, Tolstoy's War and Peace *has a far more dignified ring to it than its original name,* All's Well That Ends Well. *Luckily, these original titles didn't make it to the bookstore shelves.*

✳ ✳ ✳ ✳

✳ Long after the publication of *The Great Gatsby*, F. Scott Fitzgerald regretted not going with his preferred title, *Trimalchio in West Egg*.

✳ Peter Benchley and his editor discarded more than 100 titles for Benchley's first novel, including *Great White*, *The Shark*, *Leviathan Rising*, *The Jaws of Death*, and *A Silence in the Water*. Pressured to make a decision because the book was ready to go to press, they agreed that the only word they liked in any of the proposed titles was *Jaws*.

✳ Charles Dickens's *Tom-All-Alone's Factory that Got into Chancery and Never Got Out* was published as the more succinctly titled *Bleak House*.

✳ Stephen Crane's Civil War novel *The Red Badge of Courage* was originally titled *Private Fleming, His Various Battles*.

✳ Jane Austen thought about the implications of *First Impressions* and renamed her novel *Pride and Prejudice*.

✳ *Main Street*, Sinclair Lewis's classic exposé of small town hypocrisy, was originally called *The Village Virus*.

✳ Margaret Mitchell's *Gone with the Wind* was almost published as *Pansy*, the inappropriate name of the main character, now known as Scarlett O'Hara.

The Remarkable Erich Remarque

Despite claims by Nazi propagandists, Erich Remarque, celebrated author of All Quiet on the Western Front, *never tried to hide his heritage.*

✳ ✳ ✳ ✳

Considered one of the greatest antiwar novels ever written, *All Quiet on the Western Front* is the fictional story of German soldiers in World War I. Since it was first published in 1929, it has sold millions of copies worldwide. In 1930, it was made into a movie, which won an Academy Award for Best Picture.

Over the years, however, rumors have surfaced about the book's author, Erich Remarque. Chief among them is that his real surname was Kramer and that he spelled it backward and added a French flourish to conceal the fact that he was Jewish.

The truth is that Remarque was Catholic, not Jewish, but he did change his name. He was born Eric Paul Remark in Germany in 1898, and he legally changed his name to Erich Remarque after he served in World War I. Remarque had been the family name until his grandfather changed it in the mid-1800s.

At age 18, Remarque was drafted into the army and sent to the western front. After he was wounded in combat, he was transferred to a military hospital, where he spent the rest of the war. But his time on the front lines affected him deeply, and he used the experience as the basis for his best-selling novel.

The Nazi Party banned both the book and the movie, and in an effort to discredit Remarque's name, it spread the rumor that he was Jewish and that his real name was Kramer.

Remarque fled to Switzerland in 1931 and moved to the United States in 1939. He later returned to Switzerland, where he lived until his death in 1970.

Art by the Numbers: The Vogel Collection

So you like art, but you're operating on a shoestring budget, huh? Take a few pointers from this New York couple, who managed to amass an outstanding collection in an unlikely way.

✳ ✳ ✳ ✳

Y OU DON'T HAVE to be a millionaire to collect great art. Take it from Herb and Dorothy Vogel, two New Yorkers who, back in the mid-1960s, began collecting art that they liked, could afford, and could fit into their modest abode. And they liked a lot, could afford more than they thought, and managed to fit more than 2,000 works into their one-bedroom space.

In the Beginning, There Was Art

In 1962, Herb, a postal clerk, married Dorothy, a librarian. Their mutual love of art initially brought them together. Dorothy once said that when it came to their relationship, "It was art or nothing"—clearly a serious commitment to both one another and their shared interest. However, the Vogels were about as well off as the starving artists that they patronized. But the couple made do: By living off Herb's salary and buying art with Dorothy's, they embarked on their art odyssey.

Although the art the Vogels could afford wasn't the sort sold at Sotheby's, that didn't stop them. This was the second half of the 20th century, and New York was bursting at the seams with minimalists, conceptualists, painters, sculptors, and artists of every kind, many at the beginning of their soon-to-be-lauded careers. And the Vogels seemed to have a keen eye for the truly talented. Around 1967, the two met conceptual artist Sol LeWitt and were the first people to buy his work. This kicked off their collection, and they subsequently bought many more pieces from LeWitt, who eventually made quite a mark on the American art scene.

Their collection grew. Among the artists represented by the Vogels' collection were the likes of Christo, Andy Warhol, Chuck Close, Carl Andre, and hundreds of others. They bought often and cheaply, and never, ever sold anything.

Not So User-Friendly

Sure, the couple had an exceptional eye for artwork, but where did they store it all? The Vogels were unfazed by their typically cramped New York apartment and refused to move or rent additional space to house their collection. While almost every inch of their walls was covered in the pieces they had bought, most of the art Herb and Dorothy collected was hoarded away in closets or piled on top of shelves and boxes. As art lovers, the Vogels painstakingly maintained their collection, but accidents do happen Once, water from a fish tank splashed onto a Warhol canvas, and it had to be restored.

Bursting at the Seams

The Vogels might have been a bit preoccupied with buying art, but they weren't hermits or recluses. In fact, they were often seen out and about with the artists whose work they collected. For Herb and Dorothy, forging a connection with the people who produced the art was part of the purchasing process.

Finally, in 1992, aware that they were running out of space, the Vogels pledged their collection (all 2,000 pieces) in installments to the National Gallery of Art in Washington, D.C.—where they spent part of their honeymoon decades before. No longer stored in a coat closet, these incredible works of art can now be enjoyed by all.

The public and art elite alike lauded the Vogels, embracing the couple and their collection for its scope, intensity, and foresight. In 2008, a documentary titled *Herb and Dorothy* enlightened filmgoers about the couple and their lifelong ambition. The director, Megumi Sasaki, remarked, "I thought I was going to make a small film about beautiful, small people. But I learned . . . they are giants of the art world."

Bram Stoker and Henry Irving: Close Friendship or Unrequited Love?

Was actor Henry Irving the inspiration for Bram Stoker's famed novel Dracula, *and were their personal issues revealed in the book? Maybe, but sometimes a stake is just a stake.*

✳ ✳ ✳ ✳

BRAM STOKER, BEST known for his macabre masterpiece *Dracula*, spent much of his life as the manager of Sir Henry Irving, one of the most famous British actors of the 19th century. The two had a deep friendship, but the precise nature of their relationship is a mystery. Some say that Irving was dominant and abusive to Stoker, suggesting that the sinister Dracula character was inspired by Stoker's secret hatred for his boss. Others contend that Stoker was a homosexual who channeled his unrequited romantic feelings into sexually charged novels.

The Man Who Met His Idol

Stoker was born in Dublin, Ireland, on November 8, 1847. A sickly child, he barely left his bed until age seven. Later, he became a star athlete and president of the philosophical society at Trinity College Dublin. After he graduated, he wrote his first book, reviewed theater for local newspapers, and became obsessed with Irving.

In 1876, Stoker met his idol. Irving was acting in Dublin, and Stoker's review of his performance was so enthusiastic that Irving wanted to meet his admirer. Shortly after they met, Irving gave Stoker an autographed photo which said, "My dear friend Stoker, God bless you! God bless you!" Stoker later wrote, "Soul had looked into soul! From that hour began a friendship as profound, as close, as lasting as can be between two men."

But every spring love has its winter season. Literary historians have argued that the Irving–Stoker relationship was codependent and emotionally abusive. Irving was a world-famous actor; Stoker was an unknown novelist. Since age 18, Irving had traveled all over Great Britain, playing roles large and small. His big break came in 1874, when he played the title character in *Hamlet* at the famed Lyceum Theatre in London. His acting style was mesmeric and unique. Instead of the loud bantering and over-dramatic gestures to which theater audiences of the time were accustomed, Irving played his parts with quiet, refined dignity. In 1878, Irving became the manager of the Lyceum Theatre, and he recruited his friend Stoker as his personal assistant.

Until Irving's death in 1905, the Lyceum flourished and Stoker accompanied Irving on frequent tours abroad. While Stoker sang nothing but songs of praise for his comrade in *Personal Reminiscences of Henry Irving,* some who knew both men claimed that Irving was domineering and prone to dark moods. Stoker, however, did claim some success of his own while working for Irving: In 1897, *Dracula* was published to moderate praise. The theory that the main character is loosely based on Irving is purely speculative, although a few writers have made compelling cases. Stoker biographer Barbara Belford has claimed that *Dracula* is "bristling with repression and apprehension of homosexuality, devouring women, and rejecting mothers." Literary critic Phyllis Roth wrote that the character of Dracula "acts out the repressed fantasies of the others."

The Art of Repression

Although some may be reading a bit too much between Stoker's lines on that count, there is convincing evidence that he was a closeted homosexual. In his youth, Stoker idolized and corresponded with American poet Walt Whitman, who was famous for his liberal attitude toward homosexuality. In one letter to Whitman, Stoker declared, "I would like to call you comrade and talk to you as men who are not poets do not often talk ... I know I would not be long ashamed to be natural before you ... you have shaken off the shackles and your wings are free. I have the shackles on my shoulders still—but I have no wings."

A Wilde Situation

Some also believe that Stoker had a complex relationship with openly homosexual playwright Oscar Wilde. In 1878, Stoker married Florence Balcombe, who had recently been Wilde's girlfriend. There is little documentation of the relationship between Wilde and Stoker, but historians argue that the two men kept up icy social relations throughout their lives and that Wilde was jealous of Stoker's marriage.

And then there was the famous Wilde sodomy trial of 1895, which resulted in him spending two years in prison. Since *Dracula* was published in 1897, some historians have proposed that Stoker wrote the novel during the trial and funneled his complex feelings about homosexuality into the book. But Stoker's notes on *Dracula* reveal that he began planning the book five years before Wilde's trial.

Moreover, Stoker was one of the only artists in Wilde's circle who did not vocally come to his defense during the trials. Essays that Stoker wrote suggest that he was morally opposed to "loose" sexual codes, feeling instead that writers should censor their work so as not to propagate immorality.

Thus, while Stoker may have been in love with Irving, it seems unlikely that he ever acted on his feelings, despite conjectures about the late-night sessions that the two frequently held at the Lyceum Theatre. Whatever the inspiration for the delectably sexual *Dracula,* it is a novel that has lived for more than a century in the popular imagination.

Flubbed Headlines

✳ ✳ ✳ ✳

✳ "New Study of Obesity Looks for Larger Test Group"

✳ "Study Reveals Those Without Insurance Die More Often"

✳ "Red Tape Holds Up New Bridge"

✳ "Typhoon Rips Through Cemetery—Hundreds Dead"

✳ "Man Struck by Lightning Faces Battery Charge"

✳ "Kids Make Nutritious Snacks"

✳ "Health Officials Say Flammable Water is OK to Drink"

✳ "Bladder Control Causes Sunset Beach Flooding"

✳ "Hospitals Are Sued by 7 Foot Doctors"

✳ "Something Went Wrong in Jet Crash, Experts Say"

✳ "Police Begin Campaign to Run Down Jaywalkers"

✳ "Panda Mating Fails; Veterinarian Takes Over"

✳ "A-Rod Goes Deep, Wang Hurt"

✳ "If Strike Isn't Settled Quickly, It May Last a While"

✳ "Gasoline Stations Will Also Offer Mammograms"

✳ "Squad Helps Dog Bite Victim"

✳ "Sex Education Delayed, Teachers Request Training"

✳ "Yellow Snow Studied to Test Nutrition"

✳ "British Union Finds Dwarfs in Short Supply"

✳ "Drunks Get Nine Months in Violin Case"

✳ "Antique Stripper to Demonstrate Wares at Store"

✳ "Iraqi Head Seeks Arms"

✳ "Psychics Predict World Didn't End Yesterday"

✳ "Sewage Spill Kills Fish, but Water Safe to Drink"

✳ "Local High School Dropouts Cut in Half"

✳ "War Dims Hope for Peace"

Chapter 5

All About Animals

Bat Bombs Away!

Small, winged, nocturnal, furry—and explosive. These critters were supposed to help end the war but succeeded only in destroying an American airplane hangar and a general's car.

✳   ✳   ✳   ✳

WORLD WAR II inspired innovation and invention, as scientists and engineers from the major powers strove to develop weapons that would provide a winning edge. Many of these innovations are well known—jet engines and rockets in Germany, for example. However, those that did not work so well are scarcely remembered.

In the early days of the war, Lytle S. Adams, an American dental surgeon from Pennsylvania, conceived an idea while on vacation in the American southwest. He proposed that the United States develop a method for attaching incendiary bombs to bats and release thousands of the flying mammals over Japan. Under Adams's logic, the bats would roost in wooden buildings and explode, causing fires that would spread out of control. On paper, Adams's idea held merit—a typical bat can carry 175 percent of its body weight, and since the Japanese populace would not detect the roosting bats, the fires could spread unchecked.

In the weeks following America's entry into the war, thousands of citizens sent ideas for new weapons to the White House, and the bat-bomb proposal was one of the very few that went into development. Approved by President Roosevelt, it eventually consumed a modest $2 million of taxpayers' money.

By March 1943, a team consisting of Adams and two chemists (one from Harvard, the other from UCLA) had scoured the caves of the southwest in search of the perfect bat species for the project. Although the mastiff bat was larger and the mule-eared bat more common, the team settled on the Mexican free-tailed bat because it could carry the requisite weight and was available in large numbers (in fact, one colony of free-tailed bats near Bandera, Texas, numbered some 20 to 30 million animals).

Months of testing followed. The creatures were tricked into hibernation with ice, and then a small explosive device was surgically attached. The procedure was delicate and required lifting the bats' fragile skin, which could tear if improperly handled. The prepared bats were then loaded into cardboard cartons, which were parachuted from aircraft and opened at a preset altitude. There were numerous complications, however. Many of the containers did not open or the bats did not wake up and plummeted to their deaths. Still, the bats did succeed in burning down a mock Japanese village. On the other hand, they also managed to start a fire in an airplane hangar that destroyed a visiting general's car. Perhaps for this reason, in June 1943, after more than 6,000 bats had been used in tests, the Army handed the project to the Navy. It was renamed Project X-Ray.

The Navy eventually handed off the project to the Marine Corps, which determined that the bat bombs were capable of causing ten times the number of fires as the standard incendiary bombs being used at the time. However, when Fleet Admiral Ernest J. King learned that the bats would not be ready for deployment until mid-1945, he called off the project. Dr. Adams was bitter about the cancellation of his novel idea. He maintained that the bat bombs could have caused widespread damage and panic but without the loss of life that resulted from the use of the atomic bomb.

Back from the Brink

By the end of World War II, the population of the Japanese Akita
had dwindled to near extinction, with only 16 dogs remaining.
One man made it his mission to bring them back.

✳ ✳ ✳ ✳

IN 1944, MORIE SAWATAISHI was an engineer working in northern Japan's snow country when he heard an alarming rumor:
People were killing their Akitas, selling the pelts to Japanese
soldiers, and—as food became scarce—eating the meat. Morie
couldn't believe it. When he was a boy, the Akita had been
declared a national treasure. What had changed?

Dog Tired

Quite simply, Akitas were no longer hip. Even before the war,
the Japanese had grown tired of their native snow dogs and
wanted something novel and foreign. Akitas seemed like throw-
backs to the 19th century, when the dogs were admired for
their fearlessness (Akitas were known to corner bears).

In 1927, a preservation society was founded to protect the dogs,
but that didn't boost the pups' popularity. They did get good
publicity from Hachiko, the Akita who arrived at a Tokyo train
station every day to greet his master—even nine years after the
man's death. A bronze statue of Hachiko was erected in 1935,
and he became a symbol of loyalty. But by 1944, the statue had
been melted down for metal. So much for loyalty.

Canine Contraband

Although Morie had never been what you'd call a "dog person,"
he felt compelled to do something. He heard rumors about
a puppy for sale, the granddaughter of a prizewinning Akita.
Morie paid the breeder six times his salary for her.

Morie never named the dog, which he kept hidden in a shed
and walked only in the late night and early morning hours to
avoid detection. As his work took him across snow country, he

covertly inquired about other Akitas. By the time the war ended in 1945, Morie had put together a list—only 16 purebreds remained, and two belonged to him (he had also bought a male for breeding).

Back in the Pack

In the spring of 1946, with a litter of puppies on the way, Morie held the first postwar dog show at his home. Nearly 50 Akitas were present, but it was hard to believe that they were of the same breed. Years of malnutrition and crossbreeding with foreign dogs had given them a hodgepodge of features. What's more, no one could agree on which features were ideal. Everyone felt that ears should be erect and tails curved, but should bodies be lean or stocky? Heads round or sharp? Debates aside, Morie was just pleased to spark a dialogue and turn attention back to the breed.

Bred and Ready

"Attention" turned to "frenzy" in the early 1950s after an ad campaign featured a famous Japanese actress with her Akita (think Paris Hilton with her purse-size Chihuahua). American GIs clamored for the dogs. Suddenly, everyone in Japan was an "Akita breeder." Everyone except Morie, that is. Rather than sell his dogs for profit, he gave them away to people he knew, keeping and breeding the ones he thought had the strongest features and, most importantly, *kisho* (spirit).

For more than 60 years, Morie raised and trained Akitas— 100 in all. His last dog, Shiro, died in late 2007 at age 15; Morie followed a year later at age 92.

✳ In 1937, Helen Keller brought the first Akita to the United States. She was so moved when she heard the story of Hachiko while visiting Japan that she wanted a dog just like him.

✳ The Akita takes its name from the Akita Prefecture, a district in Japan's snow country.

✳ The statue honoring Hachiko was re-created in 1948.

Plastic Panic: What's Killing India's Cows?

Plastic shopping bags: They're not eco-friendly, but who knew they weren't e-cow-friendly? In India, where the cow is sacred, plastic bags are more than just an environmental mess—they're a holy health hazard.

✳ ✳ ✳ ✳

Sick Cows

In India, it's pretty typical to see cattle (there are more than 200 million of them in the country) roaming the streets to feed on the city's garbage. But in 2000, a police officer in the city of Lucknow noticed a strange phenomenon: skeletal cows with engorged bellies dropping dead in the street, sometimes as many as 100 a day. Other sick-looking cows were seen staggering over to dumpsters but then walking away instead of eating as usual.

A few of these sick cows were corralled to the vet, where they were anesthetized and their stomachs surgically examined. Inside, the doctors found plastic bags—up to 60 pounds of them. Apparently, the cows had eaten the bags along with whatever food was tucked inside. The cows' stomachs (they have four of them) were so full of plastic that the animals could no longer eat. The cows were starving to death as a result. What's more, India's Animal Husbandry Department found that drinking milk from the sick cows could cause cancer and tuberculosis.

Plastic Boom

Plastic bags have only been in India since the mid-1980s, when the government authorized plastic production to help the nation compete in the global market. More than 50 percent of India's plastic is used for packaging, and nearly all of it is thrown away, which creates a major waste problem in a country that's otherwise known for its recycling efforts. No joke: In India, glass bottles are used over and over again and repaired if broken; "plastics mechanics" make house calls to fix broken items with heat fusion; and anything broken or worn beyond repair is picked up by "ragpickers," a low caste of people who scavenge and sell garbage for whatever they can get.

Aye, There's the Rub(bish)

In the trash-selling game, plastic usually sells for 12 rupees a kilo (about 14 cents a pound). But plastic bags are so thin that even a bunch of them don't weigh enough to be worth much. Hence, plastic-bag pollution.

Although a few local governments tried to ban plastic bags in the late 1990s, it wasn't until 2000—with the lives of sacred cows suddenly at risk—that a national solution was sought.

Thicker Bags

Why ban the bags, asked the plastics industry, when we can just make them thicker? This way, ragpickers would take the bags, keeping them off the streets and out of the cows' bellies.

Both the government (which was making a pretty penny selling things in plastic pouches) and the plastics industry agreed this was a win-win solution. A new rule was made: All plastic bags must be thicker than 20 microns (about $^{78}/_{100,000}$ of an inch).

Unfortunately, the plastics industry also continued to make the thinner bags, which shopkeepers unwittingly continued to use—after all, who can tell the difference between 19 microns and 21 microns? Rumor had it that government officials, equipped with special bag-thickness-testing instruments, were stopping by stores to enforce the new rule. But for the most part, the situation didn't, and hasn't, changed.

Still, hope hasn't been bagged entirely: The Polythene Agony Campaign, launched in 2000, is dedicated to educating people about the dangers of plastic bags and how to dispose of them properly. As for the afflicted cows, India's Animal Husbandry Department provides rumenotomies, or "plastic-bag-ectomies," to restore them to health.

Incredible Bat Facts

It isn't easy being a bat. Thanks to Dracula, a few cases of rabies, their pointy teeth, and the fact that they hang upside down to sleep, bats inspire fear in many people. But as you'll see, bats are amazing creatures that eat bugs . . . and sometimes drink blood.

✳   ✳   ✳   ✳

Bats are the only mammals that can fly. And you thought it was the winged marmoset! Bats are exceptional in the air. Their wings are thin, giving them what is called, in flight terms, "airfoil." The power bats have to push forward is called "propulsion."

A single brown bat can catch around 1,200 mosquito-size insects in an hour. It's estimated that the 20 million Mexican free-tailed bats that live in Bracken Cave in Texas eat about 200 tons of insects . . . every night.

Vampire bats don't suck blood. They lap it up. Calm down. There are only three species of vampire bats in the whole world. If you are traveling in Central or South America, however, you might see a vampire bat bite a cow and then lick blood from the wound—no sucking involved.

Bats don't have "fat days." The metabolism of a bat is enviable—they can digest bananas, mangoes, and berries in about 20 minutes.

The average bat will probably outlive your dog. The average lifespans of bats vary depending on the species, but some types of brown bat can live to be 30 years old. Considering that many other small mammals live only two years or so, that's impressive.

Bats wash behind their ears. Bats spend more time grooming themselves than even the most image-obsessed teenager. They clean themselves and each other meticulously by licking and scratching for hours.

Bats use echolocation to get around in the dark. Bats are nocturnal, mostly because it's easier to hunt bugs and stay out of the way of predators when it's dark. But bats don't see very well, so they have to rely on navigational methods other than sight. In order to get around, bats send out beeps and listen for variations in the echoes that bounce back at them. Bats do use their eyesight to see things in the daytime, but most bat business is conducted under the blanket of darkness for convenience.

Fewer than ten people in the last 50 years have contracted rabies from North American bats. Due to television and movies, bats are thought to be germ machines, bringing disease and toxins to innocent victims. Not true. Bats avoid people. If you are bitten by a bat, go to the doctor, but don't start making funeral arrangements—you'll probably be fine.

Bats make up a quarter of all mammals. Yep, you read that right: A quarter of all mammals are bats. There are more than 1,100 species of bats in the world. That's a lot of bats!

More than 50 percent of the bat species in the United States are either in severe decline or are listed as endangered. You don't know what you've got until it's gone. Industry, deforestation, pollution, and good old-fashioned killing have wiped out many bats and their habitats. For information on how to help keep bats around, contact your local conservation society.

Cold night? Curl up next to a bat. Inside those drafty caves they like so much, bats keep warm by folding their wings around themselves, trapping air against their bodies for instant insulation.

An anticoagulant found in vampire bat saliva may soon be used to treat human cardiac patients. The same stuff that keeps blood flowing from vampire bats' prey seems to keep blood flowing in human beings too. Scientists in several countries are trying to copy the enzymes found in vampire bat saliva to help treat heart conditions and stop the effects of strokes in humans.

Presidential Pets

Nearly all of the American presidents have had pets—but don't assume that all of these presidential pals were pooches!

✳ ✳ ✳ ✳

Early First Pets

GEORGE WASHINGTON STARTED the presidential cavalcade of critters. According to the Presidential Pets Museum in Williamsburg, Virginia, George and Martha had numerous hounds with names such as Tipsy and Sweetlips; horses named Nelson and Blueskin; and a parrot.

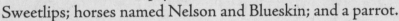

Thomas Jefferson kept two bear cubs that adventurers Lewis and Clark brought back from their journeys. The Marquis de Lafayette gave John Quincy Adams an alligator, and the Sultan of Oman gave Martin Van Buren two tiger cubs. James Buchanan was given a herd of elephants from the King of Siam.

Hail to the Chief Zoo Keepers

Theodore Roosevelt and Calvin Coolidge had so many pets that they could have hosted *Wild Kingdom*. Roosevelt's menagerie included snakes, guinea pigs, bears, kangaroo rats, owls, a flying squirrel, a lion, a hyena, and a zebra. Coolidge matched him with various breeds of dogs and cats, raccoons, canaries, a bear, a donkey, a bobcat, a wallaby, lion cubs, and a pygmy hippo.

Vote-getters

Some pets are credited with helping turn political fortunes. When Franklin Roosevelt thundered that critics couldn't pick on his "little dog" Fala, it reportedly helped him win the 1944 election. And even though he wasn't president at the time, Richard Nixon's "Checkers speech" is credited with saving his spot on the Eisenhower ticket.

Warhorses—Literally

Horses were critical on Civil War battlefields, and many sacrificed their lives for the cause.

✳ ✳ ✳ ✳

MORE THAN A million horses and mules perished during the Civil War. Horses went through rigorous training to prepare for battle. In what were essentially equine boot camps, they were taught simple commands, as well as strategies for protecting their riders under fire. Acting as a shield, a trained horse could drop to the ground to protect its rider from gunfire and provide cover from where the soldier could return shots. However, this also put horses directly in the line of fire.

Horse Heroes

Some horses achieved levels of fame on par with those of their human officers. General Lee's mount Traveler was well known for his stamina and bravery. After the war, people who saw Lee ride by would pluck hairs from Traveler's tail as a memento of the honored veteran. Union General George Meade's horse Old Baldy was a powerful steed that accompanied him in his victory at Gettysburg. Old Baldy is said to have been wounded as many as 14 times. Despite the beating he took during the war, he lived to the ripe old age of 30 and outlived his master. Today, his mounted head is on display at the Civil War and Underground Railroad Museum of Philadelphia.

Equine Value

The Union army bought horses in bulk, but in the Confederacy, many soldiers had to bring their own steeds, for which they were paid 40 cents per day. At the end of the war, Lee asked Grant to allow defeated soldiers to keep their horses in order to help plant crops at home. Grant didn't hesitate to agree.

Freaky Animal Facts

* The female boomslang, a type of African tree snake, looks so much like a tree branch that birds—its main prey—will land right on it.

* The koala, which is not actually a bear but a primitive marsupial that has existed in its present form for more than a million years, gets almost all the liquid it needs from licking dew off tree leaves.

* The emperor penguin of the Antarctic has equality of the sexes down pat: The female lays the egg, but the male has a brood pouch—a roll of skin and feathers between his legs that drops over the egg. He must then protect the egg and keep it still for two months until it hatches and the female returns to feed the chick.

* When the male snowy owl wishes to arouse a female, he dances while swinging a dead lemming from his beak.

* Roadrunners take a no-holds-barred approach to killing rattlesnakes. They jab the snakes with their sharp bills, shake them, body-slam them, and then administer a final peck in the head before devouring the prey headfirst.

* The laziest mammals are armadillos, sloths, and opossums. They spend 80 percent of their lives sleeping or dozing.

* Vultures sometimes eat so much they can't take off in flight.

* The oldest bird on record was Cocky, a cockatoo, who died in the London Zoo at the age of 82.

* Every year, approximately 3,000 birds collide with U.S. Air Force planes, causing an estimated $60 million in damages.

* Nine-banded armadillos have identical quadruplets every time they give birth.

Sweating Like a Dog

It's hot out, and you and your dog are taking a midday run. As your shirt becomes soaked with perspiration, Rover breaks a sweat in his own way—by panting. That's what you've always thought, anyway.

❊   ❊   ❊   ❊

DOGS DON'T SWEAT by panting, but they do regulate their body temperature that way. They release excess heat through their tongues while taking short, rapid breaths—sometimes as many as 300 to 400 per minute. This process expels hot air from their lungs and body cavities. But because such breathing is quick and shallow, it doesn't use up much energy, so the dog doesn't risk overheating more.

It's widely thought that dogs don't sweat at all, but that's not true either. A dog's fur prevents the release of moisture and heat. Our canine companions lack the extensive sweat glands that we have, but they do have some—and they're located in their foot pads. That's why you can see pad prints on wood floors and outdoor decks in warm weather. That's also why your pet's paws can smell kind of funky, like armpits. Dogs perspire when they get overheated or anxious, just like humans do. We get sweaty pits, they get sweaty paws.

However, the paws of a dog are too small to release the excessive body heat that's generated during a vigorous romp in the heat of summer. That's when the tongue flops out. With both cooling systems in operation, a dog is so efficient at lowering its body temperature that it can endure prolonged, high-speed chases—in pursuit of a rabbit, say—without the need to stop. On the other hand, a rabbit has no such means to cool down. Sure, it'll keep running to evade the dog, but it'll likely drop dead from heat exhaustion in the process.

The Purpose of Purring

Felines are forever mysterious to mere humans. One of our favorite kitty quirks is the act of purring.

✳ ✳ ✳ ✳

Don't cats purr because they're content? When cats purr, people are happy. But cats aren't always happy when they purr. There are actually several reasons why cats purr, and happiness is only one of them. Purring begins at birth and is a vital form of communication between mother and kitten: The kitten purrs to let its mother know it's getting enough milk, and the mother cat purrs back to reassure her kitten.

What's the significance of purring? Cats purr throughout their lives and often at times when you wouldn't expect. Cats purr when they are frightened, ill, or injured—they even purr while giving birth. Animal behaviorists believe that cats purr to comfort themselves during stressful situations and to signal their feelings to other cats. A frightened cat may purr to indicate that it is being submissive or nonthreatening, and an aggressive cat may purr to let other cats know that it will not attack. Some cats purr even when they're dying.

Domestic cats aren't the only felines that purr. Some of the big cats—lions, cougars, and cheetahs—also exhibit this endearing behavior. But it's much more soothing—not to mention safer—to stroke a purring pet cat than the king of the jungle.

✳ **Napoleon Bonaparte suffered from ailurophobia, which is the fear of cats.**

✳ **Belgium once experimented with using cats to deliver mail.**

✳ **Ancient Egyptian families mourned the death of a pet cat by shaving off their eyebrows.**

✳ **The stripes on a tiger's face are used for identification, since no two tigers sport the same stripe pattern.**

Freaky Facts: Animal Mating

* The funnel-web spider knocks his mate unconscious with pheromones before mating.

* The male *Argyrodes zonatus* spider secretes a drug that intoxicates the female, which is good because otherwise she would devour him.

* Harlequin bass and hamlet fish take turns being male and female, including releasing sperm and eggs during the mating process.

* Albatross can spend weeks courting, and their relationships can last for decades. However, the actual act of mating lasts less than a minute.

* Fruit flies perform an elaborate seven-step dance before mating. If any part of it is not completed perfectly, there will be no copulation.

* Mayflies live for only one day, during which they do nothing but mate.

* The rattlesnake has two penises. The penis of the echidnas has four heads, and a pig's is shaped like a corkscrew.

* The male swamp antechinus, a mouselike marsupial in Australia, has sex until he dies, often from starvation. Often, he's eaten because he's simply too weak after mating to escape predators.

* The idea that opossums mate through the female's nostrils is a myth. Although the male opossum has a forked penis, he mates with the female in the normal manner. Conveniently, she has two uteri, so he deposits sperm into both of them at once.

* Rattlesnakes can mate for 23 hours, and soapberry bugs can go for 11 days. But that's nothing compared to stick insects—they can go on for 2 months!

No Cats Were Harmed

What's the story with so-called "catgut," the material used in tennis and musical-instrument strings? It is really made of guts, but the guts don't come from cats.

✳ ✳ ✳ ✳

Intestinal Fortitude

THE INTESTINES OF cattle and other livestock are cleaned, stripped of fat, and prepared with chemicals before they can be made into catgut string. Historically, preparation of string from animal tissue dates back thousands of years in recorded history and likely much longer ago in reality. Intestines are uniquely suited because of their combination of strength and elasticity, even in comparison to other pretty robust naturally occurring strings like horse hair or silk. This makes sense intuitively because of the role our intestines play in our bodies, but let's not think too hard about that. One downside is that the prepared gut fibers are still very absorbent—enough so that even atmospheric humidity can warp them out of shape.

In Stitches

Historical humans realized surprisingly early that gut string was a good way to sew up wounds, and in this case the absorbency is a bonus. These humans weren't aware of issues like infection, so their ingenuity with gut string was a matter of simple craftsmanship: you should sew with the strongest material you can find, whether for clothing or shelter or for a wound.

Thousands of years later, scientists experimented and realized that gut string could dissolve in the human body. With the eventual rise of germ theory, doctors were able to create and use sterilized dissolving sutures that would be recognizable to the ancient Egyptians who first documented their sewing of wounds. In fact, most dissolvable stitches are still made with prepared animal fibers or with synthetics that were designed to mimic animal fibers in the body.

High Strung

For musical instruments, gut strings also date back thousands of years. In both Latin and Greek, the terms for strings and bowstrings (and our modern words chord and cord) were from the original Greek term meaning guts. Musicians found that gut strings made the best sound, but the strings warped, frayed, and broke quite quickly because of the effects of moisture in the air and from musicians' touch. Modern musicians can use strings with a core of gut that's surrounded by a snug winding of very fine metal. In a fine example of art imitating nature, this structure mimics the way our flexible gut fibers are arranged in our intestines, with both lengthwise fibers and circular bands.

The Gut Racket

Gut strings are still considered the gold standard by many high-level tennis players and manufacturers. In tennis, the absorbency of gut strings is counteracted with topical wax that seals the strings. Choosing and making gut strings for tennis rackets is still an artisan craft, and some cattle—most if not all tennis gut strings come from cattle—apparently produce finer quality gut strings than others, creating a Wagyu-beef-like hierarchy among cattle ranchers.

Digesting the Information

Humans have shown remarkable ingenuity since we first diverged from our most recent ancestor, but even by human standards, it's unusual to have a found material that works as both a durable tool and a nourishing food—depending on how you prepare it. Whether you're preparing for your Wimbledon debut or a period-correct Baroque chamber orchestra, consider the millennia-old tradition of the gut string.

The Great Sickness

When Western explorers, traders, and settlers made their way into the Alaskan frontier in the 18th and 19th centuries, they brought with them diseases to which the natives had never before been exposed. Maladies such as measles, tuberculosis, and influenza wiped out entire communities whose members had no natural immunity to such illnesses. So many natives died that it forever destabilized the traditional Alaskan culture. But not all of these stories had tragic endings.

✻ ✻ ✻ ✻

North to Alaska

PRACTICALLY NO ONE had heard of Nome, Alaska, before January 1925. Just two degrees south of the Arctic Circle on the Bering Sea, it was a small village that was isolated during the summer and inaccessible during the long Arctic winters.

Nome had one doctor, four nurses, a 24-bed hospital, and 8,000 units of expired diphtheria serum when it radioed for help on January 22, 1925. Doctor Curtis Welch desperately pleaded for 1.1 million units of serum to combat a diphtheria epidemic that threatened 10,000 lives. Without the serum, the mortality rate would be close to 100 percent.

The nation quickly responded to Nome's call for help. The needed serum was gathered and sent northward from Seattle by steamship. Meanwhile, 300,000 units—enough to treat 30 people—were found in Anchorage. Though not enough to control the epidemic, it was enough to contain it until the rest arrived.

But how to get the 20-pound package of serum from Anchorage to Nome? An airlift was considered but was quickly discounted: The local planes were too unreliable, and the weather was too poor for flying. The Bering Sea was already frozen over for the winter, making a delivery by water impossible. The only way to reach Nome, it was decided, was by dogsled.

In the years before the bush plane, the dogsled was the only reliable way to get around the interior of Alaska. Sled teams and their mushers traveled down the Iditarod Trail, among others, delivering mail and supplies and bringing out gold. This time, they were asked to deliver a more precious cargo.

The Great Race of Mercy

The package of serum was rushed by train from Anchorage to the town of Nenana. On January 27, 1925, "Wild Bill" Shannon and his team of 9 dogs were the first of 20 mushers and approximately 150 dogs to relay the serum 674 miles to Nome.

Known as the Great Race of Mercy, the mushers followed the Tanana River for 137 miles to the village of Tanana, where the Tanana and Yukon rivers meet. There, the route followed the Yukon for 230 miles to the village of Kaltag. From Kaltag, it was a 90-mile run to Unalakleet. Then came the most dangerous leg of the relay: 208 miles over the Seward Peninsula, which was completely exposed to the worst Arctic winter in 20 years (including –50-degree temperatures and gale-force winds). Once across the Peninsula, the last 42 miles to Nome were across the treacherous frozen ice of the Bering Sea.

At 3:00 A.M. on February 2, Gunnar Kaasen and his team of dogs, led by the now-famous Balto, made it to Nome with the serum. Eventually, another run was made to deliver the rest of the medicine. Officially, the epidemic claimed seven lives, but there were probably upwards of 100 fatalities among the natives outside Nome, who buried their children without reporting their deaths.

The mushers and their dogs became instant celebrities, receiving letters of commendation from President Calvin Coolidge, gold medals, and the princely sum of $25 for their trouble. Many participants traveled the country recounting their trip to Nome, including Kaasen and Balto, who toured the West Coast in 1925 and 1926. In 1925, a statue of Balto was placed in New York's Central Park. When he died in 1933, he was stuffed and put on display at the Cleveland Museum of Natural History.

Freaky Facts: Opossums

* The rat-tailed opossum is the only marsupial that is native to North America.

* A female opossum can bear as many as 20 babies in one litter, but because she only has about a dozen teats, every birthing is followed by an epic race through her fur to reach the safety of her pouch and find a nipple. Because each baby clamps onto the same nipple every time it feeds, those that arrive too late will perish.

* A mother opossum gestates her young for only 13 days before they are born.

* Baby opossums, which are about the size of honeybees at birth, are so undeveloped that they can't even suck milk. The mother has to use her own muscles to "pump" the milk into their tiny gullets.

* After baby opossums leave the pouch (at between two and three months), the mother carries all of them on her back for a month or so whenever she and her brood leave their den.

* Opossums do not normally hang from trees by their tails.

* Opossums have opposable, thumblike digits on all four paws and a tail that can grasp food and tree branches.

* Despite their reputation for eating anything—even rotting flesh—opossums are fastidious about hygiene. They bathe themselves frequently, including several times during meals.

* The opossum is the oldest surviving mammal family on Earth, dating back to the dinosaurs and the Cretaceous Period, some 70 million years ago.

Immortality in the Ocean

By repeating its life cycle, this creature reverses the aging process.

✳ ✳ ✳ ✳

A Handy Talent

IN THE 1990S, scientists noticed something peculiar about *Turritopsis nutricula*, a hydrozoan related to the jellyfish: It seemed to be able to live forever. The tiny creature, which can be found bobbing around blue Caribbean waters, is capable of returning to its juvenile polyp stage, even after reaching sexual maturity and mating.

The hydrozoan evolved this ability through a cell development process called "transdifferentiation." Certain other animals can do this in order to regenerate an appendage—say, an arm or a leg—but this is the only creature that can regenerate its whole body. First, it turns into a bloblike shape, which then develops into a polyp colony. *Discover* magazine described this process as being similar to a "butterfly turning back into a caterpillar." Actually, the creatures seem to only revert to this early stage when in trouble (e.g., they are starving or are in danger of getting eaten by other sea creatures). Crazily enough, they can repeat this process over and over again, whenever necessary. Thanks to this evolutionary quirk, they do not have to die, despite the ongoing threat of hunger and bullies.

Panama or Bust

Recently, scientists have noticed that, although the creatures originate in the Caribbean, some of them have been turning up in waters near Spain, Italy, and Japan. According to *Discover*, "Researchers believe the creatures are criss-crossing the oceans by hitchhiking in the ballast tanks of large ships." One thing's for sure: With its ability to turn back the clock, the *Turritopsis* is bound to become the envy of everyone in Hollywood.

Fur-Loving Führer

Did the Third Reich treat "subhumans" subhumanely? Even as Nazi Germany exterminated millions of people, it fiercely protected its animal population.

✳ ✳ ✳ ✳

THE STORY OF the Holocaust is a heart-wrenching tale: Nazis, operating under the decidedly mad precepts of Adolf Hitler and the Third Reich, systematically and diabolically destroyed those it deemed subhuman. By their twisted definitions, this number included people of non-Aryan bloodlines, such as Jews, Gypsies, and other "racially inferior" groups. Overlooking the "subhuman" stipulation entirely, however, this did not include members of the animal kingdom. Here's the story of an uneven regime that found beauty and worth in its animals while it delivered death and destruction to humankind.

Animals' Best Friend

Not many would disagree that Adolf Hitler was the Devil incarnate. Yet in the area of animal rights, he was anything but. This bizarre dichotomy also applied to Hitler's principal henchmen, Hermann Goering and Heinrich Himmler. All felt that defenseless animals deserved better treatment than they were receiving and took proactive steps to ensure it.

As early as 1927, members of the Nazi party called for action against animal cruelty. In particular, they zeroed in on the kosher butchering that was being practiced throughout the nation. (Later, in order to justify his persecution of Jews, Hitler promoted graphic films that showed them slaughtering lambs.) In early 1933, just after the Nazi organization had risen to power, its parliament passed laws regulating the slaughter of animals. Most notably, Goering enacted a ban against the act of vivisection (the dissection of animals for scientific study): "An absolute and permanent ban on vivisection is not only a necessary law to protect animals and to show sympathy with their pain, but it is also a law for humanity itself," said Hitler's right-hand man. "I have therefore announced the immediate prohibition of vivisection and have made the practice a punishable offense in Prussia [the core of the German Empire]. Until such time as punishment is pronounced, the culprit shall be lodged in a concentration camp." Goering would go on to ban commercial animal trapping, and he placed strong restrictions on hunting. He even went so far as to regulate the boiling of crabs and lobsters, allegedly because he felt distressed by their "screams" during the cooking process.

Man's Worst Enemy

It helps to understand the era's political overtones. Even as the Nazis were passing groundbreaking legislation on behalf of animals, they were drawing up decrees that would define "lesser" beings, targeting them for eventual extermination. In their demented view, these "subhumans" included anyone descended from non-Aryan parents or grandparents. Even those with one grandparent from the "wrong" bloodline were placed in jeopardy. At the same time that the Nazis passed their pro-animal laws, they opened the Dachau, Buchenwald, Ravensbrück, and Sachsenhausen concentration camps.

Although Hitler's regime would eventually exterminate an estimated six million Jews and untold numbers of other human beings, it would continue its efforts on behalf of animals. Meanwhile, the systematic killing of humans continued.

It's beyond ironic that this sort of animal-rights model should come from such a diabolical source, but that's how history played out. The Nazis viewed the issue in simplistic terms and acted accordingly. In their distorted view, Aryans ruled the hierarchy; animals were second; and subhumans, or *untermensch*, dwelled at the bottom. It was the gross misfortune of more than six million living souls that their assigned classification wasn't at least on a par with that of a dog.

Animals in Disguise

Polar bears waddling across arctic plains, chameleons turning from green to brown on a tree limb—images such as these come to mind when we think of animal camouflage. But there is far more to this survival behavior than just blending into the background of nature.

✳ ✳ ✳ ✳

✳ Polar bears are not actually white. They appear white because their fur is made up of hollow, transparent, pigment-free hairs, which scatter and reflect visible light, in much the same way snow crystals do. Underneath the fur, a polar bear's skin is black, which is how the animal appears if photographed with film that's sensitive to ultraviolet light.

✳ The Eurasian bittern, a wading bird from the heron family, nests among tall river reeds. When alarmed, it stretches its neck and either becomes motionless or, if there is a breeze, begins to sway in time with the reeds. Because of its plumage pattern—vertical dark-brown stripes on beige—the bittern becomes extremely difficult to see.

✳ The purpose of camouflage is not always to hide from predators. Sometimes the best defense is to simply look as unappetizing as possible. For this reason, the caterpillars of many species of moths have evolved to look like bird droppings.

✳ Some species of sea horse, notably the Australian leafy sea dragon (*Phycodurus eques*), mask themselves as seaweed. Not only are these creatures green, they also have long, tattered ribbons of skin that look like seaweed fronds growing from their bodies.

✳ Some varieties of what are called "geometer" caterpillars—larvae of *Geometridae* moths—

closely resemble twigs, thanks to a combination of their grayish-brown color, humps on the body that look like tree buds, a lack of abdominal legs, and the habit of resting with their bodies stretched out at an angle from the branch. Maintaining this uncomfortable pose can be taxing, so these caterpillars often spin a supportive silk thread between their head and a leaf or twig just above it.

* In Africa, one might spot what appears to be a small, densely packed cluster of dead ants shuffling by. Close inspection reveals that the ants are being transported on the back of a large predator—*Acanthaspis petax*, or the assassin bug. After killing the ants, this bug sucks their bodies dry and covers itself with their exoskeletons. This behavior enables the *Acanthaspis petax* to avoid its own predators while infiltrating anthills for an after-dinner snack.

* Eyespots on the wings of moths and butterflies serve to divert a predator's attention from the moth's vital organs to its wings, which can withstand more damage. Going a step further, the South American frog *Physalaemus nattereri* has evolved prominent black eyespots on its butt. When attacked, it heaves its backside into the air and gives the predator an "eyeful." If the predator persists, the frog treats it to a spray of unpleasant secretions.

* It isn't necessary to look terrifying to scare away predators. Hole-nesting birds, such as the European blue tit, hiss like a snake if disturbed while they incubate eggs and care for small young. A dormouse does the same if something encroaches on its dark lair, and feeding termites hiss in chorus to scare off intruders.

* When threatened, a hognose snake flattens its neck like a cobra and raises its head. If the would-be predator is not deterred by the threat, the hognose rolls over and plays dead, going so far as to let its tongue hang out of its mouth while emitting a foul scent to simulate the smell of a rotting corpse.

Laika—First Casualty of the Space Race

The first living being to be launched into orbit was a three-year-old mongrel named Laika, an unassuming stray that became an unwilling pawn in the space race. Unfortunately, contingency plans to bring the dog back to Earth were never developed, and Laika also became the first creature to die in space.

✳   ✳   ✳   ✳

THE SPACE RACE was launched on October 4, 1957, when the Soviet Union successfully sent *Sputnik 1* into orbit. And with that, the mad dash to be the first country to get a man into space was under way.

Russian Premier Nikita Khrushchev ordered members of his space program to launch a second spacecraft into orbit on the 40th anniversary of the Russian Revolution. The primary mission of *Sputnik 2* was to deliver a living passenger into orbit. The spacecraft—which was designed and built in just four short weeks—blasted into space on November 3, 1957.

Initially, three dogs were trained for the *Sputnik 2* test flight. Laika, a 13-pound mixed-breed that was rescued from a Moscow shelter, was ultimately chosen for the historic mission. Though the Soviets affectionately nicknamed the dog Little Curly, the Western press referred to her as Mutt-nik.

3, 2, 1 . . . Blast Off!

Sputnik 2's capsule was equipped with a complete life-support system, padded walls, and a harness. But during liftoff, Laika's respiration rocketed to four times its prelaunch rate, and her heartbeat tripled in speed. Sensors indicated that the cabin's temperature soared to 104 degrees Fahrenheit.

The Soviets waited until after the spacecraft was launched to reveal that Laika would never return to Earth. The mission had been so rushed that there hadn't been time to plan for the dog's return. She only had enough food and oxygen to last her ten days.

The Soviets issued conflicting information regarding Laika's death. Initial reports indicated that Laika survived four days in orbit—evidence that a living being could tolerate space. Her eventual death was attributed to either oxygen deprivation or euthanization via poisoned dog food—depending on the source.

Laika's remains were destroyed when *Sputnik 2* fell out of orbit and burned up reentering Earth's atmosphere.

Russians named postage stamps, cigarettes, and chocolate bars in Laika's honor, treating the dog as a national hero. In the meantime, there was outrage in the West over what many perceived to be the cruel and inhumane treatment of the animal.

The Aftermath

In October 2002, a Russian scientist revealed that Laika had actually died only hours into the mission due to overheating and stress.

In March 2005, a patch of soil on Mars was unofficially named in Laika's honor. Dozens of other animals—including monkeys, frogs, mice, worms, spiders, fish, and rats—have since successfully made the journey into, and back home from, space.

✳ From 1957 to 1966, the Soviets successfully sent 13 more dogs into space—and recovered most of them unharmed.

✳ Dogs were initially favored for space flight over other animals because scientists believed that they could best handle confinement in small spaces.

✳ To train for her space flight, Laika was confined to smaller and smaller boxes for 15 to 20 days at a time.

✳ In 1959, the United States successfully launched two monkeys into space. Named Able and Baker, the monkeys were the first of their species to survive spaceflight.

A New Leash on Life: Surprising Service Animals

Move over, Lassie. Step aside, Rin Tin Tin. Service animals are real-life heroes.

❊ ❊ ❊ ❊

A Friend in Need

THE TERM SERVICE *animal* is generally defined as any animal that is trained to aid a disabled handler, specifically in regards to his or her disability. While the general populace is accustomed to seeing a well-behaved dog fulfill this function, the range of animals trained for service work is far broader than just dogs being used as seeing-eye, hearing-ear, seizure-alert, or wheelchair-assistance animals.

High-Ho... *Cuddles?*

Since 1999, miniature horses have been trained as service animals for the visually impaired. Cuddles, the first guide horse, went into service two years later. Guide horses are excellent for people who are allergic to or fearful of dogs, or are simply horse enthusiasts. Guide horses, which  are comparable in size to a medium to large dog, are intelligent, docile, have excellent memories, provide stable physical support to individuals with balance problems, can be house-trained, and have a lifespan three to four times that of a service dog, on average. Some drawbacks to the miniature horse include their limited dexterity compared to a typical service dog and their more limited portability.

The Monkey Business

Since the introduction of the first service monkey in 1979, the use of Capuchin monkeys as service animals has increased with the development of a codified training system. Their longevity, portability, intelligence, and extraordinary dexterity make them superior choices for people who require assistance with a wide variety of daily activities. Not only can monkeys turn lights on and off or pick up dropped items, they can also retrieve meals from the refrigerator, take dishes from the cupboard, load CDs, and press the start button on a computer or microwave. As long as they're on a harness or leash and they're well behaved, service monkeys can even have their own seats on airplanes.

When Pigs Fly

Pigs are also occasionally used as emotional-support and therapy animals. They're extremely intelligent, but they are limited in what practical services they can perform—unless of course, their handler is a devotee of truffle hunting.

One service pig, technically a "therapeutic companion pet," caused an uproar in 2000 when US Airways transported it from Philadelphia to Seattle. Service animals don't need to be certified, nor is it permissible to ask what function the animal fulfills. Thus, the airline had to take it on good faith that the pig wasn't just a pet. The pig's owner did make prearrangements and presented a doctor's note (which was never publicly released) that allegedly indicated that the pig had a calming effect on the woman, who had a heart condition. Whether the pig had been specifically trained to calm the woman is debatable. Furthermore, the service swine's owner misrepresented its size to a significant enough degree that the pig protruded into the aisle, and she had no means of restraining or controlling it.

Although the flight was uneventful, during deplaning the pig panicked and its squealing and erratic behavior caused distress to the other 200 passengers. Not to mention the mess it left behind for the flight crew to deal with before the next flight. US Airways has since revised its policy concerning service pigs. Moral of the story: People, leave your pigs at home.

Killer Bees

As the name implies, the stings of killer bees can be fatal. Arm yourself with the knowledge to escape and you may save your life.

Killer bees are notoriously easy to provoke and hard to escape. The African honeybee, aka the killer bee, will chase humans much farther than its more common honeybee cousin (there have been reports of bees giving chase for over 600 yards). They can also stay angry for up to a full 24 hours. Here's what to do if you find yourself in the path of these bees:

❋ Don't hesitate, just run. The majority of human and animal deaths at the hands of killer bees happen because victims don't get away fast enough.

❋ Keep your hands over your eyes, nose, and mouth as you dash. Bees primarily go after the face and head, so this will give you some form of protection.

❋ Don't swat; it will only make the bees angrier.

❋ Head for anywhere indoors, and then duck into a dark corner and cover yourself. Bees usually won't follow you inside; if one does, it's likely to become confused and go toward the light at a window.

❋ If you can't find a building or a car, opt for bushes or high weeds instead. They're your next best protection.

❋ Never remain still, and never jump into water. Bees are smart enough to know that you have to come up for air sooner or later, and they'll be waiting for you.

❋ If you do get stung, use a blunt knife or the edge of a credit card to scrape the stinger from your skin as soon as possible. Don't pinch or pull on it, as this could end up increasing the flow of the toxin into the body. Get to a doctor immediately!

A New Breed of Nip and Tuck

Cropping ears and tails to meet breed standards are long-standing practices. But does Fifi really need a nose job or breast reduction?

✳ ✳ ✳ ✳

A Face-Lift for Fido

Today's pampered pets enjoy the same lavish luxuries as their humans: fancy spas, posh hotels, and upscale boutiques. So if a person can get a little work done, why can't a dog?

Two of the places leading the pack on pooch plastic surgery are Brazil and Los Angeles, which are already considered the top dogs when it comes to cosmetic enhancement. Most procedures for pets are medically motivated: face- and eye-lifts for chow chows and shar-peis, whose copious skin folds can irritate the eyes and become infected; rhinoplasty for smush-nosed breeds such as Boston terriers, pugs, and bulldogs, who often have trouble breathing; and chin-lifts for excessive droolers that are prone to mouth infections, such as mastiffs and Newfoundlands. Broken teeth from too much bone chewing can be repaired with root canals and crowns, and a pronounced overbite can be corrected with braces.

Cosmetic Canines

That said, a growing number of pet cosmetic surgeries seem to be driven by the same insecurities that compel people to go under the knife themselves. Consider Neuticles—prosthetic testicles for neutered dogs to help preserve their "self-esteem" after the goods are gone. Other popular procedures include Botox injections to fix overly droopy ears and breast reductions for female dogs after they've given birth.

U.S. dog shows prohibit surgeries that alter a dog's appearance beyond traditional breed standards. But if Miss California can have a boob job, can a poodle's nip and tuck be far behind?

Curious Creatures Among Us

* When some types of frogs vomit, their entire stomach comes out. The frogs then clean out the contents and swallow the empty stomach.

* Pacific Island robber crabs love coconuts so much that they have developed the ability to climb trees to get them.

* The water-holding frog is the greatest survivor in the animal kingdom. When it rains, the frog absorbs water through its skin. It then burrows into the sand, where it can live for up to two years off its water reserve.

* The praying mantis is the only insect that can turn its head 360 degrees.

* A chameleon can focus its eyes separately to watch two objects at once. It also has a tongue that is longer than its body.

* The eyes of the ostrich and the dragonfly are bigger than their brains.

* The electric eel can produce 350 to 550 volts of electricity up to 150 times per hour without any apparent fatigue.

* Spider silk is five times stronger than steel, but it is also highly elastic—a rare combination in materials. This silk stretches 30 percent farther than the most elastic nylon.

* A large parrot's beak can exert 500 pounds of pressure per square inch, enabling the bird to feast on such delicacies as Brazil nuts with a simple crunch.

* Wasps can make paper by mixing wood pulp with saliva to form a paste, which dries stiff.

* The egg of some species of mayfly takes three years to hatch—and then the insect lives only six hours.

Beasts of Burden

Although World War II is well-known for its tanks, airplanes, and advanced communications, both sides were surprisingly reliant on old-fashioned horses and mules to keep their armies provisioned and on the move.

✳ ✳ ✳ ✳

Horses in the German Army

DESPITE THE GERMAN Army's reputation for panzer assaults, only 1 out of 6 of its 300 divisions were armored or mechanized. A German infantry division had a strikingly low-tech component: Typically there was one horse for every three of its men. The proportion of horses was even greater for mountain and light divisions. And the use of horses grew as Allied bombers destroyed Germany's industrial plants, which were used for manufacturing tanks and for refining petroleum. Meanwhile, the extensive horse farms of Prussia and Pomerania lay untouched until they were overrun by the Soviets in 1945.

Horses were especially vital for the German and Russian armies on the Eastern Front. The region's vast expanses, primitive highways, and extreme weather put enormous strains on mechanized transports. Horses were often better suited than trucks for traversing the endless steppes or slogging through muddy or snow-covered roads.

Contrary to general belief, the German juggernaut that invaded the Soviet Union was surprisingly low-tech—it was supported by at least 750,000 horses that ate 16,350 tons of food daily. Much of the grain that was used to sustain them was taken from Russian peasants at gunpoint, adding to the horrific losses of civilians under the occupation.

The animals themselves suffered staggering losses. According to the Imperial War Museum, the Soviets used 21 million horses in the war; 14 million perished between June 1941 and

May 1945. The Germans lost 6.7 million horses in the East. When surrounded at Stalingrad, starving German soldiers ate 26,000 of them.

Relying On Mules in Burma

U.S. General Joseph Stilwell faced the seemingly impossible task of conquering Japanese-occupied Burma. He desperately needed to get supplies to China's armies. His force, "Merrill's Marauders," commanded by General Frank Merrill, had to retreat and advance across hundreds of miles of jungle, mountains, and swift-flowing streams. With mechanical equipment scarce and hard to maintain under the tropical conditions, mules proved essential to the enterprise.

During their 1942 retreat along the Chaunggyi River in Burma, many in the ragtag force keeled over from heatstroke. Then, Major Merrill succumbed to a heart attack. To survive, the soldiers had to rely on unusual means of transportation. Forestry Service officers from Great Britain commandeered 60 native men to haul equipment. Stilwell requisitioned a mule team from China that had crossed his path. He limited his exhausted men to carrying ten pounds of equipment and put everything else on the mules, including the sick and injured men who could no longer walk. Stilwell's soldiers survived, barely escaping a pursuing force of Japanese cavalry.

"Mule skinners" and "muleteers," as the animals' drivers were called, learned the peculiar habits of their animals. The mules hated bodies of water, which made crossing rivers difficult, but they despised elephants even more. That was a problem in Burma, where pachyderms had long been used to carry cargo. And, being mules, the creatures were stubborn: Their refusal to leave corrals and rest spots nearly allowed the pursuing Japanese to catch up.

In the 1944 Galahad offensive, in which Allied troops attempted to seize Burma's strategic Myitkyina airfield, the Marauders had to march over the bamboo thickets of the

Kumon mountain range, crossing one winding river some 56 times. They had to cut out steps through the passes for the mules. Some animals slipped off mud-covered passages to their deaths. Half the 3,000 humans in the force were felled by exhaustion or disease—or by kicks from the ornery beasts.

Each rifle company in the expedition was assigned 42 mules. The medics got a dozen mules, mostly to evacuate wounded. The animals, which weigh about half a ton, can carry about 200 pounds of equipment. Mortars and heavy machine guns required four mules each to haul. Other animals were obtained from Chinese troops, who captured tiny Japanese ponies and traded them to the Americans for cigarettes, flashlights, and other items. In Burma in 1944 and 1945, the British followed the Americans' lead, employing 7,000 horses and 24,000 mules, along with elephants that cleared jungle paths.

Phasing Out the Beasts

Yet World War II remained the first mostly mechanized war. The year before Pearl Harbor, the U.S. Army still had a pair of regiments with horse-drawn artillery, as well as two cavalry divisions and several regiments with combined motor and horse transport. However, horses found few uses stateside outside of patrolling the coasts for U-boats. The last U.S. cavalry unit to see action was the 26th Cavalry Regiment—the Philippine Scouts. In March 1942, at the besieged Manila Bay bastion of Corregidor, its 250 horses and 48 mules were slaughtered to supply food for the starving U.S. and Filipino defenders.

In 1943, only four new horses were brought into service. After the war, the Army's horse-breeding program was transferred to the Department of Agriculture.

Still, in at least one future conflict, beasts of burden continued to serve. During the Korean War, a mule that was captured from the Chinese Army (with brand ID 08KO) turned out to be the one that the U.S. had given the Chinese during the World War II.

Infamous Zoo Escapes

On Monday, November 9, 1874, New Yorkers awoke to shocking news. According to the New York Herald, enraged animals had broken out of the Central Park Zoo the day before, and a leopard, a cheetah, a panther, and other beasts of prey were roaming the park in search of hapless victims. Men rushed out with rifles, prepared to defend their families. But it was all for naught—the whole story was a hoax perpetuated by a reporter who was irate at what he thought were lax security measures at the zoo. As it turns out, though, that reporter may have been on to something.

✳   ✳   ✳   ✳

FAST-FORWARD TO SUNDAY, July 5, 2009, when more than 5,000 visitors were evacuated from Great Britain's Chester Zoo in Liverpool. "Chimps Gone Wild!" screamed the headlines the next morning. Apparently, 30 chimpanzees had escaped from their island enclosure. This great escape was certainly no hoax, but neither was it cause for alarm: The chimps got no further than the area where their food was kept. They gorged themselves until they had to lie down and rub their aching bellies. A short time later, zoo wardens rounded them up, and the escapees returned to their island peacefully.

When it comes to zoo escapes, primates are often among the culprits. One of the greatest of all nonhuman escape artists was the legendary Fu Manchu of Omaha's Henry Doorly Zoo. In 1968, this orangutan confounded his keepers by repeatedly escaping from his cage, no matter how well it was secured. Only when a worker noticed Fu slipping a shiny wire from his mouth did the hairy Houdini's secret come out: The orangutan had fashioned a "key" from this wire and was using it to pick the lock. Even more impressive is that he had the sense to hide it between his teeth and jaw—a place where no one was likely to look. After officials realized what the cagey animal had been up to, they stripped his cage of wires. Though Fu Manchu never escaped again, he was rewarded for his efforts with an honorary membership in the American Locksmiths Association.

Oliver, a capuchin monkey at Mississippi's Tupelo Buffalo Park and Zoo, did Fu Manchu one better. In 2007, he escaped from his cage twice in three weeks. Both times, he traveled several miles before being captured. Zookeepers suspected him of picking the lock, but they never figured out how he did it. Their solution was to secure his cage with three locks, a triple threat that has so far kept him inside. Word of Oliver's escapades drew so many visitors to the zoo that officials decided to capitalize on the capuchin culprit. A best-selling item at the zoo's gift shop is a T-shirt emblazoned with "Oliver's Great Escape," along with a map of his routes.

Evelyn, a gorilla at the Los Angeles Zoo, didn't need to pick a lock. She escaped on October 11, 2000, by climbing vines like Tarzan. After clambering over the wall of her enclosure, she strolled around the zoo for about an hour. Patrons were cleared from the area, and Evelyn's brief attempt to experience the zoo from a visitor's point of view ended when she was tranquilized and returned to her enclosure without incident.

Juan, a 294-pound Andean spectacled bear at Germany's Berlin Zoo, had a much more amazing adventure on August 30, 2004. It began when he paddled a log across the moat surrounding his habitat. He then scaled the wall and wandered off to the zoo playground. There, he acted just like a kid, taking a spin on the merry-go-round and trying out the slide. When he left in search of further amusement, clever animal handlers decided to distract him with a bicycle. Sure enough, Juan stopped to examine the two-wheeler as if he were contemplating a ride. Before he could mount it, however, an officer shot him with a tranquilizer dart, thus ending his excellent adventure.

Cher Ami: Heroic Pigeon

A carrier pigeon named Cher Ami braved brutal enemy fire to save nearly 200 trapped American servicemen.

✳ ✳ ✳ ✳

DURING WORLD WAR I, the U.S. Army Signal Corps put nearly 600 carrier pigeons into service in France. Among them was Cher Ami, a genuine hero among heroes.

During the Meuse-Argonne offensive in October 1918, more than 500 members of the 77th Infantry, led by Major Charles Whittlesey, found themselves trapped behind enemy lines. Worse, their location began taking artillery fire from fellow American forces that were unaware of their predicament. By the second day, nearly 300 members of the 77th had been killed.

Whittlesey dispatched messages to division headquarters via carrier pigeon. The first birds were quickly shot down, leaving only a pigeon named Cher Ami to finish the job. A barrage of bullets whizzed past the little bird as he rose from the bushes; for a moment, it seemed that he too would go down. Cher Ami took a bullet in the chest, and another shot nearly severed the leg holding his message canister, but he managed to make it to headquarters. Within hours, the surviving members of Whittlesey's "Lost Battalion" were safely behind American lines.

Cher Ami was hailed as a hero. Army medics worked to save his life, going so far as to carve a tiny wooden leg to replace the one that had been injured by enemy fire. The pigeon was awarded the French Croix de Guerre with Palm medal for his service and later became the mascot of the Department of Service.

Cher Ami died on June 13, 1919. He is now on display at the National Museum of American History in Washington, D.C., in an exhibition titled "The Price of Freedom: Americans at War."

The World's Most Dangerous Animals

The world's most dangerous animal has a legacy of death that stains history, and it seems incapable of ever being tamed. What is this ravenous, bloodthirsty beast? Humans, of course. After humans, here are the most dangerous animals in the world.

✳ ✳ ✳ ✳

Mosquitoes

THEY BITE 270 MILLION people each year, killing as many as 2 million. The seemingly innocuous mosquito is, in fact, the deadliest non-human on the planet, spreading up to a dozen diseases, including malaria. With more than 3,500 different species, mosquitoes inhabit every corner of the globe, but only a handful carry diseases. The species that live in Africa, Asia, and the Americas are the deadliest.

Venomous Snakes

Snake attacks account for close to 125,000 deaths per year, making them the deadliest reptiles on the planet. Only about 450 of the 3,000 known snake species are venomous, and the deadliest live in Africa, Asia, and Australia. Depending on the species and the severity of the bite, death can occur in a matter of hours if no treatment is administered.

Scorpions

The scorpion is a highly compact killing machine that takes more than 1,500 human lives every year. Just 25 of the 1,500 species of scorpions are deadly to humans, and the most venomous live in Africa, the Americas, and Central Asia. The most lethal is the fat-tailed scorpion of North Africa.

Lions

Responsible for several hundred human deaths each year—mostly in African and Indian villages—the king of the jungle is certainly worthy of our respect and fear. Lions have been known to kill animals weighing more than 3,000 pounds. They can reach speeds of 36 miles per hour, but they cannot sustain that speed for long, so they stalk-and-ambush their prey instead.

Crocodiles

The Nile crocodile is the most dangerous croc to humans, followed by those in Australia. With more than 60 teeth of various sizes, crocodiles are active, vicious, and treacherous brutes that kill hundreds of people each year. But crocs don't just kill prey with their lethal chompers: With an incredibly swift twist of its tail, a croc can capture prey by swatting it off the shore and into the water, where it is seized and devoured.

Elephants

We generally think of elephants as friendly, approachable, even cute animals. But don't let that image fool you—each year they crush more than 500 humans. Most elephant attacks occur in Africa and India, but attacks in zoos and circuses around the world are becoming more common. Elephants can be very unpredictable, and even so-called tame elephants have turned on and attacked trainers who have known them for years. The number of elephant attacks continues to rise as humans decimate the habitats of this majestic beast. Although elephants are herbivores, when humans get in the way of their food supply, they will use their six tons of bulk and their menacing tusks to trample and gore a person to death in a matter of seconds.

Box Jellyfish

Nearly a hundred swimmers and sunbathers are fatally stung every year by box jellyfish, which live in the tropical waters off Australia, the Philippines, Papua New Guinea, Malaysia, Indonesia, and Vietnam. A few people have died within seconds of being touched by the jellyfish's cluster of long tentacles, which can number between 40 and 60 in large specimens. When the tentacles come into contact with human skin, they react immediately—clinging to the skin and releasing their venom. If the victim runs or thrashes about, absorption of the venom accelerates. And if the victim attempts to remove the tentacles, even more venom is released. Fortunately, the sting of the box jellyfish is usually not fatal, but death by cardiac arrest can occur if antivenom is not given immediately.

Hippos

Found around rivers and lakes throughout Africa, no animal exudes more raw power than the mighty hippo. They might appear lazy and slow, but a hippo can run faster than a person and can maneuver its bulky head and jaws with deadly efficiency. The hippopotamus is an herbivore, but it won't hesitate to attack a human if it feels threatened. Hippos are responsible for approximately 150 human deaths per year, and, like elephants, they charge, trample, and gore their victims to death.

Cape Buffalo

Found in Africa south of the Sahara, Cape buffalo are imposing creatures, standing about five feet tall and weighing nearly a ton. Tip to tip, their horns can measure as much as 58 inches across. Although they are herbivores, Cape buffalo consider humans viable predators, and they won't hesitate to charge and put those horns to work. It's been estimated that each year, Cape buffalo are responsible for about 40 human fatalities.

Leopards

The most adaptable of the big cats, leopards can be found in deserts and forests, on mountains, at sea level, and in lands as diverse as China, India, and Kenya. Leopards have been called the most physically perfect of the big cats. Weighing 125 pounds or more and averaging two feet tall at the shoulder and seven feet long from nose to tail, they are among the most powerful animals in the world. Left unchallenged, leopards tend to be shy and reserved, and they will avoid confrontation with humans. But when challenged, an angry leopard is ferocious, capable of concentrating all its energy into a short-range attack of lightning speed, resulting in about 30 human deaths each year.

It's a Dog's Life

Ask a proud dog owner about the age of his or her beloved pooch, and you'll likely hear this response: "Well, Fido is five in human years, so multiply that by seven, and he's 35 in canine years." But doggone it, it's just not that simple.

* * * *

Calculating Canine Years

A NUMBER OF FACTORS are taken into account when one attempts to equate a dog's developmental level with that of a human being. Breed, size, heredity, nutrition, and training can affect and influence the development (and, therefore, the "human age") of a dog. Generally speaking, human beings experience developmental stages at these approximate ages: infant: 1 year; toddler: 3 years; youngster: 6 years; adolescent: 11 years; teen: 15 years; adult: 20 years; mature adult: 65 years; old coot: 85 years. Note that the rate of development is faster and closer together in the early years and spreads out as time goes on.

Dogs experience similar developmental stages—they just don't hit at the same points and in the same time frames as humans do. For example, an infant gains certain motor and communication skills, as well as knowledge, in his or her first year. In the same time period, most dogs—regardless of breed or size—reach a physical and "emotional" maturity similar to that of a teenager.

Almost all dogs develop at the same rate in their first 5 years. Relative to human years, they generally follow this range:

> 1 human year = 15 dog years
>
> 2 human years = 24 dog years
>
> 3 human years = 28 dog years
>
> 4 human years = 32 dog years
>
> 5 human years = 36 dog years

A Breed Apart

Relatively speaking, a larger breed will "age" at a faster rate than a smaller breed. At 7 human years, a Chihuahua is considered 44 in dog years, whereas a Labrador has hit the big 5–0. This span will continue to grow as the years go by. At 10 years old, the Chihuahua is 52 in dog years, whereas the Lab is ready to collect social security at 66.

Part of this growth disparity is due to the fact that larger breeds tend to have shorter life spans than smaller breeds. Larger breeds—such as Great Danes, St. Bernards, and Irish Wolfhounds—are susceptible to disabling ailments such as arthritis and hip dysplasia. Other concerns include nutrition and healthcare. If the family dog receives a balanced meal, as well as regular checkups and shots, it should live a long life. Also, just as humans do, dogs start to experience problems with sight, hearing, mobility, teeth, gums, and digestion as they near the "senior citizen" status of 60 or 70 dog years—a mere 12 or 13 human years. But the good news, according to veterinarians, is that a dog's average life span has increased from 7 to 12 years in the past eight decades.

More Myths About Dogs

A wagging tail indicates a friendly dog. It depends on the position of the tail when it's wagging. A loose, mid-level wag usually indicates an approachable dog, but tails held high or low could signal an aggressive or defensive demeanor.

A dog is sick if its nose is warm. The assumption is that a warm-nosed dog has a fever, but the fact is that a thermometer is the only reliable way to measure a dog's temperature. The normal body temperature of a dog is 100.5 to 102.5 degrees Fahrenheit.

Holy Guano, Batman!

The location of the Bat Cave was always one of Batman's best-kept secrets, but the Caped Crusader would need actual superpowers to enter the real bat cave near Fredericksburg, Texas. The ammonia fumes produced by the guano of the millions of bats that live there (normally around two million; more than three million in July and August) can be deadly for humans. The smell alone is certainly enough to keep most mortals at bay.

✳ ✳ ✳ ✳

Batman would also likely be annoyed by the hordes of tourists who flock to this abandoned railroad tunnel each evening to witness one of nature's most awe-inspiring spectacles. One plus, though: The place is virtually insect-free.

Like Clockwork

Natural phenomena are often difficult to predict, making it hard for the average person to schedule a time to experience them firsthand, but not so with the Fredericksburg bats. Although they'll return separately throughout the night, they emerge, more or less en masse, each and every evening about an hour before or after sundown.

First, two or three small black specks can be spotted circling the tunnel's mouth. Perhaps these harbingers are sent to check things out, for in less than a minute, the entire sky fills with a flapping frenzy of wings from more than two million furry, pointy-eared, Mexican free-tailed bats. The bats circle the surrounding area, gaining momentum and consuming any insects that may be lingering before flying off into the deepening twilight sky in search of more fertile feasting grounds. Because each bat consumes its own weight in insects nightly, it's no wonder they need to go out in search of food.

Nothing to Fear

It takes about an hour from the emergence of the first sentries before the last stragglers join the night's hunt, but for that short time, observers are treated to one of the most bizarre and awe-inspiring sights they are ever likely to behold. Of course, some folks might be a bit timid about being in such close proximity to so many bats. In reality, the bat-phobic have nothing to worry about. The old wives' tales about bats getting caught in people's hair are just myths. True, bats can't see well with their eyes, but their radar systems are so finely tuned that they can pick up matter as fine as a single strand of hair, so there's no way they would purposely entangle themselves. Park rangers confirm that bats want to avoid contact with people as much as people want to avoid contact with bats. That same extraordinary radar also prevents bats from ever colliding with each other, even within the cramped quarters of the tunnel, and allows a mother bat to immediately find her baby among all the other bats living in the abandoned tunnel tenement.

A Sight to Behold

Visitors stand at viewing areas that are far enough away from the tunnel's entrance to avoid the full force of the pungent aromas emanating from inside. Although its unpleasant stench might indicate otherwise, the guano is actually a good thing. Regularly mined for use as fertilizer, bat guano serves as its own little ecosystem, supporting a variety of life forms. The park rangers there are happy to wax poetic about the gory but fascinating details of all the species that survive and thrive in guano—but be warned, their tales are not for the squeamish.

Folks in the nearby capital of Austin can also view bats—albeit a mere 1.5 million of them—as they emerge each evening from under the city's Congress Avenue Bridge.

The Dogs of War

Even "man's best friend" was drafted into military service during World War II. The four-legged soldiers protected their two-legged comrades; in fact, no patrol led by a war dog was ever ambushed or fired upon without warning.

✳ ✳ ✳ ✳

THE USE OF dogs in war was not new. In ancient times, Romans and Gauls drove packs of semi-wild dogs onto battlefields to attack and terrify their enemies. During World War I, both sides used them as watchdogs and light beasts of burden, among other duties. Then, between 1939 and 1945, dogs became sophisticated, highly trained assistants to their masters in uniform. Using their endurance, speed, and extraordinary senses of hearing and smell, they saved many lives during combat.

Canine Guards, Scouts, and Soldiers

Germany had the largest prewar canine program of any of the belligerents and trained as many as 200,000 dogs. Each concentration camp had an SS dog unit, and the animals were trained to attack prisoners, instilling fear. Other canines were trained as patrols or scouts and used in combat. The SS valued dogs, particularly German shepherds, because they were fast, intelligent, low to the ground, and, with proper training, almost fearless. The Wehrmacht used dogs on all fronts, and even provided some 25,000 dogs to its Japanese allies to use in China.

On the Eastern Front, the Soviet Union conscripted breeds such as the reliable Samoyed, a hardy Siberian sled dog, to pull light equipment and bear wounded men. Eventually, the Soviet Army formed 168 canine units from a variety of breeds and developed specialized guard breeds, such as the black Russian terrier, at its Red Star Kennels outside Moscow. Some of the Red Star's less-fortunate graduates included "suicide dogs," which were conditioned to locate food underneath tanks while carrying backpacks loaded with high explosives.

To the west, Britain and France also recruited dogs for auxiliary duty. In May 1940, as the Wehrmacht swept across France, nearly 200 dogs were among the host that was evacuated from Dunkirk across the English Channel.

Dogs for Defense—Training Canines in the United States

In the United States, patriotic members of the American Kennel Club started an organization called "Dogs for Defense" and called on citizens to donate dogs to the war effort. The War Department began inducting dogs as service animals in March 1942, assigning them to the Quartermaster Corps. During the war, the U.S. Army, Marines and Coast Guard trained more than 10,000 German shepherds, Doberman pinschers, Belgian sheepdogs, and other "acceptable" breeds for what became popularly known as the "K-9 Corps."

U.S. military dogs were trained at several special camps, including one that was established on Cat Island near Gulfport, Mississippi. The 8- to 12-week training course went well beyond teaching Private Rover to "sit," "come," and "stay" on command: Dogs acclimated to the sounds and smells of battle, practiced riding in military vehicles, and even learned to wear gas masks.

On the front lines and in the rear, U.S. war dogs were assigned a variety of duties, depending on their breed, training, and personality. More than 9,000 dogs pulled sentry and reconnaissance duty. They were especially effective in detecting enemy soldiers hiding in foliage or sneaking up on a camp. About 150 dogs were specially trained for messenger duty, running dispatches under fire between the front lines and military headquarters. Others were assigned to the Medical Corps and performed outstanding service locating wounded GIs. Still others, known as "M-Dogs," sniffed out enemy land mines.

Army war dogs proved particularly effective in the Pacific Theater, where dense eye-level vegetation obscured enemy soldiers from the average GI. In an age before infrared sensors and satellite reconnaissance, a dog's senses of sight, smell, and sound were the best tools a foot soldier had for picking out an enemy hiding in the bushes. In just one of many such instances, the Quartermaster Department reported that during the marine landings at Bougainville Island in November 1943, "D-Day dogs" ran messages, pointed out snipers in trees, and sniffed out dug-in Japanese defenders at ranges of over 100 yards, giving marines time to find cover before the enemy opened fire.

Canine Heroes

GIs learned to value their dogs, and several valiant animals were awarded company citations for heroism. A few dogs even earned combat medals such as the Purple Heart before the War Department changed its policy to restrict combat awards to humans. One famous German shepherd named Chips faithfully served the 3rd Infantry Division in all of its major European and African operations. He wore eight battle stars bestowed by his company's men. In one instance while fighting in Sicily, Chips broke away from his handler and attacked an enemy pillbox by himself, mauling one enemy soldier and forcing the machine-gun crew to surrender.

At war's end, the dogs, like their human companions, were repatriated to civilian life. The War Department reconditioned the four-legged warriors to view humans as their best friends and thoroughly tested them for docility before returning them to their original owners.

✳ In 1993, Disney produced a television movie about Chips called *Chips the War Dog.*

✳ In recognition of his service, Chips was awarded the Distinguished Service Cross, Silver Star, and Purple Heart, but all were later revoked when the War Department decided that animals shouldn't receive medals.

The Lowdown on Rats

* There is a National Fancy Rat Society.

* The average life span of a rat is less than three years, but one pair can produce 2,000 offspring in a year.

* Chinese health officials give feral rats flavored birth-control pills.

* A group of rats is called a mischief.

* One pair of rats sheds more than a million body hairs each year, and a single rat can produce 25,000 droppings in a year.

* Some rats can enter an opening as small as half an inch wide.

* Each year, rats cause more than $1 billion in damages in the United States alone.

* Some rats can swim up to a half a mile in open water, dive through water-plumbing traps, travel in sewer lines against strong currents, and stay underwater for up to three minutes.

* In the mid-14th century, the Black Death swept across Europe, killing an estimated 25 million people—about a third of Europe's population (60 percent of Venice died within 18 months). No one made the connection with the hordes of flea-laden rats that infested European cities.

* Rats constantly gnaw anything softer than their teeth, including bricks, wood, and aluminum sheeting.

* A rat can fall 50 feet without injury. What's more, rats can jump 36 inches vertically and 48 inches horizontally.

* Rats use their tails to regulate their body temperature, to communicate, and to balance.

* Rats are color blind and cannot vomit or burp.

Chapter 6

Money and Business

Where Did That Name Come From?

Ever wonder where some everyday products got their names? Read on to find out.

※　※　※　※

WD-40: In 1953, the Rocket Chemical Company began developing a rust-prevention solvent for the aerospace industry. The name WD-40 indicates what the product does (water displacement) and how many attempts it took to perfect it.

Starbucks: *Moby-Dick* was the favorite book of one of the three founders of this coffee empire. He wanted to name the company after the story's fabled ship the *Pequod*, but instead, he and his partners settled on the name of the first mate, Starbuck.

Google: In the 1930s, mathematician Edwin Kasner asked his nephew to think of a word that could represent 1 followed by 100 zeros. The boy, Milton Sirotta, came up with *googol*. The creators of the popular search engine varied the spelling and adopted the word to represent an infinite amount of information.

M&Ms: Chocolate pellets coated in sugar were popular in Britain for decades under the brand name Smarties. When Forrest Mars (the son of the founder of the Mars candy company) saw soldiers eating them during the Spanish Civil War, he and his business partner, R. Bruce Murrie, bought the U.S. rights. But there was already an American candy product called Smarties, so Mars and Murrie used their initials to form a new brand name.

GAP: Don and Doris Fisher opened their first store in 1969 to meet the unique clothing demands of customers between childhood and adulthood, which was identified and popularized as "the generation gap."

Aspirin: In 1899, the German company Bayer trademarked the word *aspirin* as a composite of the scientific name of the drug. "A" indicates that it comes from the acetyl group, "-spir" represents its derivation from the plant genus *spiraea*, and "-in" was a common ending for drug names in the 19th century.

Nike: In 1971, the founders of a small sports-shoe business in Beaverton, Oregon, were searching for a catchy company name. Designer Jeff Johnson suggested Nike, for the Greek goddess of victory. Nike is now the world's largest sportswear manufacturer.

Jeep: Eugene the Jeep, a character in a 1936 *Popeye* comic strip, was actually a dog that could walk through walls, climb trees, and fly. When U.S. soldiers were given a new all-terrain vehicle in the early 1940s, they were so impressed that they may have named it after the superdog.

Scotch Tape: When the purportedly penny-pinching executives, or "Scotch bosses," at the Minnesota Mining and Manufacturing Company (3M) didn't put enough adhesive on their tape, people complained. The company responded by putting better adhesive on its new product. The tape stuck—and so did the name.

Rubik's Cube: This brain-teasing toy is named after its creator, Hungarian architect Erno Rubik. First introduced in 1977, the perplexing puzzle was popular in the 1980s, and more than 100 million of the cubes were sold. It sparked a trend, and similar puzzles were created in various shapes. The Rubik's Cube has seen a recent resurgence in popularity and retains a place of honor on many desktops.

The Birth of the Good Humor Man

The controversy behind a beloved slice of Americana is revealed.

✳ ✳ ✳ ✳

AMERICAN SUBURBIA: LAUGHING children, lemonade stands, sprinklers, and, of course, that ice cream truck emitting its soothing jingle.

In 1920, Harry Burt, an ice cream shop owner in Ohio, invented the first ice cream confection on a stick. His store was selling a lollipop called the Jolly Boy Sucker, as well as an ice cream bar covered in chocolate. Inserting a wooden stick into the ice cream bar made it "the new, clean, convenient way to eat ice cream."

Burt quickly patented his manufacturing process and started promoting the Good Humor bar, in accordance with the popular belief that one's palate affects one's mood. He then sent out a fleet of shiny white trucks, each stocked with a friendly Good Humor Man and all the ice cream bars kids could eat. By 1961, 200 Good Humor trucks wound their way through suburbia.

Not Without a Little Bad Humor

Ice cream on a stick quickly found itself facing solid competition from other be-sticked frozen treats, including the Popsicle. Good Humor took the makers of these treats to court, alleging that its patent gave them exclusive rights to frozen snacks on a stick. Popsicle countered that Good Humor only owned the rights to ice cream on a stick, whereas Popsicles were flavored water. When Popsicle attempted to add milk to its recipe, a decade-long court battle ensued, centering on what precise percentage of milk constitutes ice cream, sherbet, and flavored water.

Popsicle and Good Humor have settled their differences and are now owned by the same corporation. As for the Good Humor Man, he's endangered but not extinct. Good Humor stopped making its ice cream trucks in 1976, but they are still owned by smaller distributors.

Mall of America:
Consumer's Delight

For fans of "retail therapy," Mall of America in Bloomington, Minnesota, is a little slice of heaven. Boasting 2.5 million square feet of retail space and more than 520 stores, it's the largest enclosed mall in the United States and the second-largest mall in North America, after the West Edmonton Mall in Edmonton, Canada. Here are a few more fascinating facts about the shopping center.

✳ ✳ ✳ ✳

✳ More than 40 million people visit Mall of America each year—that's more than the populations of North Dakota, South Dakota, Iowa, and Canada combined.

✳ Mall of America is so large it could hold 258 Statues of Liberty, 7 Yankee Stadiums, or 32 Boeing 747s.

✳ More than 5,000 weddings have been performed at Mall of America since it opened in 1992.

✳ The mall generates nearly $2 billion in economic activity each year.

✳ On any given day, four out of ten visitors to the mall are tourists.

✳ Mall of America employs 11,000 year-round workers— 13,000 during peak periods.

✳ The mall's roof contains eight acres of skylights.

✳ More than 32,000 tons of trash are recycled by the mall each year.

✳ The mall's 100-member security force includes three bomb-sniffing dogs.

"But Wait...There's More!"

Before there was QVC, there was Ron Popeil—an inventor who became a multimillionaire by pitching labor-saving devices on TV. Here are some of his famous and infamous products.

✳   ✳   ✳   ✳

1. **Veg-O-Matic:** Popeil learned to be a pitchman from his father, Samuel, who was also an inventor and salesman of kitchen gadgets. Samuel's Chop-O-Matic was introduced in the mid-1950s at the amazing low price of $3.98. Ron renamed it the Veg-O-Matic, pitched it as "the greatest kitchen appliance ever made...", and the two items sold more than 11 million units.

2. **Pocket Fisherman:** Popeil touted this device as "the biggest fishing invention since the hook...and still only $19.95." The handle was a mini tackle box that contained a hook, line, and sinker. Samuel came up with the idea for the Pocket Fisherman in 1963, after he was nearly injured by the tip of a fishing pole. Today it sells for $29.99 and has a double-flex rod hinge that unfolds to a fully extendable position. Since 1963, more than two million units have sold.

3. **Mr. Microphone:** Launched in 1978, Mr. Microphone was a device that turned radios into annoying precursors of the karaoke machine. The TV commercial featured Popeil's daughter and her boyfriend riding in a convertible and using Mr. Microphone, which made one wonder why he didn't follow this invention up with Mr. Earplugs. Nevertheless, he sold more than a million of them at $19.95 a pop.

4. **Electric Pasta Maker:** This gadget allows you to make 12 different shapes of preservative-free homemade pasta in just five minutes. You can even use it to make homemade sausage. Since 1993, more than one million people have purchased the Electric Pasta Maker.

5. Solid Flavor Injector: Resembling a syringe with a large plastic "needle," this gadget was used to inject fillings—such as dried fruit, small vegetables, nuts, chocolate chips, and candy—into foods such as hams, roasts, cupcakes, and pastries. At just $14.95, it didn't cost a fortune to add a bit of flair and pizzazz to your food.

6. Electric Food Dehydrator: Introduced in 1965 at $59.95, Ron Popeil called it "the most famous food dehydrator in the world!" The sun might disagree with that claim, but the Electric Food Dehydrator, which brought Popeil back from semiretirement, currently sells for $39.98.

7. Smokeless Ashtray: The Smokeless Ashtray promised to suck up the smoke from cigars and cigarettes before it filled the room, and in the 1970s, Ron sold more than a million of these contraptions at $19.95 each. With most smoking being done outside today, the Smokeless Ashtray has been replaced by something less expensive—wind.

8. GLH Formula Number 9 Hair System: Got a bald spot? Ron Popeil can fix it with the GLH Formula Number 9 Hair System. GLH isn't real hair, but rather a spray that matches your hair color, thickens thinning hair, and covers bald spots. More than one million cans have sold for only $9.95 for the spray or $20 for the spray, shampoo, and finishing shield.

9. Showtime Rotisserie and BBQ Oven: Introduced in 1998, the Showtime Rotisserie and BBQ Oven is by far Ron Popeil's most successful product to date. He has sold seven million units of this oven in three different models: the $99.95 Compact Rotisserie, the $159.80 Standard Rotisserie, and the $209.75 Pro Rotisserie. Popeil's pitch for the Showtime Rotisserie is "Set it and forget it!," which has been repeated so many times in infomercials that it's impossible to forget.

Copyright Clarifications

What is the link between Jaws, Starlight Express, *and* Strawberry Fields? *They are all titles of creative works—and they cannot be copyrighted.*

✳ ✳ ✳ ✳

L IKE A PATENT or a trademark, a copyright gives the creator of an original work the exclusive right to control and profit from his or her efforts. Although literary works, movies, songs, paintings, and video games can all be copyrighted, titles of books, plays, songs, and films cannot.

Copy Cats

According to the U.S. Copyright Office, title 17, section 102 of the copyright code extends only to "original works of authorship" and clearly states that ideas, concepts, and titles are not subject to copyright protection. The copyright for a novel such as *The Da Vinci Code*, for example, is designed to restrict others from creating derivative works based on Dan Brown's story of a Harvard symbologist deciphering a code that's hidden in the works of Leonardo da Vinci. It does not, however, prevent someone else from writing a book or a play or making a movie about codes hidden in famous paintings, as long as it is not judged to be a copy of Brown's work. There is also nothing to prevent someone from using *The Da Vinci Code* as a title for his or her work. This is one reason why so many movies have the same or similar titles. The other reason, of course, is that Hollywood has never been known for its originality.

In some instances, titles that fall into the category of brand names may be entitled to protection under trademark laws, but for the most part, there is nothing to stop you from publishing a book with the same title as a best seller. As long as the actual content of the book is not copied or closely adapted, no one can claim infringement. That's not to say someone won't try, though, and copyright lawyers tend to be a particularly litigious bunch.

Curious Classifieds

✳ ✳ ✳ ✳

✳ "Sheer stockings. Designed for fancy dress, but so serviceable that lots of women wear nothing else."

✳ "And now, the Superstore—unequaled in size, unmatched in variety, un-rivaled inconvenience."

✳ "Don't stand there and be hungry . . . come in and get fed up."

✳ "Open house—Body Shapers Toning Salon—free coffee & donuts"

✳ "Golden, ripe, boneless bananas, 39 cents a pound"

✳ "Semi-Annual after-Christmas Sale"

✳ "Mother's helper—peasant working conditions"

✳ "Wanted: A boy who can take care of horses who can speak German"

✳ "Dinner Special—Turkey $2.35; Chicken or Beef $2.25; Children $2.00"

✳ "Teeth extracted by the latest Methodists"

✳ "Stock up and save. Limit: one"

✳ "Widows made to order. Send us your specifications."

✳ "Tattoos . . . While you wait"

✳ "For you alone! The bridal bed set . . ."

✳ "Braille dictionary for sale. Must see to appreciate!"

✳ "Attorney at law; 10% off free consultation"

✳ " 'I love you only' Valentine cards: Now available in multipacks"

✳ "For sale: an antique desk suitable for lady with thick legs and large drawers."

✳ "We do not tear your clothing with machinery. We do it carefully by hand."

✳ "For Sale—Amana washer $100. Owned by clean bachelor who seldom washed."

Origins of 12 Modern Icons

Who knows why some images endure, while others slip through our consciousness quicker than 50 bucks in the gas tank. In any case, you'll be surprised to learn how some of our most endearing "friends" made their ways into our lives.

✳   ✳   ✳   ✳

1. **The Aflac Duck:** A duck pitching insurance? Art director Eric David stumbled upon the idea to use a web-footed mascot one day when he continuously uttered, "Aflac... Aflac... Aflac." It didn't take him long to realize how much the company's name sounded like a duck's quack. There are many fans of the campaign, but actor Ben Affleck is not one of them. Not surprisingly, he fields many comments that associate his name with the duck, and he is reportedly none too pleased.

2. **Alfred E. Neuman, the face of *Mad* magazine:** Chances are you're picturing a freckle-faced, jug-eared kid, right? The character's likeness, which was created by portrait artist Norman Mingo, was first adopted by *Mad* in 1954 as a border on the cover. Two years later, the magazine used a full-size version of the image as a write-in candidate for the 1956 presidential election. Since then, several real people have been said to be "separated at birth" from Mr. Neuman, including Ted Koppel, Jimmy Carter, and George W. Bush.

3. **Betty Crocker:** Thousands of letters were sent to General Mills in the 1920s, all asking for answers to baking questions. To give the responses a personal touch, the company's managers created a fictional character. The surname Crocker was chosen to honor a retired executive, and "Betty" was selected because it seemed "warm and friendly." In 1936, artist Neysa McMein blended the faces of several female employees to create a likeness. Crocker's face has changed many times over the years: She's been made to look younger

and more professional, and now she has a more multicultural look. At one point, a public opinion poll rating famous women placed Betty Crocker second to Eleanor Roosevelt.

4. **Duke, the Bush's Baked Beans Dog:** Who better to trust with a secret recipe than the faithful family pooch? Bush Brothers & Company was founded by A. J. Bush and his two sons in 1908. In 1995, the advertising agency working for Bush's Baked Beans decided that Jay Bush (A. J.'s great-grandson) and his golden retriever, Duke, were the perfect team to represent the brand. The only problem was that the real Duke was camera shy, so a stunt double was hired to portray him and handle all the gigs on the road with Jay. In any case, both dogs were sworn to secrecy.

5. **The California Raisins:** Sometimes advertising concepts can lead to marketing delirium. In 1987, a frustrated copywriter at Foote, Cone & Belding was working on the California Raisin Advisory Board campaign and said, "We have tried everything but dancing raisins singing 'I Heard It Through the Grapevine.'" With vocals by Buddy Miles and design by Michael Brunsfeld, the idea was pitched to the client. The characters plumped up the sales of raisins by 20 percent, and the rest is Claymation history!

6. **Joe Camel:** Looking for a way to revamp Camel's image from an "old-man's cigarette" in the late 1980s, the R.J. Reynolds marketing team uncovered illustrations of "Old Joe," who was originally conceived for an ad campaign in France in the 1950s. In 1991, the new Joe Camel angered children's advocacy groups when a study revealed that more kids under the age of eight recognized Joe than Mickey Mouse or Fred Flintstone.

7. **The Coppertone Girl:** In 1959, an ad for Coppertone first showed a suntanned little girl's white buttocks being exposed by a puppy. "Don't be a paleface!" was the slogan, and it reflected the common belief of the time that a suntan

was healthy. Artist Joyce Ballantyne Brand created the pig-tailed little girl in the image of her three-year-old daughter Cheri. When the campaign leapt off the printed page and into the world of television, the role of the little girl became Jodie Foster's acting debut. As the 21st century dawned—and along with it changing views on sun exposure and nudity—Coppertone revised the drawing to reveal only the girl's lower back.

8. **Juan Valdez:** This coffee lover and his trusty donkey have been ensuring the quality of coffee beans since 1959. Back then, the National Federation of Coffee Growers of Colombia wanted to put a face on its thousands of coffee growers. The Doyle Dane Bernback ad agency found one alright! By 1981, Valdez's image was so well known that it was incorporated into the Federation's logo. Originally played by Jose Duval, the role was taken over by Carlos Sanchez from 1969 to 2006. In his spare time, Sanchez manages his very own small coffee farm in Colombia.

9. **Mr. Whipple:** The expression "Do as I say, not as I do" took on a persona in the mid-1960s—Mr. Whipple, to be specific. This fussy supermarket manager (played by actor Dick Wilson) was famous for admonishing his shoppers by saying, "Ladies, please don't squeeze the Charmin!" The people at Benton & Bowles Advertising figured that if Mr. Whipple was a habitual offender of his own rule, Charmin toilet paper would be considered the cushiest on the market. The campaign included a total of 504 ads and ran from 1965 until 1989, landing it a coveted spot in the *Guinness Book of World Records* as the planet's longest-running ad campaign. A 1979 poll listed Mr. Whipple as the third-most-recognized American behind Richard Nixon and Billy Graham.

10. **The Gerber Baby:** Contrary to some popular beliefs, it's not Humphrey Bogart, Elizabeth Taylor, or Bob Dole who so sweetly looks up from the labels of Gerber products. In fact, the face that appears on all Gerber baby packaging belongs to mystery novelist Ann Turner Cook. In 1928, when Gerber began its search for a cherubic face to help promote its new brand of baby food, Dorothy Hope Smith submitted a simple charcoal sketch of the tot—promising to complete it if it was chosen. As it turned out, that wasn't necessary because the powers that be at Gerber liked it just the way it was. In 1996, Gerber updated its look, but the new label design still incorporates Cook's baby face.

11. **The Pillsbury Doughboy:** Who can resist poking the chubby belly of this giggling icon? This cheery little kitchen dweller was "born" in 1965 when the Leo Burnett advertising agency dreamt him up to help Pillsbury sell its refrigerated dinner rolls. The original vision was for an animated character, but instead, agency producers borrowed a unique stop-action technique that had been used on *The Dinah Shore Show*. After beating out more than 50 other actors, Paul Frees lent his voice to the Doughboy. So if you've ever craved Pillsbury rolls while watching *The Adventures of Rocky and Bullwinkle*, it's no wonder: Frees was also the voice for Boris Badenov and Dudley Do-Right.

12. **Ronald McDonald:** Perhaps the most recognizable advertising icon in the world, this beloved clown made his television debut in 1963 and was originally played by future *Today* weatherman Willard Scott. Nicknamed the "hamburger-happy clown," Ronald's look was a bit different back then: He had curly blond hair, a fast-food tray for a hat, a magic belt, and a paper cup for a nose. Ronald's makeover must have been a hit because McDonald's now serves more than 52 million customers every day around the globe.

The Power of the Wish Book

Every day, hundreds of thousands of people shop at Sears but few customers realize that this retail institution ushered in the era of mass consumption that marked the 20th century.

✳   ✳   ✳   ✳

SEARS, ROEBUCK AND CO. began in 1886 when railroad-station agent Richard Sears founded the R. W. Sears Watch Company in Minneapolis, Minnesota, to sell watches through the mail.

In 1887, Sears moved his business to Chicago, where he hired young Alvah Roebuck to be the company's official repairman. The following year, Sears printed the company's first catalog, which was sent to rural communities around the country. The catalog offered watches, diamonds, and jewelry, all with a money-back guarantee. The latter was important to Sears's sales tactics because it gained the trust of farmers and rural residents who had been swindled by unscrupulous traveling salesmen.

In 1889, Sears sold his successful business to become a banker. But he missed the mail-order business that he helped to pioneer, so he returned to Minnesota and repartnered with Roebuck. By 1893, the pair had moved to Chicago as Sears, Roebuck and Co., and their catalog had grown to 196 pages. In addition to jewelry and watches, the revamped catalog—which expanded to 507 pages in 1895—offered sewing machines, furniture, clothing, tools, saddles, shoes, musical instruments, and much more.

The Wish Book

Having worked on the railroad, Richard Sears knew what rural communities were like, and during this era, roughly 65 percent of Americans lived in nonurban areas. He knew that they had only the local general store and the occasional traveling salesman to fill their needs. The Sears catalog, which customers began calling the "Wish Book," allowed country folk a taste of the urban life. It transformed the looks, tastes, and attitudes of rural residents faster than anything prior to the automobile. Sears wrote the descriptions for the catalog items in a folksy style, addressing each of his customers as "Kind Friend." The initiation of free rural mail delivery in 1896 and parcel post in 1913 enabled the company to send merchandise to even the most isolated customers. By the turn of the century, Americans rated the Sears catalog as their favorite book after the Bible. Most importantly, it allowed the rural population to take part in the mass consumption that defined the 20th century.

In 1895, Roebuck sold his interest in the company to Julius Rosenwald, who made the company stronger with his administrative skills. In 1906, the company constructed a new $5 million mail-order plant on Chicago's West Side; with more than three million square feet of floor space, it was the largest business building in the world at the time.

Due to poor health, Sears retired from active participation in the company in 1908. When he died in 1914, the catalog had evolved into a slick-looking publication with factual descriptions instead of the folksy-sounding tone preferred by Sears.

Sears Stores

Initially, Sears, Roebuck and Co. was solely a catalog business, and its success was dependent on rural America. However, by the early 1920s, the country's urban population outnumbered its rural residents for the first time. Sears's new vice president, Robert E. Wood, lobbied for retail expansion, and in 1925, the first Sears store opened in Chicago. By 1933, the retail operation had grown to 400 stores, and store sales had topped mail-order revenue for more than two years. Additionally, some of the mail-order merchandise was sold under the Sears brand name; this was the beginning of the Sears-associated lines such as Craftsman, Kenmore, and Die Hard.

The Sears "Big Book" catalog was discontinued in 1993.

Dr Pepper: "The Friendly Pepper-Upper"

✳ ✳ ✳ ✳

When someone in Texas asks for a Coke, they're probably asking for a Dr Pepper, a formula born at Waco's Old Corner Drugstore.

L ONG BEFORE CARBONATED beverages were bottled and branded, the only way that people could enjoy a sip of soda was to visit the neighborhood soda fountain. During the 1800s, soda fountains, which were often located inside pharmacies, were ornate, imposing edifices that required maintenance and tweaking, and drinks had to be mixed on the spot.

Pulling Double Duty

In Texas, most small towns had a drugstore where the pharmacist also served as a "soda jerk," a name owing to the jerking motion exhibited when he pulled forward on the soda fountain's spigot. The druggist's specialty, of course, was to compound and dispense medicines, so mixing up the sweet syrups, roots, herbs, and berries that flavored carbonated water was a natural offshoot for his talents. In fact, many of the earliest soft drinks were actually bitter concoctions, geared toward settling the stomach or treating a particular ailment.

In Waco, a local flavor blend attained a certain degree of fame during these halcyon days. The story began with pharmacist Charles Alderton of Waco's Old Corner Drugstore. He was fascinated with the idea of the fountain and enjoyed devising new flavors to mix with carbonated water. He also noticed that customers had a difficult time choosing one taste over another and saw the opportunity to create something special.

Alderton experimented with a range of ingredients stocked at the store, and he mixed them in myriad combinations. Drop by drop, sip by sip, through a painstaking method of trial and error, he succeeded in creating a particularly tasty syrup in 1885.

A New Taste Is Born

Further refinements were made to the formula, which was willingly taste-tested by shop owner Wade Morrison and local patrons. To Morrison and Alderton's delight, Waco residents soon began to request the fountain specialty by name. At the time, however, it was known only as a "Waco." Soon, druggists east and west of the Brazos were hounding Morrison about purchasing jugs of the syrup for their own shops.

To satisfy the demand, Morrison and Alderton began to mix up large batches of the syrup. But that wasn't enough. As the number of soda parlors that served the Waco swelled, the two realized that they needed a more efficient setup. Enter bottler R. S. Lazenby, who packaged the drink exclusively for the growing retail market; by 1891, he transformed what had been a pharmacist's experiment into a sweet gold mine.

A Now-Familiar Name

By then, the Waco nickname was deemed too local, so the drink was renamed "Dr. Pepper's Phos-Ferrates." Allegedly, the name was inspired by a Dr. Charles Pepper, who operated a pharmacy in Rural Retreat, Virginia, and had once been Morrison's boss. As a tribute, Morrison suggested the name for the new beverage. Some claim that Morrison was in love with Pepper's teenage daughter, but her father didn't approve. The story, which is disputed by Waco's Dr Pepper Museum and the current company, claims that he named the drink in honor of his lost love.

Regardless of the name's origin, there's no confusion when it comes to how successful Dr Pepper became. From its inauspicious beginnings, the drink joined the pantheon of carbonated beverages to become one of the most adored brands of soda pop.

Chapter 7

Intriguing Individuals

Queen of the Stalingrad Skies

Lily Litvak, the "White Rose of Stalingrad," terrorized German pilots and helped inspire a nation to a monumental victory.

✳   ✳   ✳   ✳

IN THE UKRAINE city of Krasy Luch stands a tall marble statue with 12 gold stars on its column, capped with a large, striking bust of a woman crowned in an aviator's cap and glasses.

The monument is dedicated to Lily Litvak, the celebrated Soviet fighter ace who struck fear into German pilots in the skies over Stalingrad. The 12 gold stars on her statue commemorate the 12 solo kills she recorded in her brief but illustrious career.

A Deadly Flower

Litvak was affectionately called the "White Rose of Stalingrad" for the white lily (which was mistaken for a rose) painted on the sides of the cockpit of her Yak-1 fighter plane. Her skill and tenacity became so well-known that German pilots would allegedly peel away when they saw the white flowers coming.

A licensed flying instructor by her 18th birthday, Litvak volunteered for an aviation unit after Germany invaded the Soviet Union, but she was rejected because of her lack of flying time. After embellishing her experience, she joined the famed all-female 586th Fighter Regiment, where she honed her skills.

Initially, Litvak faced more of a challenge from chauvinistic attitudes than from enemy pilots. When she transferred to an all-male air unit at Stalingrad in September 1942, her commander refused to let her fly. After continuous pleading, she was finally given a plane and quickly made believers of her male comrades after scoring her first two kills on her second combat mission.

A Timely Inspiration for a Beleaguered Nation

For the next 11 months, the stunning beauty with golden blonde hair and captivating gray eyes outwitted and outfought German pilots, who often couldn't accept that they had been shot down by a woman. A German fighter ace who was shot down by Litvak refused to believe that he had been bested by a woman until they were introduced and she described to him minute details of their dogfight that only the two of them could have known.

Through the course of 168 combat missions, Litvak was shot down two or three times, once sustaining serious injury to her legs. She bounced back each time with a fiercer will to fight that was further hardened following the death of her fighter-pilot husband. Her luck finally ran out on August 1, 1943, when eight German fighters ganged up against her and sent her crashing to her death 17 days shy of her 22nd birthday.

Although her remains were not found until 1979, Litvak's heroics in the skies over Stalingrad were never forgotten by the Soviet people. Her bravery and achievements provided a timely inspiration to a nation that was facing defeat and was desperate for something from which it could draw hope. In 1990, in recognition of her contribution to her nation's monumental victory, Litvak was posthumously awarded its highest honor, the Hero of the Soviet Union.

The Woman Called "Moses"

Harriet Tubman, one of America's most celebrated heroes, was responsible for ensuring the freedom of many slaves, both before and during the Civil War.

✳ ✳ ✳ ✳

IN THE 1800s, the "Underground Railroad" was a euphemism for any type of assisted escape made by slaves to the Northern states, Canada, or the British West Indies, where slavery was illegal. It was a loosely organized system of white abolitionists and free blacks who secretly worked together to move slaves through the countryside, into safe houses, and on to freedom. Significant risks were attached to any work that was undertaken on behalf of escaping slaves on the Underground Railroad, as American law permitted slave owners to reclaim fugitive slaves anywhere they might be found within the United States.

Meet Moses

Harriet Tubman was often likened to the biblical character Moses for leading more than 300 slaves along the Underground Railroad to freedom from 1849 to 1860. What many people don't know is that Tubman's involvement didn't stop when the Civil War began—she continued to be active throughout the conflict. Despite the fact that she had no formal military training, Tubman researched, planned, and led a successful Union military operation with Colonel James Montgomery along the Combahee River in South Carolina on June 2, 1863.

In 1849, Tubman escaped her life as a slave in Maryland by stealing away into the surrounding forests, ultimately making her way to the free state of Pennsylvania. When the Civil War began in 1861, the 40-year-old Tubman was already the most famous black woman in America. Her diminutive appearance was rendered more unusual by her missing front teeth and a large dent in the back of her skull—the result of a blow delivered by a slave master in her youth.

During the years that Tubman led slaves to freedom, she did not work for reward or commercial gain. Her burning desire to assist her family and other slaves in the South was her sole motivation. Tubman kept virtually no records of her secret missions to the South, and there is no precise accounting of the number of slaves whom she assisted to freedom. Some evidence suggests that Tubman made at least 20 trips down south.

Although she was offering freedom to her Underground Railroad passengers, she was ruthless with her human cargo. If escaping slaves seemed overly fearful or were inclined to return to their owners, Tubman would brandish a loaded revolver to change their minds. She was proud that every slave she assisted was delivered to freedom.

In 1851, abolitionist John Brown declared that "slavery is war," and Tubman likely believed the same thing. She conducted her raids like a guerilla fighter—often in disguise, using the fields and forests as her cover. She used code phrases taken from the Old Testament to communicate to those awaiting rescue. "When the good ship Zion comes along...be ready to step on board" was one favorite.

Tubman's fame and success in infiltrating Southern slave holdings during the early 1850s caused blacks to revere her and the plantation aristocracy to hate her. One Maryland slave owner offered a reward of $40,000—about $1 million by today's standards—for her capture. This development prompted Tubman, somewhat wisely, to move to the small Canadian city of St. Catherines, which stands just across Niagara Falls from Buffalo, New York, and was a popular northern terminus for the Underground Railroad. She lived there from 1851 to 1857.

Harriet the Spy

Tubman's knowledge of the South led her to a variety of wartime assignments for the Union army. After working as an army nurse, Tubman was asked in 1862 to conduct a number of scouting and spying missions in the Southern states. Her experiences with the Underground Railroad were great training for this military spy and reconnaissance work.

Official Union records list Tubman as an "advisor" to Colonel James Montgomery in a combined gunboat and land mission against Confederate positions near South Carolina's Combahee River on June 2, 1863. In fact, the mission was the first American military engagement of any significance to be planned and directed by a woman.

As Tubman and a small Union force tracked enemy positions along the Combahee, she noted the locations of deadly Confederate torpedoes, or mines, placed in the river. Tubman later led a Union company of 300 black soldiers, who were supported by river gunboats, toward the Confederate targets. Those gunboat crews and land forces then went on a rampage, destroying Confederate supplies and confiscating valuable farm animals. The soldiers also succeeded in freeing more than 700 slaves from nearby plantations.

Tubman was never a commissioned officer, nor did she hold a rank in the Union army, but she continued to work as a spy and military scout until the war's end. She often disguised herself as a simple middle-age slave woman as she made her military observations.

Harriet Tubman's contributions to both the freedom of slaves and the Union war effort were profound. After the war, she remained busy. During the Reconstruction period, she established schools to educate former slaves. She also helped found an old age home in Auburn, New York. Despite her achievements, however, Tubman received little public acclaim and lived a long life of relative poverty. Her memoir, *Scenes in the Life of Harriet Tubman,* only attracted significant attention after she died in 1913 at age 93.

The Giants of Seville

Known as the "Kentucky Giant," Martin Van Buren Bates served in the Civil War, and after a stint as a circus performer, he attempted to settle down with his giantess wife and raise a family in the town of Seville, Ohio.

✳ ✳ ✳ ✳

MARTIN VAN BUREN Bates was born in Kentucky in 1837, the youngest of 11 children. He was more than six feet tall by the time he was 13 and ultimately grew to be seven feet, nine inches tall. He weighed about 470 pounds. Bates enlisted in the Confederate army during the Civil War and was captured by the Yankees in 1863. In prison camp, he was known as the "Kentucky Giant." He joined a circus in Cincinnati after the war.

Male Giant Seeks Female Giant

While touring in New Jersey in 1869, Bates met Anna Swan, a giantess from Halifax, Nova Scotia. Anna had been a staple of P. T. Barnum's circus since she was 16. She joined Bates's troupe, and the two married in London in 1871 during a tour of Europe. Following the stillbirth of a daughter and a bout of consumption, the couple returned to America and bought a 120-acre farm in Seville, where they became active members of the community. They built a mansion suited to their size, with 8-feet-high doorways and 14-feet-high ceilings.

Anna gave birth to a 23-pound, 12-ounce son in 1879, but the couple's dream of starting a family ended 11 hours later when their baby died. Depression and deteriorating health hastened Anna's death in 1888, and Martin sold the house they had built. He ultimately remarried, to a woman of regular stature before he passed away 31 years later. Both Martin and Anna are buried with their son in Seville's Mound Hill Cemetery; their grave site is marked with a statue of Anna. Their mansion was torn down in 1948, but their barn remains. The Seville Historical Society Museum displays mementos from their life.

Birth Names of 18 Vintage Film Stars

How many of these film stars can you pair with their birth name?

1. Jack Benny	A. Archibald Leach
2. George Burns	B. Asa Yoelson
3. Eddie Cantor	C. Benjamin Kubelsky
4. Joan Crawford	D. Bernard Schwartz
5. Tony Curtis	E. Dino Crocetti
6. Dale Evans	F. Edward Israel Iskowitz
7. Judy Garland	G. Frances Gumm
8. Cary Grant	H. Frances Smith
9. Rita Hayworth	I. Joseph Levitch
10. William Holden	J. Julia Jean Turner
11. Al Jolson	K. Leonard Slye
12. Boris Karloff	L. Lucille Fay LeSueur
13. Jerry Lewis	M. Margarita Cansino
14. Dean Martin	N. Nathan Birnbaum
15. Marilyn Monroe	O. Norma Jean Baker
16. Roy Rogers	P. Ruby Stevens
17. Barbara Stanwyck	Q. William Beedle
18. Lana Turner	R. William Henry Pratt

Answer Key: 1. C; 2. N; 3. F; 4. L; 5. D; 6. H; 7. G; 8. A; 9. M; 10. Q; 11. B; 12. R; 13. I; 14. E; 15. O; 16. K; 17. P; 18. J.

The Collyer Brothers: Pack Rats Extraordinaire

It all started out so well. The Collyer boys were born into a fairly prominent New York family—Homer in 1881 and Langley in 1885. Their father, Herman, was a doctor; their mother was an educated woman who worked occasionally as an opera singer. The boys attended Columbia University and earned degrees in law and engineering, respectively.

❋ ❋ ❋ ❋

In 1909, the family broke apart when Dr. Collyer left for unknown reasons. Homer, Langley, and their mother stayed in the family house at 2078 Fifth Avenue, smack in the middle of Harlem, which was an affluent white neighborhood at the time. But as Harlem changed, so did the boys. When their mother died in 1929, they were left to fend for themselves, and that's when things got really bizarre.

What's Up with the Quirky Neighbors?

The Collyer boys weren't very good with details like paying bills, so they had no telephone, electricity, or running water in the house. Paranoid about burglars, the eccentric brothers boarded up all the windows and put iron gates over the doors. Kerosene lamps lit the house, and a kerosene stove provided nominal heat during the frigid New York winters. Water was retrieved from a pump at a nearby park—but only under the cover of darkness. This is also when they did their junk collecting.

In 1933, Homer went blind, and he would later be crippled by a battle with rheumatism. Langley prescribed his ailing brother a treatment: He was to eat 100 oranges per week, supplemented with a few peanut butter sandwiches for good measure. It's no surprise that Homer never regained his eyesight.

On top of the reclusiveness, home remedies, and water fetching, the Collyer brothers were constantly hoarding. Individuals who suffer from this "pack rat syndrome" (also known as syllogomania) save everything. This is never done in a neat, organized way—one of the hallmarks of a compulsive hoarder is a totally chaotic, stuffed-to-the-rafters home. Hoarders have so many possessions that they are usually rendered incapable of carrying out basic living functions such as washing dishes and cleaning the house.

Such was the case with Homer and Langley. Hoping that one day his brother would regain his eyesight, Langley saved every New York newspaper for several decades—by the end, he had several tons of newspapers. The brothers also amassed a collection that included sewing machines, baby carriages, rusted car parts, chandeliers, mannequins, old bicycles, thousands of books, and five pianos. If one of the two happened upon something, it went into the house, piled on top of everything else. Over the years, they created a palace of junk.

Knock, Knock

In 1942, after the Collyers had neglected to pay their mortgage for some time, their bank set eviction proceedings in motion, and a cleanup crew was sent to 2078 Fifth Avenue. They were met by an irate Langley, and the police were summoned.

The police eventually entered the fortress but not without a struggle: All entrances to the house were blocked with what the officers identified as "refuse" and "garbage" that was "neck-deep." When they finally found Langley, he wrote a check for the remainder of the mortgage and sent the authorities on their merry way. For the next five years, the Collyer brothers lived in an increasingly reclusive manner, and sightings of the eccentric men became less frequent.

Then in 1947, the police received a phone call from a man identifying himself as Charles Smith, who claimed that there was a dead body in the Collyer house. When police arrived, they found it more impenetrable than before. The only way into the house was through an upper-story window, and even then, gaining access involved removing huge chunks of junk and throwing them to the street below.

When one of the officers finally got inside the cavernous house, he went through the labyrinthine rooms searching for the source of a nasty odor. Between several piles of trash, the officer found Homer. Police reports stated that he'd died from a combination of malnutrition and cardiac arrest and had only been dead for a few hours. Langley, however, was nowhere to be found.

A manhunt was undertaken in New York City, but most people figured that Langley was still inside the house, waiting to catch one of the officials with his homemade booby traps. As it turned out, they were right—sort of. Langley was indeed in the house, but he wasn't about to catch anyone. His body was discovered about ten feet from his brother's and had been providing lunch for the neighborhood rats for a couple of weeks by the time he was found. It appeared that Langley was trying to bring food to Homer when he was caught in one of his own traps.

The Numbers

In the end, more than 100 tons of junk were removed from the Collyer house. That's more than 200,000 pounds of shoes, medical equipment, suitcases, phonebooks, newspapers, animal parts, tapestries, etc.

The house was razed, and Collyer Brothers Park now stands in its place. Over the years, the obsessive brothers have been the subject of several books and plays, and even a comic book, though no one's made a movie of their lives . . . at least not yet.

❋ Each year, more than 25 million trees are used to make newspapers. By dispensing with a single run of just one major metropolitan Sunday paper, more than 75,000 trees could be saved.

❋ Recycling just one ton of paper saves 17 trees.

❋ The average U.S. household discards more than 13,000 pieces of junk mail per year—44 percent of which goes into the trash without ever being opened.

Cary Grant's Acid Test

In the 1950s and '60s, Cary Grant was one of Hollywood's top box-office draws, starring in popular movies such as An Affair to Remember *(1957) and* Charade *(1963). In sharp contrast to the debonair persona that he cultivated on-screen, Grant also participated in an experimental psychotherapy program in which he underwent more than 100 trips on the hallucinogenic drug LSD.*

✳ ✳ ✳ ✳

IT'S NOT EASY to picture Hollywood's quintessential leading man as an acid eater. On-screen, Cary Grant's matinee-idol looks and sophisticated charm made him one of the most popular stars ever. He was ranked No. 2 on the American Film Institute's list of Greatest Screen Legends and was a favorite of Alfred Hitchcock, appearing in some of the acclaimed director's best films, including *Suspicion* (1941), *Notorious* (1946), *To Catch a Thief* (1955), and *North by Northwest* (1959). Author Ian Fleming even partially modeled his James Bond character on the British-born Grant.

Offscreen, however, Grant was somewhat insecure and suffered a turbulent personal life that saw him married five times. In fact, it was Grant's third wife, Betsy Drake, who, around 1956, introduced him to doctors prescribing the experimental drug lysergic acid diethylamide, better known as LSD. By the late 1950s, Grant was a regular patient of Los Angeles psychiatrists Dr. Mortimer Hartman and Dr. Arthur Chandler, who supervised the actor's frequent acid trips.

He Turned On, Tuned In, but Didn't Drop Out

Prior to 1966, LSD was available legally in the United States as an experimental psychiatric drug. Researchers such as Hartman, Chandler, and Dr. Oscar Janiger—cousin of the famed Beat poet Allen Ginsberg—recruited hundreds of L.A. residents as human volunteers to gauge the therapeutic potential of LSD in treating neuroses in people who are unresponsive to conventional therapies. Before the drug became associated with the hippie movement and before Timothy Leary urged users to "turn on, tune in, and drop out," the use of psychedelics in scientific research drew little attention from the public or media.

Although Grant's acid trips took place outside of the usual clinical environment, Dr. Hartman or Dr. Chandler always designated a monitor to constantly observe him and talk him down should he experience a bad trip.

Grant publicly discussed the therapeutic value of LSD. He likened the hallucinations he experienced during those LSD trips to the act of dreaming and felt that the drug helped him come to terms with his star image and reconcile his past, particularly unresolved conflicts with his parents. He also credited LSD therapy with helping him gain control over his drinking.

LSD and Alcohol

Dr. Albert Hofmann first formulated LSD by accident in 1938 at a Swiss pharmaceutical company. Five years later, he accidentally discovered its effects when he undertook the world's inaugural acid trip. The CIA and U.S. military later tried using the drug's disorienting effects as a nonlethal warfare tactic before researchers began to explore its psychiatric potential. Grant was not the only Hollywood celebrity to participate in these studies: Other volunteers included actors Jack Nicholson, James Coburn, and Dennis Hopper; author Aldous Huxley; and musician/conductor Andre Previn.

While Grant spoke positively about his LSD use, he also recognized the inherent dangers of the drug. "I found it a very enlightening experience," he once said. "But it's like alcohol in one respect: A shot of brandy can save your life, but a bottle of brandy can kill you. And that's what happened when a lot of young people started taking LSD, which is why it became necessary to make it illegal."

Fast Facts: Marilyn Monroe

✳ Marilyn Monroe was born in Los Angeles on June 1, 1926. Her birth name was Norma Jean Mortenson, but she was baptized Norma Jean Baker—with her mother's last name—because the identity of her father was considered undetermined.

✳ Monroe's mother suffered from bipolar disorder and spent much of her life in mental institutions. As a result, Marilyn spent two years at the Los Angeles Orphans Home. She also lived in many foster homes as a child.

✳ Marilyn Monroe was married three times: In 1942, she wed 21-year-old Jim Dougherty. The two had only dated for six months before their wedding, but Marilyn was anxious to escape her unhappy childhood, and her foster parents encouraged her to get married. Monroe and Dougherty divorced in 1946 when Marilyn began to focus on her modeling career. Next, in 1954, she famously married Joe DiMaggio after two years of dating; they divorced nine months later. Finally, Monroe married playwright Arthur Miller in 1956; they stayed together until 1961.

✳ Monroe signed her first acting contract in August 1946. The deal with 20th Century Fox paid her $125 per week.

✳ By the time she was a teenager, Marilyn Monroe had very dark blonde hair, which sometimes looked brunette in black-and-white photographs. While working as a model in her late teens, she gradually lightened her hair. With the change in hair color, Monroe got more modeling jobs.

✳ Though she used the name Marilyn Monroe as early as 1946, she didn't legally take on the moniker until a decade later. (Monroe was her mother's maiden name.)

✳ Monroe has been featured in numerous tributes in American culture. Elton John's song "Candle in the Wind" was written in her honor. Her image also graced a commemorative U.S. postage stamp that was created in 1995.

* Monroe won the Golden Globe for Best Motion Picture Actress for her role in *Some Like It Hot* (1959). She was also named the female World Film Favorite at the Golden Globes in 1954 and 1962, but she was never nominated for an Oscar.

* Contrary to her on-screen persona, Marilyn Monroe loved to read Tolstoy and other classic writers. In fact, while she was under contract at Fox, she was enrolled at UCLA, studying literature.

* *Playboy* magazine has shown plenty of love for Marilyn over the years. The men's magazine first featured Monroe in 1953 as "Sweetheart of the Month" in its debut issue. (Hugh Hefner reportedly paid $500 for the rights to her nude photo—a shot taken seven years prior for which Monroe herself had only received $50.)

* Fed up with being typecast as a dumb blonde, Marilyn moved to New York City to study with the legendary Lee Strasberg at the Actors Studio. Strasberg had enormous respect for her talent as an actress—not her image as a sex goddess.

* Hollywood folklore says that during the production of *Some Like It Hot* (1959), Monroe's notorious habit of being late or not showing up escalated and that when she did show up, she often needed dozens of takes to get even the simplest lines correct. Few realize that part of the reason for her behavior was due to a miscarriage that occurred during the film's shooting schedule. Monroe was battling a profound depression and was relying on prescription drugs to help her sleep and to wake her up.

* Marilyn Monroe completed 29 films during her career. She was working on her last—*Something's Got to Give*—when she died on August 5, 1962. The film was never released.

Becoming *Black Like Me*

Journalist John Howard Griffin used deception to force America to take a closer look at racism.

❋ ❋ ❋ ❋

IN 1959, JOHN Howard Griffin—a white man from Texas—pondered the issues of race in the United States. The Civil Rights Movement had started only four years prior; and Griffin decided that the best way to find out how a black man lived was to transform himself into one and travel through the segregated Deep South. Griffin detailed his odyssey in a series of articles for *Sepia* magazine, a monthly publication for black audiences.

Before Griffin assumed his new persona, he visited the Deep South as a white man; he reported meeting many friendly, hospitable people along the way. Next, with the aid of a dermatologist, Griffin ingested oral pigmentation medication and underwent treatments that darkened his skin.

When he returned to the same towns he had previously visited, now as a black man, he received a very negative reaction. He later reported that the daily difficulties of living as a black man outweighed even those moments of overt racism. Seemingly small tasks that white people took for granted—getting a bite to eat, requesting a glass of water, using a restroom—took up a large amount of his time. Often he would be turned away from establishments and directed toward black-only diners and shops, which were usually located far away.

Griffin's story spread like wildfire. In 1961, he wrote *Black Like Me*, a best seller that told of his journeys. Television interviews and magazine features followed. White Americans finally realized what life was really like for blacks in the South.

Griffin received criticism and death threats, so he moved his family to Mexico. He had offered an eye-opening look at the racism prevelant in the South, but he paid a steep price for it.

What Is a Cowboy?

It seems that there are as many ideas of what a cowboy really was as there are John Wayne movies in which he played one.

✳ ✳ ✳ ✳

THE COWBOY IS probably the most recognized and beloved symbol of the Old West. But what did cowboys really do? What were their responsibilities? What made them cowboys?

An Unexpected Origin

The first surprise may be that the American cowboy can trace his roots to the young men who herded cows on the haciendas, or ranches, in Spain. The task of herding cattle began in that country in medieval times. The Spanish word for cow is *vaca*, and those who herded them on horseback were called *vaqueros*, a name that eventually evolved into the American term *buckaroo*.

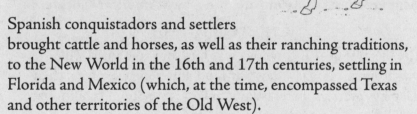

Spanish conquistadors and settlers brought cattle and horses, as well as their ranching traditions, to the New World in the 16th and 17th centuries, settling in Florida and Mexico (which, at the time, encompassed Texas and other territories of the Old West).

American Traditions

Many customs of the vaqueros became parts of the American cowboy tradition. They used lariats or lassos to rope cattle. The vaqueros put saddles on their horses, although early American cowboys added the horn in the middle so they'd have something to tie the lariat to. Branding a cow's hide with a hot iron as a means of identifying the owner also came from Spain, and the Mexican sombrero was an ancestor of the cowboy hat. Spurs (for prodding a horse to run faster), chaps (leather over-pants to protect the legs), and the rodeo (Spanish for *roundup*) are also vaquero contributions.

In the early 1800s, settlers from the newly formed United States began moving west in greater numbers, bringing with them English riding and ranching customs that would eventually blend with those of the vaqueros. The real heyday of the American cowboy came after the end of the Civil War, when former soldiers from both sides moved west in search of opportunities. At the same time, demand for beef in the East increased. Ranchers who took advantage of the huge open ranges in Texas needed scores of cowboys to tend their herds, round up cattle, and drive them to the markets in rail centers such as Abilene, Kansas, to be loaded onto trains.

Because of their experience, many Mexican cowboys were hired by the Texas ranchers. In addition, because the huge demand for ranch workers generally won out over racism and discrimination, a large number of African Americans who had been freed from slavery were hired as cowboys. Native Americans were also sometimes hired as cowboys, often as part of government programs to "assimilate" them into American culture.

The Daily Grind

The lifestyle of real cowboys was far from romantic. Their pay was around a dollar a day, but food and beds were free; their sleeping quarters were usually in large one-room bunkhouses. Besides herding cattle, cowboys also had to herd horses, both tame and wild, to keep themselves supplied with the three to four fresh mounts each needed every day. Wild horses were "broken" by riding them until they settled down, a practice known as "bronco-busting." Before the annual cattle drives, herds had to be rounded up for branding and castration, which were tasks performed by the cowboys. Cattle were herded using special horse skills known as "cutting." Many of these activities and skills can be seen today in rodeo competitions.

Life on a cattle drive gives us the image of what we might call the "lonesome cowboy." The Chisholm Trail was 1,000 miles long, and cattle would only be driven 15 miles a day, which meant that a drive could last as long as two months. Singing, telling stories, and other campfire activities were the cowboys' primary forms of entertainment. While Mexican cowboys carried lariats and knives, American cowboys who were Civil War vets preferred rifles and handguns for hunting, scaring off varmints, and defending themselves. Due to cowboy demand, the Colt company developed the famous Colt .45 six-shot revolver, another symbol of the Old West.

Cowboy Fashion

The fancy clothes worn by popular cowboy singers of today and the past had no place in real life. Working cowboys wore wide-brimmed hats that combined features of the Mexican sombrero and the hats worn by both sides in the Civil War. Cowboys wore cotton bandannas to wipe off sweat and protect their eyes and nose from dust. Leather cowboy boots had pointed toes to help them slip easily into stirrups, high heels to keep them in the stirrups, and high tops to guard the cowboys' calves while riding. The standard cowboy pants were jeans with a smooth seam on the inside to protect from blistering while riding. Leather chaps worn over the jeans gave added protection.

There's usually not much mention of "cowgirls" in cowboy history. However, some women, usually wives and daughters of ranchers, worked on the ranches and occasionally owned them.

The life of the true Texas cowboy—herding cattle and working the trail drives in the mid- to late-1800s—was much tougher, much more boring, and much less romantic than the version generally seen in the movies. Modern cowboys are better paid and still perform some of the traditional duties, although horses have been replaced by pickup trucks, all-terrain vehicles, and sometimes even helicopters. Helicopters? That sounds like something that cowboy John Wayne might even like.

10 Famous People Who Were Adopted

The image of the typical or "normal" American family—with a father, a mother, 2.5 kids, and a dog—has become less and less familiar over time. These days, families are "blended" and "progressive" and more than a little creative in terms of structure. Below are a few well-known celebrities who were ahead of the curve. Each famous figure listed below was orphaned, fostered, or adopted at a young age—and clearly didn't let that set them back.

✳ ✳ ✳ ✳

1. **Babe Ruth:** George Herman Ruth, Jr., born in 1895 in Maryland, lost six of his seven siblings in childhood due to disease and poverty. Babe's tavern-owning parents placed him and his sister in orphanages, sending Babe to St. Mary's Industrial School for Boys. It was there that Babe met Brother Matthias, who taught him how to play baseball. The rest is history—Babe Ruth was one of the greatest and most beloved players to ever set foot on a baseball field.

2. **Bo Diddley:** In 1928, one of America's most influential blues musicians, Ellas Bates—better known as Bo Diddley—was born to a desperately poor couple in rural Mississippi. At a young age, he was adopted, along with three cousins, by his mother's cousin, who moved the family to Chicago in the mid-1930s. Nicknamed "The Originator," Diddley would go on to record nearly 40 records and was inducted into the Rock and Roll Hall of Fame in 1987.

3. **Dave Thomas:** Dave Thomas, the founder of fast-food restaurant giant Wendy's, was given up for adoption at birth. Sadly, his adoptive mother died when he was five. Thomas left high school in the tenth grade to work full-time at a restaurant. After a stint in the army, Thomas

moved to Columbus, Ohio, where he opened his first Wendy's restaurant in 1969. He would later found the Dave Thomas Foundation for Adoption to promote adoption-law simplification and reduce adoption costs in the United States.

4. **Deborah Harry:** Best known as the lead singer of Blondie, the New Wave innovator that produced hits such as "Call Me" and "Heart of Glass," Deborah Harry was given up at three months and adopted by a couple from New Jersey. Harry led the typical rock-star lifestyle, and she lived to tell the tale. Blondie was inducted into the Rock and Roll Hall of Fame in 2006, and Harry continues to tour and act.

5. **Malcolm X:** The childhood of the man who would become "black power" leader Malcolm X was not a happy one. His father, Earl Little, was a Christian minister who was killed in 1931 when Malcolm was a small boy. Following his father's death, his mother had a nervous breakdown and was committed to a mental hospital. Malcolm and his siblings were put into an orphanage and later fostered by various families. Malcolm would later convert to the Nation of Islam and emerge as one of the most influential civil rights activists of the modern era.

6. **Steve Jobs:** The eventual cofounder of Apple Computers and the brain behind the iPhone, iPod, and iPad, Steven Paul was adopted as an infant by Paul and Clara Jobs in February 1955. Jobs held an internship with Hewlett-Packard and did a stint at Atari, Inc., before he and Stephen Wozniak developed the first Apple computer. These days, the iPhone is ubiquitous, and Apple computers are synonymous with style and technical savvy.

7. **Scott Hamilton:** Dorothy and Ernest Hamilton adopted Scott in 1958 when he was just six weeks old. In 1984, Hamilton won an Olympic gold medal in men's figure skating, making him the first American male to medal in the

sport since 1960. In 1986, Hamilton created Stars on Ice, a professional ice show that visits cities around the world.

8. **Harry Caray:** During his decades-long career, baseball announcer Harry Caray called the shots for the St. Louis Cardinals, Chicago White Sox, and Chicago Cubs. Harry Christopher Carabina was born in 1914 in one of the poorest sections of St. Louis and was still an infant when his father died. By the time he was ten, his mother had died, too, so an aunt raised him from that point. In 1989, Caray was inducted into the Baseball Hall of Fame as a broadcaster, and, in 1990, he joined the Radio Hall of Fame. A statue of him stands outside legendary Wrigley Field on Chicago's north side.

9. **Melissa Gilbert:** Best known for her portrayal of Laura Ingalls on *Little House on the Prairie*, Melissa Gilbert was adopted at birth by Ed Gilbert and Barbara Crane, both Hollywood actors. Melissa's dad died when she was 11, and Michael Landon, who played her father on television, became a surrogate father to her. Melissa's siblings include adopted brother Jonathan Gilbert, who portrayed Willie Oleson on *Little House,* and sister Sara (who is not adopted), who played Darlene on *Roseanne.* Melissa continues to act, mostly in made-for-TV movies, and she served as president of the Screen Actors Guild from 2001 to 2005.

10. **Faith Hill:** Adopted when she was only a few days old, Audrey Faith Perry was raised in Star, Mississippi, by Ted and Edna Perry. The country music superstar was the only adopted kid in the family and formed a good relationship with her biological mother later in life. Faith always knew that she was adopted and refers to her childhood as "amazing."

Franz Mesmer Transfixes Europe

The Age of Enlightenment saw an explosion of new ideas. One of these was the possibility of tapping into people's subconsciouses, causing them to enter a dreamlike state where they might find relief from various ailments, whether through actual effect or merely by the power of a hypnotist's suggestion. One early practitioner of this technique became so famous that his very name is now synonymous with the ability to send patients into a trance—the art of mesmerism.

※　※　※　※

FRANZ ANTON MESMER was a late bloomer. Born in Germany in 1734, Mesmer had difficulty finding a direction in life. He first studied for the priesthood, and then drifted into astronomy and law before finally graduating from the University of Vienna at age 32 with a degree in medicine. He married a well-to-do widow and set up practice in Vienna, becoming a doctor to the rich and famous and using his connections to cater to an upper-crust clientele. He lived comfortably on a Viennese estate and counted among his friends Wolfgang Amadeus Mozart, who wrote a piece for Mesmer to play on the glass harmonica, an instrument that had just arrived from America.

At first, Mesmer's medical prescriptions were unremarkable; bleeding and purgatives were the order of the day, and Mesmer followed accepted medical convention. But Mesmer's attention was also drawn to the practice of using magnets to induce responses in patients, a technique that was much in vogue at the time. Mesmer experimented with magnets to some effect and came to believe that he was successfully manipulating tides, or energy flows, within the human body. He theorized that illness was caused by the disruption of these flows and that health could be restored by a practitioner who could put them back in order. He also decided that magnets themselves were unnecessary props and that he could perform the manipulation of the tides himself, because of what he termed his "animal magnetism"—the word *animal* merely stemming from the Latin term for *breath* or *soul*. He would stir the tides by sitting in a chair opposite a patient with their knees touching, gazing unblinkingly into their eyes, making passes with his hands, and massaging the areas of complaint, often continuing the treatment for hours until the patient felt the magnetic flows moving inside his or her body.

Europe Becomes Mesmerized

Mesmer gained notoriety as a healer, his fame growing to the point where he was invited to consult on famous cases of the day. He investigated claims of unusual cures and traveled around Europe, holding demonstrations at which he induced symptoms and their cures merely by pointing at people, much to the amazement of his audience. But after a 1777 scandal—he temporarily restored a blind piano player's sight, which caused her to lose her audience because the novelty of watching her play was now gone—Mesmer decided to move to Paris.

France would prove to be a good fit for Mesmer. He resumed seeing patients, while at the same time seeking approval from the scientific community of Paris for his techniques. The respect and acknowledgment he felt he deserved from his peers was never to come, but his popular reputation soared; Marie Antoinette once wrote to Mesmer and begged him to reconsider after he announced that he intended to give up his practice. His services were in such demand that he could no longer treat patients individually; he resorted to treating groups of patients with a device he called a *baquet*—a wooden tub bristling with iron rods around which patients would hold hands and collectively seek to manipulate their magnetic tides. Mesmer himself would stride back and forth through the incense-laden room, reaching out and tapping patients with a staff or finger. For a complete cure, Mesmer believed patients needed to undergo a convulsive crisis—literally, an experience wherein they would enter a trancelike state, shake and moan uncontrollably, and then be carried to a special padded chamber until they recovered. The treatment proved particularly popular with women, who outnumbered men 8–1 as patients of Mesmer. This statistic did not go unnoticed by the monitors of public decency, who drew the obvious conclusion that something immoral was taking place, though they were unable to produce much more than innuendo in support of their accusations.

When I Snap My Fingers...

Mesmer's incredible popularity also made him an easy target for detractors. Mesmerism became such a fad that the wealthy even set up baquets in their homes. But like many fads, once it was over, it was ridiculed. As a result, Mesmer's client base declined, and he was even mocked in popular theater.

Copycats emerged to the extent that in 1784, the king set up a commission—including representatives from both the Faculty of Medicine and the Royal Academy of Science—to investigate all claims of healing involving animal magnetism. Benjamin Franklin, in Paris as an ambassador at the time, was one of the investigators. In the end, the commission determined that any treatment benefits derived from Mesmerism were imagined. This rejection by the scientific community, combined with the erosion of his medical practice, drove Mesmer from Paris in 1785. He kept an understandably low profile after that, spending some time in Switzerland, where he wrote and kept in touch with a few patients until he died in 1815.

Mesmer's legacy remains unresolved. Some still view him as a charlatan of the first order. Others see in his techniques the foundation of modern hypnotherapy, which has become a well-recognized practice in psychiatry. Regardless, it is indisputable that Franz Anton Mesmer's personal animal magnetism continues to capture our imagination even today.

A Doll of a Woman: Cindy Jackson

Cindy Jackson is living proof that any woman can achieve a little girl's fantasy of resembling a Barbie doll—provided she is willing to put forth the time, money, and determination to undergo dozens of procedures on her face and body.

✳ ✳ ✳ ✳

Coming Up Barbie

JACKSON'S PLATINUM-HAIRED PERFECTION may invite comparison to stereotypes, but she is no dumb-blonde-joke punch line. On the contrary, the Ohio-farm-girl-turned-world-celebrity is a member of MENSA. However, Jackson has refused to accept the idea that a brainiac can't also be a beauty, so she has spent around "the cost of a mid-size family car" on dozens of cosmetic procedures since 1988. She says that it all goes back to the Barbie doll she received from her parents at age six. "Through Barbie I could glimpse an alternative destiny," Jackson says on her website.

Although photos of Jackson at age eight reveal a relatively attractive girl, Jackson felt homely compared to her beautiful mother and sister. She also felt unloved by an undemonstrative father and misunderstood by schoolmates, who took her habit of daydreaming for standoffishness. At home, she would pore over her mother's issues of *Vogue* magazine and imagine a life filled with glamour and excitement . . . a life just like Barbie's.

Jackson studied art and photography after graduating from high school and then worked several jobs until she could buy a one-way ticket from Ohio to London. She arrived in 1977 and spent ten years rocking with British punk bands. When her father died in 1988 and left her an inheritance, a lightbulb went on: She could now afford to lift her heavy eyelids, reduce her chin and nose, and add breast implants. At last she could become Barbie—or at least a reasonable facsimile.

Globetrotter Barbie

That year, Jackson methodically set about going under the knife and soon had more plastic than the original Barbie. Her official "wish list" included a smaller, more feminine nose, fuller lips, whiter teeth, a smaller jaw, less "tired-looking" eyes, a more defined waistline and flatter stomach, higher cheekbones, slimmer thighs, and larger breasts. She's undergone 14 surgeries, often with multiple procedures during each operation. She's also had numerous minor procedures, such as dermabrasion. But not every operation went well—her first breast implants solidified into cementlike bags and had to be removed.

After only a few years, she achieved Barbie doppelganger status to the extent that TV talk shows began to invite her on as a guest. The BBC even paid for a live breast implant procedure, which was viewed by millions.

She has written an autobiography, *Living Doll*, and another book on cosmetic-surgery tips, which she offers along with her own beauty products on her website. She models designer fashions, makes appearances around the world, champions a number of animal-protection charities, and writes articles for *Cosmopolitan*, *The Daily Telegraph*, and other publications. Yet Jackson somehow finds several days a week to devote to researching cosmetic surgery. Now in her fifties, she looks decades younger and says that she is still toying with additional improvements. And though it may seem as though her ideas of beauty are only skin-deep, she contributes a portion of her book earnings to the prevention of cruelty to animals.

✳ According to the American Society of Plastic Surgeons, there were more than 1.6 million cosmetic surgeries performed in the United States in 2010.

✳ In 2010, approximately $10.7 billion was spent on cosmetic procedures, which include surgeries and nonsurgical treatments such as Botox injections, laser hair removal, and chemical peels.

Before They Were Billionaires

Warren Buffett may seem like he's made out of money, but he wasn't born that way. Check out what some of the world's richest people were doing before they were billionaires.

✳ ✳ ✳ ✳

Warren Buffett

WITH CLOSE TO $40 billion in his back pocket, Warren Buffett probably doesn't dine at many buffets. His empire has long landed him in the top couple of spots on the *Forbes* list of the world's wealthiest folks. His fortune is entirely self-made too: Buffett grew up delivering newspapers and chasing lost balls at a golf course in Nebraska. He started toying with the stock market at the ripe old age of 11.

By high school, Buffett was making $50 per week with a pinball business that he'd started. Buffett's big fortune, however, came in the early '60s, when he bought 5 percent of American Express's stock despite the company's near-bankrupt state. He may have looked like a fool at the time, but after American Express rebounded and his stock value skyrocketed, Buffett was the one laughing—all the way to the bank.

Sam Walton

Wal-Mart founder Sam Walton used a small-town approach to create what's now one of America's biggest retail chains. Walton started his career in sales at JCPenney and later opened his own Ben Franklin store franchise; soon, he oversaw 15 stores.

Walton opted to strike out on his own and cater to the rural market. The first Wal-Mart opened in Rogers, Arkansas, in 1962. As the chain grew, so did Walton's worth: When he died in 1992, his net worth was approximately $22 billion.

Ted Turner

Mr. Television wasn't always a wealthy chap, either. Ted Turner worked for his father's outdoor advertising company as a young man, dutifully mowing around billboards maintaining ad-housing properties. He worked his way through college and got kicked out of the Coast Guard before taking over the family business.

In 1970, Turner's fortunes really turned around when he turned down everyone's advice and decided to buy a tiny TV station in Atlanta. That station eventually expanded into a broadcasting empire, with networks such as TBS and CNN helping Turner to earn billions.

Michael Dell

Michael Dell clearly did something right. He has billions of dollars and one of the world's largest computer companies in his portfolio. And to think, Dell started out working at a Chinese restaurant as a kid. He used his earnings to buy valuable postage stamps from traders. He sold his stamp collection to buy a computer, used his computer to start a newspaper subscription business, and used the profits from that to buy a BMW.

In college, Dell continued his entrepreneurial streak by going door-to-door selling computers that he'd refurbished. Once he had made $180,000, he dropped out of school and started Dell.

Bill Gates

Wrestling with Warren Buffett for the title of *Forbes's* richest man, Bill Gates had an estimated $53 billion in 2010, down from $58 billion in 2008. (Hey, even filthy rich people aren't immune to the slumping economy!) Microsoft's main man was making major money while still a teenager: Gates formed a programming club in high school, and he helped build a traffic-monitoring program called Traf-O-Data. The program brought in around $20,000 and gave Gates his first real taste of success.

Gates started to study at Harvard as a prelaw student, but then he and programming pal Paul Allen saw a "window" of opportunity, you might say, and they left the scholastic world behind. The duo formed a little venture known as Microsoft and never looked back.

Houdini Unbound

Magic still thrives, thanks to the mystifying antics of entertainers such as David Blaine and Criss Angel. But no one holds a candle to the great Harry Houdini, master magician and escape artist extraordinaire.

✳ ✳ ✳ ✳

The Early Years

HOUDINI WAS BORN Ehrich Weisz on March 24, 1874, in Budapest, Hungary. His family emigrated to America when Ehrich was about four years old, settling in Wisconsin. Ehrich started performing magic at age 12, billing himself as Eric the Great. He ran away from home to entertain at fairs and circuses, but he later joined his family at their new home in New York City.

At 15, Ehrich read a biography of famed French magician Jean Robert-Houdin; the book changed his life. In honor of his hero, he took the stage name Harry Houdini. He performed solo for a while, and then, in 1892, he teamed up with his brother Theo. As the Houdini Brothers, the duo performed at venues, such as Coney Island and the 1893 Chicago World's Fair. In 1894, Harry married Wilhelmina Rahner, who replaced Theo in the act.

The Escape Artist

Houdini was skilled at magic, card tricks, and escape artistry, but widespread fame eluded him until he took the advice of renowned booking agent Martin Beck, who encouraged him to eschew the small stuff and concentrate on illusions and escapes. Beck put Houdini on the vaudeville circuit, which took him around the country. To generate publicity in each town he visited, Houdini would ask the local police to lock him in their sturdiest cell—from which he would promptly escape.

Houdini traveled to Europe in 1900, and it was there that he really gained a reputation, by routinely escaping from the seemingly inescapable. Soon he was the highest-paid entertainer on the continent, raking in $2,000 a week. When he returned to the States, he set out to prove that there was nothing from which he could not escape. Several of Houdini's stunts were literally death-defying: Once he came frighteningly close to suffocating while escaping from a buried coffin.

In 1918, Houdini created a magic act that would become a staple of later magicians—he made a live elephant disappear at the famed Hippodrome in New York City. He also introduced another fan favorite, swallowing several needles and a piece of string, and then pulling the string from his throat with the needles threaded. "My professional life has been a constant record of disillusion," said Houdini, "and many things that seem wonderful to most men are the everyday commonplaces of my business."

More than a Magician

In the 1920s, Houdini established himself as a debunker of fake spiritualists, testifying before a congressional committee on the subject in 1926. As a magician, he knew the tricks of the trade, and would often don disguises to visit "mediums" who claimed to be able to talk to the dead. Once they'd gone through their act, Houdini would reveal how the tricks were done. It was Houdini's way of giving back—he hated charlatans who preyed on grieving families.

Houdini died on October 31, 1926—not in a failed escape attempt, as some legends have it, but from complications from a ruptured appendix. There is some debate regarding the events leading up to his death, which almost certainly involved an incident in which he was punched in the abdomen by a student at McGill University in Montreal. Houdini's funeral was held on November 4, 1926, in New York; it was attended by more than 2,000 mourners.

The Jerry Springer Story

It's not every day that a mayor from the Midwest becomes famous for chatting with cheating transvestites. Behold the odd story of the one and only Jerry Springer.

✳ ✳ ✳ ✳

HIS NAME'S BEEN chanted more times than some spiritual mantras (usually by mullet-headed audience members rooting for a fistfight), but Jerry Springer wasn't always the mic-holding talk show referee that we know him as today. Springer started his career as a politician, making it all the way to Cincinnati's mayor's office before moving into his more notorious trash-talkin', daddy-drama-filled domain.

The Political Years

After graduating from law school in 1968, Springer worked as a campaign aide to Robert F. Kennedy. By the early 1970s, he'd landed a seat on the Cincinnati City Council. Perhaps taking a cue from his future talk-show guests, Springer soon found himself in the center of a sex-filled scandal: In 1974, he was forced to resign from his post after word broke that he'd visited a prostitute at a less-than-legit "massage parlor." (He evidently paid by check, which led to his public outing.)

Within a few years, however, Springer was able to regain the public's trust. He never denied paying the prostitute, and he used his honesty as the centerpiece of a second campaign for council in 1975. Springer was elected once more as a city councilman and was even appointed mayor of Cincinnati for a one-year stint during that second term.

Television Transition

While serving as mayor, Springer started delivering regular commentaries on a local radio station. That led to a gig with NBC station WLWT, where Springer parlayed a role as political reporter into a job as main anchor and a managing-editor position. It was there that he developed the catchphrase that would later become his daily show-closing statement: "Take care of yourself and each other."

While working at WLWT, Springer was approached about doing a nationally syndicated political talk show. In 1991, *The Jerry Springer Show* went on the air, and oddly enough, it was originally designed to emulate the serious tone of Phil Donahue's early programs. Looking for more pizzazz (and higher ratings), Springer soon decided to throw in some twists—namely, ditching the serious political focus and instead bringing on outrageous, low-class guests who'd gladly go bonkers on his stage.

It worked: The revamped show became a massive hit, and the more extreme the guests became, the higher the ratings climbed. Controversy was the formula, and it paid off big time. Audiences came to expect the outrageous, and Springer delivered: Guests would tear off clothes, attack each other, and do just about anything in front of the cameras. The now-infamous chant of Springer's name ("Jer-ry! Jer-ry! Jer-ry!") became synonymous with sleazy fighting and out-of-control antics. A cast of off-duty cops was on constant standby, poised and ready to break up any argument that neared the point of danger.

Springer Realities

So how much of *The Jerry Springer Show* is actually real? Producers claim they work to maintain accuracy, although some sources suggest that they encourage guests to "emphasize certain plot points" to achieve higher levels of drama.

As for the guests who are about to be surprised—the guys who find out that their girlfriends are actually men married to goats—producers do let them know in advance that some sort of secret is going to be revealed on the show. They use a standard list of "stock secrets" to provide four possible twists, one of which is the actual truth. For Springer, there's one truth that can't be denied: Love him or hate him, he's carved out his own niche in American pop culture.

The Odd Odyssey of George Foreman

Part shaman, part huckster, George Foreman went from back-alley thug to Olympic gold-medal winner to heavyweight champion of the world before he found his true calling at the pulpit of the people. Whether he's praising the rewards of religion or promoting the glory of grilling, Foreman is formidable, forceful, and full of fun.

✳   ✳   ✳   ✳

ONE OF THE most feared fighters to ever stride inside the roped ring, George Edward Foreman was blessed with a hammer for a fist, troubled with a tempestuous temper, and burdened by a Texas-size chip on his shoulder. Born in the tiny hamlet of Marshall, Texas, and raised on the hard streets and back alleys of Houston, big George spent most of his youth looking for trouble—and finding it.

Foreman's father left the family home and his seven children when George was young, leaving George's mother Nancy to raise the entire brood by herself. A hardened street fighter, petty thief, and common criminal, the troublesome teenager dropped out of high school at age 16. After one extremely intense encounter with the law that forced him to hide inside a sewer pipe to escape the police dogs nipping at his heels, Foreman decided to turn his life around. Enticed by an advertisement for President Lyndon Johnson's Jobs Corps and its slogan ("If you have dropped out of school and want a second chance in life, then the Job Corps is for you"), Foreman signed up for the program. He eventually relocated to California, where he met counselor and boxing coach Doc Broaddus, who fine-tuned Foreman's raw brawling skills and helped turn the criminal chump into a heavyweight champ.

The Sweet Science

Once he stepped inside the squared circle, it was clear that George had the potential to become a formidable fighter. In addition to his natural instincts, pure power, and athletic prowess, he had an intimidating stare and an aggressive ring presence that often poleaxed his opponents before a single punch was thrown.

Foreman shot up the amateur rankings and earned a spot on the 1968 Olympic boxing team. He accompanied the squad to Mexico, where he easily pummeled his way to the gold medal. When he accepted his award, he wrapped himself in an American flag, a display of patriotism that stood in stark contrast to the raised black fists of protest by fellow African American medal winners John Carlos and Tommie Smith. He returned home a hero.

Within four years of turning pro, Foreman was the top-ranked contender for the heavyweight title with 37 consecutive victories, most of them by knockout. On January 22, 1973, Foreman decked champion Joe Frazier in two rounds to capture the crown. He successfully defended his title twice before agreeing to meet former champ Muhammad Ali in Zaire in what would become known as "The Rumble in the Jungle," perhaps the most famous fight in the history of the pugilistic pursuit.

Foreman had been able to use his brute strength, menacing mug, and pistonlike jab to neutralize every opponent he had faced. But Ali—a shrewd, skillful, and strategic showman—could not be bullied. The former champ allowed Foreman to pound him at will in the early rounds, carefully protecting his head while letting his body absorb the blows. The maneuver, dubbed the "rope-a-dope," fatigued Foreman, which allowed Ali to reclaim his crown by knocking out the champion.

Waking Up

That loss changed Foreman forever. Although he continued to fight, it was clear that his spirit was exhausted. After retiring from the ring in 1977, he became an ordained minister and began preaching on the same streets that he'd prowled as a kid in the poverty-ridden haunts of Houston. In 1984, he founded the George Foreman Youth and Community Center, a non-denominational facility that provides kids with the safe haven that Foreman himself had needed but couldn't find. In 1987, with his ministry and youth center in need of a monetary boost, Foreman returned to the ring, shocking scribes of the sweet science by winning 24 consecutive bouts. In 1994, at age 45, Foreman floored Michael Moorer to win the IBF and WBA heavyweight titles, becoming the oldest boxer in history to capture the crown and the only fighter in history to go a full two decades between title victories.

Retirement for Real

Big George hung up the gloves for good in 1998. By this time, the once-scowling scrapper had become a smiling symbol of success. His transformation has been miraculous. Foreman has gone from a monosyllabic menace to a well-spoken and highly regarded humanitarian, Christian, and caregiver. He is also a very wealthy entrepreneur and TV marketer, using the airwaves to hawk everything from mufflers to clothing to his world-famous George Foreman grill. He also tried his hand at situation comedy, starring in the short-lived TV series *George*. Unfortunately, like many of the opponents who stood toe-to-toe with him, it flopped.

Foreman has also published a series of best-selling cookbooks and spiritual testaments. He authored an inspirational guide to fatherhood, another subject on which he is an expert. The father of ten children, he is renowned for naming all five of his sons George, although to avoid obvious confusion in the Foreman household, they are known by their nicknames: Monk, Red, Joe, Big Wheel, and Little George.

The Man Behind the Mouse

Walt Disney is arguably the most famous moviemaker in the world—an icon adored by millions. But how much do you really know about him? Maybe less than you think.

✳ ✳ ✳ ✳

✳ As a youngster, Disney sold his drawings to his neighbors.

✳ At age 16, Disney tried to join the military but was rejected for being too young. He joined the Red Cross instead. He was sent to France, where he drove an ambulance.

✳ Disney grew his trademark mustache at age 25.

✳ The first commercially released Mickey Mouse cartoon, *Steamboat Willie,* was also the first Disney cartoon to feature synchronized sound. It premiered in New York City on November 18, 1928.

✳ Disney provided the voices of both Mickey and Minnie Mouse for nearly 20 years.

✳ Disney's first animated feature film, *Snow White and the Seven Dwarfs,* cost nearly $1.5 million to produce. It was a huge gamble for the Disney studio but went on to tremendous financial success and critical acclaim.

✳ Disney won more Academy Awards than any other individual: 32 total.

✳ Following the success of *Snow White,* Disney and his brother, Roy, gifted their parents with a new house that was close to their studios. A month later, their mother died from asphyxiation caused by a broken furnace in the new home. It was a tragedy from which Disney never recovered.

Howard Hughes: Oddball!

Here are a few fascinating facts about the wacky recluse.

✳ ✳ ✳ ✳

WHEN IT COMES to bizarre behavior, few people can match the antics of billionaire Howard Hughes. Born in Houston in 1905, Hughes was a shrewd businessman whose personal interests ranged from aviation to motion pictures. But Hughes also had a dark side, which ultimately overshadowed his many important accomplishments. Here's a glimpse:

Associates reported that Hughes was obsessed with the size of peas, and he even created a special fork so he could sort them on his plate.

During the making of *The Outlaw* (1943), a biography of Billy the Kid that Hughes produced, Hughes became fixated on a perceived flaw in Jane Russell's bras, which he insisted made it look as if she had two nipples on each breast. This problem so vexed him that he designed a steel underwire bra for her to wear during the shoot.

During World War II, the U.S. military gave Hughes $18 million to build three "flying boats" for use in transporting troops and supplies. The contract stipulated that the planes be ready within three months, but Hughes saw that as merely a suggestion. He ultimately delivered one plane—the "Spruce Goose"— in 1947, two years after the war had ended. The plane was flown only once, by Hughes himself, and then mothballed.

Becoming More and More Eccentric

Hughes was germophobic and went to extraordinary lengths to protect himself from contamination. Everything he touched had to go through a rigorous cleansing process that involved four separate scrubbings with soap and water. The item then had to be wrapped in a tissue or a paper towel and handed to him by someone wearing white cotton gloves. Plus, Hughes hired only Mormons to serve on his personal staff because he believed that they were more hygienic than non-Mormons.

In 1966, Hughes traveled to Las Vegas and booked two floors at the Desert Inn for ten days. And then he refused to leave. He bought the hotel and turned the ninth floor into his private home.

Television was a like a drug to Hughes, who, in his later years, kept the TV on at high volume 24 hours a day. He bought the Las Vegas CBS affiliate and dictated which shows would run and when. However, Hughes often took so long to make up his mind that the station was unable to release a reliable schedule to the public.

It's the Little Things

Over the years, Hughes wrote hundreds, perhaps thousands of memos to his highly paid team of lackeys. But rather than focus on the intricacies of his business empire, many of these missives dealt with the ridiculously mundane. One three-page memo detailed how canned fruit was to be prepared for him.

He was also a not-so-secret racist. He intended to purchase the ABC television network for $200 million, but he canceled that plan after viewing an episode of *The Dating Game* in which a black man went on a date with a white woman.

Hughes refused to get rid of anything that belonged to him—including his own urine, which he collected and kept in jars.

At the time of his death on April 5, 1976, Hughes was virtually unrecognizable. His frame was skeletal, his hair long and matted, and his fingernails and toenails grotesquely long. He had lived as a recluse for so much time that officials had to rely on fingerprints to confirm his identity.

Canada's Cryptic Castaway

This mute amputee has a foothold in Nova Scotian folklore— nearly a century after his death.

✳ ✳ ✳ ✳

Who Is This Man?

ON SEPTEMBER 8, 1863, two fishermen in Sandy Cove, Nova Scotia, discovered an unusual treasure washed ashore: a lone man in his twenties with newly amputated legs, who had been left with just a loaf of bread and jug of water.

There were a few clues, such as his manner of dress, that led the townspeople to speculate that the fellow was an aristocrat. But there was no point in asking him—he didn't speak. In fact, he was said to have uttered only three words after being found: "Jerome" (which the villagers came to call him), "Columbo" (perhaps the name of his ship), and "Trieste," an Italian village.

Based on these three words, the villagers concocted various romantic stories about him: that he was an Italian nobleman captured and mutilated by pirates (or perhaps a pirate himself), a seaman punished for threatening mutiny, or maybe an heir to a fortune, crippled and cast away by a jealous rival.

Charity Case

Jerome went to live with Jean and Juliette Nicholas. Jean was fluent in five languages, none of which proved successful in communicating with Jerome.

In 1870, when the Nicholases moved away, the town rallied together and paid the Comeau family $140 a year to take him in. On Sundays after mass, locals would stop by and pay a few cents for a look at the maimed mute. Jerome lived with the Comeaus for the next 52 years until he died on April 19, 1912.

Records suggest that Jerome was no cool-headed castaway. Though he never spoke intelligibly, hearing certain words (specifically *pirate*) would send him into a rage. It's also been said that he was not a fan of cold weather, so he spent winters with his leg stumps shoved under a stove for warmth. In his younger days, he enjoyed sitting in the sun, but he spent the last 20 years of his life as a complete shut-in, huddled by the stove.

Mystery Revealed

Jerome's panic about the cold makes sense if the latest hypotheses about him are true. One group of scholars uncovered a story in New Brunswick about a man who was behaving erratically and couldn't (or wouldn't) speak. To rid themselves of the eccentric, members of his community put him on a boat to New England— but not without first chopping off his legs. The man may have never made it to New England but instead wound up on the beach at Sandy Cove.

Another theory suggests that Jerome was a man who was found in 1861, pinned under a fallen tree in Chipman, New Brunswick, with frozen legs that had to be amputated. Without a doctor nearby, the man was sent down the St. John River to Gagetown and then shipped back to Chipman, where he was supported for two years by the town's parish and nicknamed "Gamby" by the locals (which means "legs" in Italian). At that point, the parish got tired of taking care of him and paid a captain to drop him across the bay in Nova Scotia. Another account suggests that after the surgery, the man wasn't returned to Gagetown but put directly on a boat that landed at Sandy Cove.

Regardless of which theory is more accurate, both suggest that the reason for Jerome's arrival in Nova Scotia is that an entire town disowned him.

But New Brunswick's loss has been Nova Scotia's gain. There, Jerome is a local legend. He has been the subject of a movie (1994's *Le secret de Jérôme*), and a home for the disabled bears his name. Tourists can even stop by his grave for a quick snapshot of the headstone, which reads, quite simply, "Jerome."

Real Names of 16 Famous People

"A rose by any other name" is supposed to smell as sweet, but while that may be true in love, it isn't always so in show business. Below are the real names of some famous folks—see if you can match the real name to the alias.

✳ ✳ ✳ ✳

1. Reginald Kenneth Dwight
2. Paul Hewson
3. Mark Vincent
4. David Robert Jones
5. Caryn Elaine Johnson
6. Eleanor Gow
7. Samuel Langhorne Clemens
8. Tara Patrick
9. McKinley Morganfield
10. Farrokh Bulsara
11. Robert Allen Zimmerman
12. Demetria Gene Guynes
13. Marion Morrison
14. Allen Konigsberg
15. Georgios Panayiotou
16. Jay Scott Greenspan

A. John Wayne
B. Woody Allen
C. Elton John
D. Mark Twain
E. Freddie Mercury
F. Bob Dylan
G. George Michael
H. Whoopi Goldberg
I. Bono
J. Jason Alexander
K. Demi Moore
L. Vin Diesel
M. David Bowie
N. Muddy Waters
O. Elle MacPherson
P. Carmen Electra

Answer Key: 1. C; 2. I; 3. L; 4. M; 5. H; 6. O; 7. D; 8. P; 9. N; 10. E; 11. F; 12. K; 13. A; 14. B; 15. G; 16. J

Chapter 8

Earth and The Universe

Strange Weather Facts

✳ The fastest winds ever recorded on Earth blew through the suburbs of Oklahoma City, Oklahoma, on May 3, 1999. The 318-mile-per-hour gusts were recorded during an F5 tornado that destroyed hundreds of homes.

✳ The most rain in a three-day period fell on the tiny island of La Reunion in the South Indian Ocean. In March 2007, close to 13 feet of rain fell as Tropical Cyclone Gamede passed within 120 miles of the island.

✳ Cloud-to-cloud lightning can stretch over amazing distances. Radar has recorded at least one of these "crawlers" that was more than 75 miles long.

✳ On February 18, 1979, a snowstorm blanketed southern Algeria; it was the only time it has ever snowed in the Sahara.

✳ May 2003 featured the most tornadoes recorded in one month in the United States. Experts confirmed 543 twisters, far more than the 399 recorded in June 1992.

✳ Most people evacuate when a hurricane is on the way, but some deliberately fly into the storms. The National Oceanic and Atmospheric Association's Hurricane Hunters pilot C-130 aircrafts straight into a hurricane's eye to measure data such as wind speed and direction, which are critical in predicting where a hurricane will go and how strong it is.

The Dirt on Green Architecture

In the beginning, all architecture was environmentally friendly, or "green." Adobe houses, mud huts, dugouts, and igloos are all natural homes. Synthetic materials last longer than natural ones, but now we know how such construction affects the ecosystem.

✱　✱　✱　✱

ENVIRONMENTAL DEGRADATION, RUN-OFF, and pollution are just a few problems caused by modern development. To help preserve the planet, there is a growing movement to restore harmony between architecture and the environment.

Tools of the Trade

It's not just solar panels anymore. Green architecture uses a wide variety of natural, sustainable materials, including:

✱ recycled-rubber roofing: works just as well as shingles, no tar necessary

✱ insulation made of cotton or newspaper: recycled and non-hazardous

✱ cork flooring: recycled and inexpensive

✱ fiber-cement siding: recycled, durable

✱ photovoltaic (PV) cells: heating option that uses little energy

✱ wood-and-plastic-composite decking: looks the same as the real thing and is weather-resistant

✱ natural paints: beautiful, nonpolluting

✱ recycled glass: looks good, feels good

These materials, once difficult to find, are becoming more easily accessible to architects and builders. Many consumers realize that going eco-friendly doesn't mean sacrificing quality. In fact, many green materials are more durable than their synthetic counterparts.

Compressed earth blocks (CEB), for example, are composite bricks made from clay, sand, and a concrete stabilizer. These bricks are uniform, economical, largely locally produced, nontoxic, nonflammable, and bug-resistant.

Location, Location, Location

If you want to build an environmentally sensitive home, location is just as important as the materials you use. An adobe-brick home that's built in the middle of a delicate ecosystem will still affect wildlife and compromise the groundwater.

Some eco-minded architects feel that the kindest thing a person can do for the planet is not build anything at all. Instead, they recommend "repurposing" an existing structure, thereby using less energy and fewer resources.

If you are building from scratch, you have several "green" considerations. If you are building a business, choose a location that allows employees to bike to work or use public transportation. Building a home with a woodstove? Close proximity to a sustainable forest is prudent; burning fossil fuels to drive miles for firewood defeats the purpose.

Give Green a Go

There are lots of simple ways you can "green up" your lifestyle.

* Buy products made from recycled materials.

* Turn off computers at the power strip before you go to bed.

* Use rechargeable batteries or solar-powered devices.

* Turn off the lights when you leave a room.

* Clean your house with natural products, such as vinegar and baking soda.

* Take advantage of natural light by using energy-efficient glazed skylights and lightbulbs.

* Maintain an organic garden.

The World's Most Toxic Places

Toxic chemicals and radioactive materials are putting humans in serious danger in many parts of the world.

✳ ✳ ✳ ✳

Environmentalists have tracked down the ten most toxic sites on the planet, and what they've found paints a startling picture. Experts say that living in one of these cities is like "living under a death sentence." This is no exaggeration.

Sumgayit, Azerbaijan

This former Soviet industrial center is now home to countless contaminants. Untreated sewage and mercury-laden sludge are among the chief concerns and lead to unusually high cancer and death rates. Scientists have also found a large number of premature births and babies born with defects and deformities.

La Oroya, Peru

In this city, toxic emissions from mining result in food that's filled with high levels of lead. In fact, inspectors found that only 1 percent of children have normal amounts of lead in their blood. Local hospitals say that many babies are never even born because of prenatal damage.

Sukinda, India

Chromium is the issue in this mine-heavy region of India: Its untreated water has been found to have more than twice the recommended level. Side effects range from internal bleeding to widespread tuberculosis and infertility.

Vapi, India

In Vapi, chemical manufacturing plants produce pesticides, pharmaceuticals, and fertilizers, but with no safe disposal system, the waste runs right into the groundwater. The pollution there is so severe that some areas are now devoid of biological life.

Sukinda, India

Chromium is the issue in this mine-heavy region of India: Its untreated water has been found to have more than twice the recommended level. Side effects range from internal bleeding to widespread tuberculosis and infertility.

Vapi, India

In Vapi, chemical manufacturing plants produce pesticides, pharmaceuticals, and fertilizers, but with no safe disposal system, the waste runs right into the groundwater. The pollution there is so severe that some areas are now devoid of biological life.

Dzerzhinsk, Russia

The *Guinness Book of World Records* once named Dzerzhinsk the most chemically polluted city in the world. The average life expectancy there is only 44 years.

Norilsk, Russia

Russia's second-largest city above the Arctic Circle has hundreds of tons of copper and nickel oxides in the air, thanks to metal mining and processing plants in the area. Scientists report that the snow there is black and the air tastes of sulfur. Life expectancy in Norilsk is low, and the rate of illness among children is high.

Chernobyl, Ukraine

More than 20 years after the world's worst nuclear disaster, much of Chernobyl is still unlivable. The meltdown of a reactor's core sent unfathomable amounts of radiation into the city. Thousands of cases of cancer have been detected in young adults from the area, and millions still suffer from various health-related problems.

Kabwe, Zambia

Children bathe in contaminated water in this African nation, which was once home to massive lead-mining operations. Lead saturates the city's water and soil, and there are no health standards to keep the community safe. Many children have blood-lead levels just barely under an amount that's considered deadly.

Beach Squeaks and Dune Booms

There are two types of sonorous sands, which create sounds when stepped on or moved. Marco Polo attributed their strange songs to musical evil spirits, but scientists now know better.

✳ ✳ ✳ ✳

Sand Symphonies

"SINGING SAND" is generally classified as either "squeaking" or "booming." Squeaking sands are found near shorelines and, when compressed with a wave or footstep, emit a high-pitched chirp or croak that lasts only a fraction of a second. The real songsters of the silica world are the booming dunes, of which only 35 exist worldwide. Their sound is described as whistling or roaring, and has been compared to a kettledrum, foghorn, or pipe organ. Their pitch is low, but their sound is surprisingly loud, carrying up to six miles away and comparable to a rock concert or a revving motorcycle.

We Have Ways of Making You Talk

Booming dunes are made of highly polished, perfectly spherical sand that is extremely dry (as little as 0.1 percent moisture can impede the sand's ability to sing).

The dunes are quite tall—at least 150 feet—and when the slopes reach approximately 35 degrees, the sand is prone to avalanche. The grains collide around 100 times per second, slipping off each other and emitting sound. The waves caused by these collisions synchronize them, so the grains collide in tempo, creating a unified pitch. This vibrating exterior layer of sand, which is looser and less dense than the inner layer, causes the air around the dune to vibrate, amplifying the sound like a loudspeaker.

Although scientists have long been interested in singing sands, recording artists are now incorporating this strange music into their work. The most beautiful music comes from dunes in the Arabian country of Oman.

If the Moon Were Made of Green Cheese...

Ever wonder how many slices of cheese the moon would make if the moon were made of green cheese? Of course you have!

✳ ✳ ✳ ✳

21,900,000,000 km³ : volume of moon

0.0002835 km³ : volume of one Kraft single

74,074,074,100,000: number of Kraft singles that could be produced with moon-size wheel of cheese

3,741: number of years the population of American students—from kindergarten through high school—could be fed grilled cheese sandwiches made from this cheese for school lunch

The idea of the moon being made of green cheese dates back to John Heywood's 1546 collection *Proverbes*. But when Heywood wrote, "The moon is made of a greene cheese," he was not referring to the color green, but green as in "new" or "unaged." However, there are several cheeses that are actually green in hue. Here are five of the more popular ones:

1. Basiron Pesto: A bright-green Dutch gouda-style cheese with a strong pesto flavor.

2. Sage Derby: The practice of adding sage to this cheese to impart a green color dates back to the 17th century.

3. Green Thunder: This Welsh cheese is infused with garlic and green herbs.

4. Vermont Sage: This Vermont classic is flavored with bits of sage that give it a greenish hue.

5. Schabziger: Produced only in one canton of Switzerland, the practice of adding fenugreek to this Swiss cheese dates back to at least the 8th century.

The Big Shakedown

In 1811 and 1812, Missouri and surrounding states were rocked by devastating earthquakes.

✳ ✳ ✳ ✳

IN LATE SEPTEMBER 1811, Shawnee Chief Tecumseh made an appeal for peace with the whites—he also vaunted himself a force of the highest order. "You do not believe the Great Spirit has sent me. You shall know," he reportedly said. "I . . . shall go straight to Detroit. When I arrive there, I will stamp on the ground . . . and shake down every house in Tuckhabatchee."

On December 16, the day Tecumseh was thought to have arrived in Detroit, the first of a series of powerful earthquakes occurred near New Madrid, Missouri, destroying the small town of Tuckhabatchee.

In 1811, New Madrid was a popular destination for boatmen traveling the Mississippi River. It also sat directly on an active seismic fault zone three miles below the Earth's surface. By December, pressure that had been building up for centuries reached a breaking point.

December Jolt

The first quake struck on December 16 at around 2:15 A.M. Suddenly, there was a sound like distant thunder. Then, the earth began to shake violently; people were thrown from their beds, furniture flew across rooms, and trees snapped like twigs.

There was a "complete saturation of the atmosphere, with sulphurious vapor, causing total darkness," recalled an eyewitness. Lightning bathed the scene in eerie light. Ducks, geese, and other birds flew overhead, screeching loudly. Cattle stampeded. Chimneys crashed to the ground.

New Madrid was not the only area that was hit hard. The town of Little Prairie, 30 miles downriver, was completely destroyed by the quake and its numerous aftershocks. One witness saw the ground "rolling in waves of a few feet in height, with a visible depression between. By and by those swells burst, throwing up large volumes of water, sand and a species of charcoal."

On the water it was even worse. Boats were tossed around by the tremors. One boat was lifted 30 feet and propelled upriver for more than a mile—the Mississippi was running backward.

"The Devil Has Come Here"

For the next month, the earth continued its restless movement. On January 23, 1812, another quake struck. George Henrich Crist, who lived in north-central Kentucky, recalled, "The earth quake or what ever it is come again today. It was as bad or worse than the one in December. A lot of people thinks that the devil has come here. Some thinks that this is the beginning of the world coming to a end."

Final Shot...?

Like a seasoned stage performer, Mother Nature saved her biggest bang for the finale. The quake on February 7, which struck at around 3:15 A.M., was as big or worse than the ones before it. It completely destroyed New Madrid. A boatman anchored there wrote of "the agitated water all around us, full of trees and branches." He decided to keep his craft where it was as nearly two dozen other boaters frantically cut their mooring lines and headed for open water; they were seen again. A trapper hunting in Tennessee saw the ground "sink, sink, sink, carrying down a great park of trees."

The New Madrid earthquakes were felt as far east as New York City (over 1,000 miles away) and as far south as Milledgeville, Georgia (575 miles away). Only the area's scarce population prevented a greater loss of life.

Since then, the population of the area has grown enormously. And underneath them lies the New Madrid fault, like a ticking time bomb waiting to go off.

Fast Facts About Space

* Skylab, the first American space station, fell to Earth in thousands of pieces in 1979. Thankfully, most of them landed in the ocean.

* Skylab astronauts grew 1½ to 2¼ inches due to spinal lengthening and straightening as a result of zero gravity.

* The cosmos contains approximately 50 billion galaxies.

* Since 1959, more than 6,000 pieces of "space junk" (abandoned rocket and satellite parts) have fallen out of orbit, and many have hit Earth's surface.

* The surface gravity of Jupiter is more than two-and-a-half times greater than that of Earth.

* Uranus is unique among the planets in that its equatorial plane is almost perpendicular to its orbital plane.

* If you could fly across our galaxy from one side to the other at light speed, it would take 100,000 years to make the trip.

* Every year, the Sun loses 360 million tons.

* If you attempted to count the stars in a galaxy at a rate of one per second, it would take about 3,000 years to finish.

* Earth is the only planet in our solar system that's not named after a god.

* Neptune takes 165 Earth years to orbit the Sun. It appears blue because it is made of methane gas. Winds on Neptune can reach 1,200 miles per hour. Neptune has eight moons.

* Objects weigh slightly less at the equator than at the poles.

* In 2007, a small meteorite from Siberia sold for $122,750 at an auction in New York City.

Flubbed Headlines

✳ "Enraged Cow Injures Farmer with Ax"

✳ "Killer Sentenced to Die for Second Time in 10 Years"

✳ "Juvenile Court to Try Shooting Defendant"

✳ "Soviet Virgin Lands Short of Goal Again"

✳ "Mayor Says D.C. Is Safe Except for Murders"

✳ "Check With Doctors Before Getting Sick"

✳ "Fisherman Arrested for Using Wife as Shark Bait"

✳ "Statistics Show Teen Pregnancy Drops Off Significantly After Age 25"

✳ "Specialist: Electric Chair Can Be 'Extremely Painful'"

✳ "Students Cook & Serve Grandparents"

✳ "County to Pay $250,000 to Advertise Lack of Funds"

✳ "Volunteers Search for Old Civil War Planes"

✳ "One-Armed Man Applauds the Kindness of Strangers"

✳ "Nicaragua Sets Goal to Wipe Out Literacy"

✳ "Man Shoots Neighbor with Machete"

✳ "Gunfire in Sarajevo Threatens Cease-fire"

✳ "Prisoner Serving 2,000-year Sentence Could Face More Time"

✳ "Thugs Eat Then Rob Proprietor"

✳ "Dealers Will Hear Car Talk at Noon"

✳ "Kicking Baby Considered to Be Healthy"

✳ "Local Man Has Longest Horns in Texas"

✳ "Lawmen from Mexico Barbecue Guests"

✳ "Traffic Dead Rise Slowly"

✳ "Scientists Note Progress in Herpes Battle; Ear Plugs Recommended"

✳ "British Left Waffles on Falkland Islands"

✳ "Two Convicts Evade Noose, Jury Hung"

The Hollow Man

John Symmes was convinced that Earth was hollow. All he needed was someone to fund his expedition, then he could show the world the marvels that lay beneath its surface. He found that someone in Ohio.

✳ ✳ ✳ ✳

AS A CHILD, John Symmes was drawn to the sciences. After serving in the War of 1812 and retiring four years later, he moved his family to St. Louis, Missouri, and opened a business selling provisions to troops. It was successful enough that Symmes could print up a massive amount of pamphlets explaining his Hollow Earth theory. Dated April 10, 1818, the pamphlet began, "I declare the Earth is hollow and habitable within." The pamphlet went on to discuss Symmes's belief that there were openings, or "hollows," at both poles, through which the interior portion of Earth could be accessed.

In 1819, Symmes and his family moved to Newport, Kentucky. It was there that he began giving lectures on his Hollow Earth theory. Symmes went into great detail about his idea: The inside of Earth was made up of "concentric spheres," one inside the other. Furthermore, the spheres were positioned in such a way that large holes leading to the hollow portion were located at both the North and South Poles. The North "hole" was almost 4,000 miles in diameter, while the one at the South Pole was even bigger—a whopping 6,000 miles in diameter. Obviously, holes that large shouldn't be difficult to find, which is why Symmes was looking to raise money to fund an expedition in search of one of them.

Friends in High Places

His goal was to find the north opening, which he thought was located "one degree northward of latitude 82." Symmes also expected to find plant life, animals, and perhaps even humans living inside the hollow. If all went well, the expedition would return the following spring. Of course, nobody could go anywhere until they raised enough money to fund the expedition.

Symmes was often joined at his lectures by friend Jeremiah Reynolds, who was a firm believer in Symmes's theories and planned to accompany him on his expedition, which he hoped to kick off in Siberia in the fall. His team would consist of 100 men, who would set off for the pole using reindeer and sleighs.

In 1822, Symmes convinced Ohio millionaire James McBride, who reached out to Kentucky Representative Richard M. Johnson (who would later become vice president under Martin Van Buren), who in turn petitioned Congress to finance the expedition. Incredibly, the petition made it to the floor, although it was defeated by a margin of 56–46.

A Dream Unfulfilled

The following year, Symmes himself petitioned both houses of Congress, but that too was rejected. Undaunted, he asked the General Assembly of the State of Ohio to recommend him to Congress so that he could address them directly. After the assembly briefly discussed the matter, a motion was granted to "indefinitely postpone" the issue.

Even though he spent years touring the United States trying to raise money to fund his expedition, Symmes would never get the chance to search for the entrance to the Hollow Earth. He passed away on May 29, 1829, without ever raising the necessary funds. In 1873, Symmes's son Americus erected a stone monument over his father's grave; it was topped with a replica of the Earth with a hollow center. The area has since been renamed Symmes Park.

As for Symmes's partner, Jeremiah Reynolds successfully raised funds for his own expedition. Using money from a New York investor, Reynolds purchased a ship, the SS *Annawan*, and set sail for Antarctica to find the opening that Symmes believed was located there. He never discovered it.

Chapter 9

Waging War

The Bounty Jumpers

During the Civil War, when the North and South started offering bounties to encourage civilians to enlist, a new breed of con artist arose to get while the gettin' was good.

✳ ✳ ✳ ✳

WHILE PATRIOTISM CAN be a very persuasive tool in the recruitment of soldiers, the governments involved in the Civil War quickly discovered that they needed something a bit more tangible to convince new recruits to don uniforms and fight for their respective causes. To sweeten the pot, both the North and South offered bounties to soldiers who signed up. In fact, the North increased cash bonuses as the need for fresh fighters grew. What began as less than $100 soon became quite lucrative for the average person. In 1864, a soldier wrote home: "I receive for re-enlisting nearly eight hundred dollars which I shall devote to straightening things at home."

A New Kind of Scam

Considering that the average pay for Northern infantry privates was $13 a month, such a large bounty was very enticing. But these bonuses also attracted a new group of con artists called bounty jumpers, who would register for the armed forces, pocket their bounties, and then never appear for duty. They'd often use false names and addresses so they could move on to another community or state and repeat the offense. One bounty jumper enlisted 32 different times before he was caught and sentenced to four years in prison.

Bidding Wars

The amounts of money in play rose as the army had an increasingly difficult time recruiting new soldiers. While the draft was in place, communities were required to send a certain number of recruits to war. If a community didn't want to send its own residents, it could simply find other people to send to the army in their place. Districts would collect cash (which would be given in addition to any bounties the army itself was offering) and engage in bidding wars, while soldiers and bounty jumpers moved from area to area in search of the best offer. When President Lincoln called for 500,000 more recruits in 1864, bounties skyrocketed. It was not uncommon to see offers as high as $1,000 during these recruitment drives.

On to the Con

Eventually, the method of paying bounties changed in order to fight the jumpers: Instead of a huge lump sum, the bonuses were paid in installments during service or came in the form of a deposit upon enlisting, with the balance paid when and if the soldier survived the war. Even with this change in payment, it is estimated that the North spent between $500 million and $700 million on bounties alone.

✳ **Confederate Lieutenant George Dixon was saved from a bullet by a $20 gold piece in his trouser pocket at the Battle of Shiloh. The coin stopped the bullet from entering his leg and possibly saved his life. The coin had been given to him by his sweetheart as a good luck charm. He later served on the submarine *H. L. Hunley*. The wreckage of the *Hunley* was discovered in 2000. When an archaeologist was preparing to lift Dixon's skeleton from the sub, she found a gleaming gold piece engraved "Shiloh. April 6th, 1862. My life Preserver. G.E.D."**

The Plot to Kill Hitler

A failed assassination attempt resulted in the execution or imprisonment of "conspirators," many of whom were not even involved in the plot.

❋ ❋ ❋ ❋

ALTHOUGH THE ATTEMPT on Hitler's life on July 20, 1944, was not the first against the Nazi leader, the consequences of the failed coup were far-reaching. The Gestapo used the attempt as justification for the murders or incarcerations of thousands of dissenters who had no connection to the plot whatsoever. Meanwhile, the failed plan pushed Hitler into a form of self-exile at his Prussian headquarters and his mountain retreat near the village of Berchtesgaden. He never again trusted the German military, which brought him great victories at the start of the war.

A Marked Man

Several plans to kill Hitler were developed as early as 1938, but were aborted when German Army leaders wavered. Various attempts on Hitler's life were made in 1943. In one instance, a bomb disguised as a bottle of brandy was brought aboard Hitler's private airplane, but the explosives failed to detonate. One week later, a colonel planned to blow up both Hitler and himself with bombs carried in each pocket of his overcoat; he changed his mind when Hitler left a weapons exhibit in Berlin early. A third attempt involved a captain who planned to detonate a hand grenade while embracing the Führer; this plot failed when the event Hitler was scheduled to attend was canceled.

As the fortunes of war turned against Germany in mid-1943, army plotters became convinced that Hitler had to be killed in order for peace to be restored with the western Allies, a move they felt would also prevent a Russian invasion of Germany. The most important part of the coup fell to a conspirator who was both a colonel and a count—Claus von Stauffenberg. He was a staff officer whose bloodlines descended from a Prussian general in the Napoleonic War and the former chamberlain to the last king of Wurttemberg.

In August 1943, Stauffenberg met Henning von Tresckow, an officer with Army Group Central who'd had a hand in organizing two of the previous attempts on Hitler's life. Although he had been an early supporter of Hitler's National Socialist movement, Stauffenberg became disenchanted with the regime after a debacle at Stalingrad that saw tens of thousands of German soldiers killed or captured.

In April 1943, Stauffenberg was critically wounded in North Africa when he was struck by bullets from a strafing plane. The attack cost the career soldier the sight in his left eye, part of his right arm, and two fingers on his left hand. While recuperating in a hospital, he decided to take an active role in the conspiracy to kill Hitler.

Operation Valkyrie

In September 1943, coconspirator Ludwig Beck, the former chief of the Army General Staff, asked Stauffenberg to formulate a plan for the seizure of power. The plan, code-named Operation Valkyrie, called for Reserve Army units to arrest Nazi leaders and sympathizers and seize control of media outlets and key government institutions after Hitler was assassinated.

By July 1944, Stauffenberg had recovered sufficiently from his wounds to be appointed chief of staff to the General of the Reserve Army. The position gave him access to Hitler's military meetings, including one that was scheduled for mid-July at Hitler's so-called "Wolf's Lair"—Rastenburg in East Prussia.

On July 20, 1944, Stauffenberg flew to Rastenburg carrying a briefcase filled with British-made plastic explosives. After a brief meeting with the Supreme Commander of the Armed Forces (OKW), Chief Field Marshal Wilhelm Keitel,

Stauffenberg excused himself for several minutes, at which time he armed the bomb. The explosives were set to detonate in ten minutes.

Stauffenberg accompanied Keitel into the meeting, which had just started as the men arrived. Stauffenberg managed to place himself one position down from Hitler at the large rectangular oak table that dominated the conference room. He placed his briefcase on the floor against one of two thick supports, which ran down the width of the table. As the military men listened to a situation report about the Russian Front, Stauffenberg slipped out of the room.

Foiled Again

In a move that would have fateful consequences, an officer attending the meeting accidentally kicked and knocked over the briefcase while leaning over the table to get a better look at a map. The officer bent over, picked up the briefcase, and then moved it to the other side of the oak support beam, away from Hitler.

When the bomb detonated at 12:42 P.M., the explosion caused the building's roof to collapse and the windows to shatter. As flames and smoke rose from the demolished mass of rubble, Stauffenberg stepped into his car, bluffed his way through two checkpoints, and boarded his airplane for the three-hour journey to Berlin, confident that he had killed Hitler.

Four people died in the blast, but Hitler was not one of them. The Nazi dictator was thrown across the room by the force of the explosion but managed to stagger out of the rubble with help from Keitel. Although he was far from dead, the dictator suffered ruptured eardrums, burns to his legs, a severely bruised right arm, and singed hair.

One of Stauffenberg's coconspirators, General Fellgiebel, witnessed Hitler leaving the conference room and immediately phoned his contact in Berlin with news of the failed attempt.

The situation became even more confusing when Stauffenberg phoned after landing at the Berlin airport to say that Hitler was dead.

Believing the assassination attempt was successful, others involved in the plot began carrying out their assigned roles. As Keitel began contacting key members of the military to alert them of the assassination attempt, several of the less-resolute conspirators faltered and changed sides. When Berlin radio announced that Hitler would address the nation that evening, the final bloody chapter of Operation Valkyrie began.

Payback Time

Stauffenberg was quickly arrested at the Benderblock, the headquarters of the Reserve Army. In the ensuing confusion, he tried to escape and was wounded in the process. Shortly after midnight, Stauffenberg and three other conspirators were taken to a courtyard and executed.

The Gestapo immediately began rounding up suspects. More than 5,000 people were eventually arrested. Those who survived the brutal interrogations were given show trials before the SS-controlled People's Court and its bullying judge, Roland Freisler. Under SS Chief Himmler's new "blood laws," the families of the conspirators were also punished. Eight of the conspirators were hung from meat hooks at Berlin's Plotzensee Prison (the executions were filmed for Hitler's benefit). An estimated 100 to 200 people who were implicated in the plot were either executed or—as in the case of Field Marshal Erwin Rommel—committed (or were "allowed" to commit) suicide.

As many of the conspirators belonged to the German Army's officer ranks, Hitler never again trusted his generals and punished them by lavishing favor on the ranks of the Waffen-SS. The failed assassination attempt had further implications for the army, whose members were now forced to recite a pledge of loyalty to the Führer and use the stiff-armed Nazi salute.

A House Divided

Although her loyalty was to her husband and the Union government he served, Mary Todd Lincoln couldn't ignore her ties to the Confederacy.

❋ ❋ ❋ ❋

MARY TODD LINCOLN was a Northerner by marriage but a Southerner by birth. Raised in a slaveholding family in Kentucky, Mary Todd moved to Illinois at age 20 to live with her sister Elizabeth. She was a hit within Northern social circles, and various men—including Stephen A. Douglas—courted her. Abraham Lincoln was the new legal partner of her cousin, who spoke highly of the lanky lawyer and helped make the introduction between Mary and Abe. Despite her sister's objections, Mary Todd married Lincoln three years later.

A Family Apart

Like many families in the nation, the Todd family was torn apart by the Civil War. Most of the family supported the Confederacy—one of Mary's brothers, three half-brothers, and three brothers-in-law actively fought for the South. Mary's stepmother was related to John C. Breckenridge, the former U.S. vice president and Confederate general. Only Mary and a few other family members supported the Union. Her sister Elizabeth, however, had opposed her marriage to Lincoln and thus offered her no political support. Of the Kentucky Todds, only Mary's older brother Levi was a staunch Unionist. But his health was poor and he never served in the Union army.

Mary Lincoln remained loyal to the North and to her husband, of course, but newspapers and Washington society regularly used her Confederate family ties against her. Southerners called her a traitor to her roots, and Northerners suspected her tangled allegiances. Though she did sometimes arrange for her family to get travel passes to cross enemy lines in both directions, there is no evidence that she ever aided the Confederacy. Mary

Lincoln was caught between a rock and a hard place—she couldn't publicly mourn her Confederate family members without seeming disloyal to the Union, but she couldn't ignore them without seeming disloyal to her family.

Death Tolls in the Todd Family

In such a large family, tragedies were sure to be plentiful. Levi died of illness in 1864. Mary's half-brother Samuel died at Shiloh. Another half-brother, David, was wounded at Vicksburg and died a few years after the war. A third, Alexander, was killed at Baton Rouge.

Most newsworthy of all was the death of her brother-in-law, Benjamin Hardin Helm, a West Pointer who was married to her sister Emilie. He was offered a high-ranking paymaster post by Lincoln but refused it, instead accepting a field position in the Confederate army. He rose to the rank of general before his death at the Battle of Chickamauga.

When Benjamin Helm died, Emilie had trouble going to visit Mary in Washington, D.C., and her mother in Kentucky. Because she had gone to the South with her husband, traveling to either place involved crossing Union lines, and because she was the widow of a Confederate officer, local commanders would not give her permission to do so. To make matters more difficult, Emilie refused to take an oath of loyalty to the Union, saying that it would be an insult to her late husband. Lincoln gave her a special pass to permit her and her young daughter Katherine to travel to Washington, where she stayed with the Lincolns at the White House. Despite the fact that Emilie was a family member, the Lincolns were strongly criticized for "giving aid to the enemy" in this fashion.

The family conflict even filtered down to succeeding generations. Katherine Helm and the Lincolns' son Tad would argue about the war. As they played together at the White House, Katherine would shout, "Hurrah for Jeff Davis!" Tad would yell in reply, "Hurrah for Abe Lincoln!"

"Mister" Lindbergh: Fighter Pilot

Franklin Delano Roosevelt tried to prevent American hero and controversial prewar isolationist Charles Lindbergh from serving in the Army Air Corps. Nonetheless, Lindbergh flew combat missions in the Pacific—and even tagged a kill.

✳ ✳ ✳ ✳

To MANY, HE was "Lucky Lindy," the heroic pilot who had flown solo across the Atlantic. To others, he was a jaded recluse who fled his native country for Europe after the intense media attention following the kidnapping and murder of his son. Still others knew him as the man to whom the Nazis awarded the Service Cross of the German Eagle, which he received during a visit to Germany in 1938. During that visit, Charles Lindbergh was impressed by the energetic, disciplined drive of the German people and their burgeoning war industry. He returned to the United States convinced that Depression-era America could not stand against Germany. Lindbergh used radio spots to plead his isolationist viewpoint. As a member of the America First Committee, Lindbergh delivered speeches that many considered at best naive and at worst anti-Semitic. Then Japan attacked Pearl Harbor, and any ambivalence that Americans had about the war dissipated overnight. Lindbergh also shed his isolationist stance and volunteered to resume his commission in the Air Corps, but President Roosevelt, whom Lindbergh had publicly chastised, refused the airman's services.

Eventually, Lindbergh was permitted to ply his trade as an adviser and spent the first years of the war test-flying virtually every model of American aircraft that was used in the conflict. Not content with stateside service, however, Lindbergh managed to get assigned as a technical assistant in the Pacific Theater, where he helped pilots acclimate to Chance Vought's very fast F4U Corsair aka "the bent-winged bird" and clandestinely participated in several combat missions.

Lindbergh soon got himself attached to the 475th Fighter Group and its gregarious commander, Colonel Charles MacDonald, who took an immediate liking to Lindbergh and agreed to let him fly combat missions with the "Satan's Angels." In deference to Lindbergh's civilian status, the pilots were ordered to refer to the celebrity pilot as "Mister Lindbergh." During his time with the 475th, Lindbergh used flying techniques that he'd first perfected during his transatlantic trip to dramatically increase the fuel efficiency of the group's P-38 Lightnings.

On July 28, 1944, Lindbergh's squadron encountered two Mitsubishi 51 Sonias—armed reconnaissance planes—one of which was piloted by Captain Saburo Shimada, a skilled and experienced flyer. The vastly outnumbered enemy planes managed to slip out of the Americans' sights time after time. During the dogfight, Shimada and Lindbergh raced toward each other head-to-head with guns blazing. Despite the fact that Lindbergh had never fought another aircraft, he bore down on the Sonia and let loose a burst of gunfire that stopped the Sonia's engine. At the last possible moment, Lindbergh veered off, and the stricken enemy plane swept by mere feet from him. This was Lindbergh's only confirmed kill of the war.

On August 1, Lindbergh, MacDonald, and several other pilots, who were bored by routine missions escorting bombers and strafing buildings, used Lindbergh's fuel-saving techniques to fly to the far-flung Palau islands in search of enemy aircraft; Lindbergh was nearly killed in a resulting dogfight. Word of the incident spread to the U.S. high command, and MacDonald was sent stateside on disciplinary leave. Lindbergh was also ordered home. He later claimed that the reason MacDonald was punished was because the flight to Palau had been deemed impossible, and the Air Corps didn't like being proved wrong.

American Women in Uniform Break Barriers

They flew Flying Fortresses from factories to airfields, translated secret radio transmissions from the French underground, and helped build the atomic bomb. American women who joined military auxiliary units during the war shattered many stereotypes, but not without their share of opposition.

✳ ✳ ✳ ✳

AFTER PEARL HARBOR, every branch of the military established auxiliary branches for women to serve, freeing many men to fight abroad. Hundreds of thousands of women served in the Women's Army Auxiliary Corps (WAAC), the Women Accepted for Volunteer Emergency Service (WAVES), the Women Airforce Service Pilots (WASP), the Marine Corps Women's Reserve (MCWR), and the Coast Guard's SPARS.

The Navy's First Female Officer

In 1942, Mildred H. McAfee became the first female commissioned officer in U.S. Navy history when she was sworn in as a Naval Reserve Lieutenant Commander to serve as the first director of the WAVES. A year later, 27,000 WAVES were at work. Most held clerical positions, but WAVES also worked in aviation, the Judge Advocate General's Corps, and intelligence.

Danger in the Skies

More than 1,000 women earned their WASP wings, logging over 60 million miles of operational flights. Though their jobs were risky, the WASPs were considered civilian employees and didn't receive military benefits. Piloting aircraft from factories to military bases or towing aerial targets for training, the female pilots freed up experienced male pilots for combat duty. Thirty-eight WASP pilots lost their lives in service. The program was disbanded in December 1944, after a Congressional bill to give WASPs full military status and benefits failed to win approval.

Serving with Distinction, Overseas and at Home

Founded on May 14, 1942, the U.S. Army's WAAC was the largest military branch in which women served, with 150,000 enlisting by the end of the war. At first, some U.S. politicians and the American public had difficulty accepting women in army uniforms. "Who will then do the cooking, the washing, the mending, the humble homey tasks to which every woman has devoted herself; who will nurture the children?" asked one Congressional opponent of the corps. Even after the WAAC was formed, the women did not earn as much as men, nor did they receive any military benefits until November 1942, when Congress passed a second bill giving the corps full military status. Simultaneously, "Auxiliary" was dropped from the corps's name.

Though many WAACs worked in office support roles, others served as weather forecasters, cryptographers, radio operators, mechanics, parachute riggers, intelligence analysts, bombsight maintenance specialists, and control-tower operators. Overseas positions were highly coveted by WAAC personnel. Third Officers Martha Rogers, Mattie Pinette, Ruth Briggs, Alene Drezmal, and Louise Anderson became the first WAACs deployed overseas when they joined General Eisenhower's headquarters in Algiers in 1943; they followed his command until V-E Day two years later. "Their contributions in efficiency, skill, spirit and determination are immeasurable," said Eisenhower of WAACs under his command in 1945.

Despite exemplary service records, gossip and bad publicity severely curtailed WAAC recruitment in 1943. Enlisted men ridiculed the idea that females could be soldiers, and many forbade their wives or girlfriends to join. Rumors of promiscuity were widely reported, but a War Department investigation found nearly all of the gossip to be unsubstantiated. However, an early WAAC slogan, "Release a Man for Combat," was reconsidered as a bad choice of words and changed to "Replace a Man for Combat." In the Pacific, a theater where many U.S. servicemen had not seen an American woman in 18 months, WAACs were escorted by armed guards and lived inside compounds surrounded by barbed wire. Despite the tribulations, Generals Marshall and Eisenhower remained supportive of the WAAC, and many servicemen who worked with WAACs were impressed with their service.

Several WAACs worked on the Manhattan Project. At the Los Alamos atomic laboratory, Master Sergeant Elizabeth Wilson ran a cyclotron and Jane Heydorn worked as an electronics technician. In Italy, the 6669th Headquarters Platoon of General Mark Clark's 5th Army became one of the most famous WAAC units, often serving within a few miles of the front line. In Great Britain, WAAC stenographer Ruth Blanton worked in army intelligence, translating radio reports from the French underground that specified bridge locations and German-troop strengths in preparation for D-Day. In February 1945, a battalion of 800 African American WAACs was stationed in Birmingham, England, and later Paris; they were responsible for directing mail to all U.S. personnel in the European Theater. In the Southwestern Pacific, many WAACs who had been sent as clerks and typists were retrained as drivers and mechanics. In the Philippines, WAAC units moved into Manila three days after the city's occupation.

After the war, WAAC director Oveta Culp was awarded the Distinguished Service Medal, the second-highest U.S. military decoration. In 1948, the service's new director, Colonel Mary A. Hallaren, became the first commissioned female officer in the U.S. Army. WAACs went on to serve in the Korean and Vietnam wars. The WAAC was disbanded in 1978, when women were integrated into regular units of the U.S. Army.

The Man Who Tried to Outrun War

Wilmer McLean knew all too well that "you can run, but you can't hide."

✳ ✳ ✳ ✳

Unwelcome Guests

IN 1854, WILMER McLean bought a Virginia estate along the Bull Run Creek near Manassas Junction in Prince William County. He worked hard and made major improvements to the property, including building a large stone barn.

When the Civil War erupted in April 1861, Virginia became a focal point of the conflict. In July, the first major engagement of the war, the First Battle of Bull Run (or First Manassas), was fought by Union and Confederate troops who trod all over McLean's land. At one point, an artillery shell went down his chimney, fell into the kitchen fireplace, and exploded in a pot of stew, scattering food all over the room. During the battle, wounded Confederates were placed in McLean's large barn, which the Union then shelled and destroyed.

That was enough for McLean. After the armies moved on to other battles, he bought a farm in an isolated part of southern Virginia. It seemed a good choice—over the next three years, two more battles occurred near Bull Run.

Parlor Trick

But there's no escaping fate. On Sunday, April 9, 1865, the war found McLean once again. He was walking in the village of Appomattox Court House when a Confederate officer looking for a meeting place for generals Ulysses S. Grant and Robert E. Lee approached him. The first building he suggested was rejected, so McLean offered his home. There, in McLean's front parlor, the two generals negotiated an end to the Civil War.

The irony was not lost on McLean, who reportedly remarked, "The [Civil] War began in my front yard and ended in my parlor."

Shoichi Yokoi: Lost Soldier, Found

Shoichi Yokoi fled from the Americans invading Guam in 1944 and was not captured until 1972. Repatriated to Japan, he quickly became a sensation—for better or worse.

✳ ✳ ✳ ✳

ON JULY 21, 1944, Sergeant Shoichi Yokoi of the Japanese Imperial Army was engaged in a desperate fight with American troops whose tanks were ripping apart his regiment. As the situation became increasingly dire, Yokoi chose to flee rather than be killed—or worse, captured alive. He was not alone: More than 1,000 Japanese soldiers were hiding in the jungles of Guam when Americans secured the island. Only Yokoi and eight other soldiers remained undiscovered.

By 1964, his companions had either surrendered or died, and Yokoi was alone. He knew that the war was over, but he chose to remain hidden. "We Japanese soldiers were told to prefer death to the disgrace of getting captured alive," he later said. But he also didn't want to be killed by the Americans, so he hid and survived alone for another eight years—a remarkable 28 years total—before he was discovered by hunters, captured, and taken into custody.

How Did He Do It?

Before the war, Yokoi had been a tailor's apprentice. His skills were of no help against American tanks, they proved immensely useful in the jungle. By pounding the bark of the native pago tree, Yokoi was able to make a durable fiber that he used to make clothing. He reworked a piece of brass to fashion a needle and repurposed plastic from a flashlight to make buttons.

While in hiding, Yokoi ate snails, rats, eels, pigeons, mangoes, nuts, crabs, prawns, and occasionally, wild hog. Although he boiled all the water that he drank and cooked the meat thoroughly, he once became ill for a month after eating a cow.

Surviving in the jungle *undetected* for 28 years is quite a feat. As such, Yokoi went to great lengths to disguise his shelters. His most permanent dwelling was a tunnel-like cave, painstakingly hand-dug using a piece of artillery shell. At one end of the three-foot-high shelter, a latrine emptied down an embankment into a river; at the other end, a small kitchen contained some shelves, a cooking pot, and a lantern made out of a coconut shell.

"It is with much embarrassment that I have returned alive"

The Japanese word *ganbaru* refers to the positive character traits associated with sticking to one's task during tough times. For many Japanese who survived the war, Yokoi was the living embodiment of ganbaru. However, for the young people of Japan's increasingly Western postwar culture, the 56-year-old Yokoi was a reminder of the previous generations' blind loyalty to the Emperor, which had caused Japan's disgrace. Though he longed to see the Emperor and wrote a letter to apologize for having survived the war, Yokoi was never granted his wish.

Reassimilation

Millions of television viewers across Japan watched Yokoi's return trip to his native village, where a gravestone listed his date of death as September 1944. From those who considered him a hero, Yokoi received gifts of money totaling more than $80,000 and many marriage offers. He purchased a modest home and married Mihoko Hatashin, who he described as a "nice, old-fashioned" girl, unlike the modern Japanese women whom he described as "monsters whose virtue is all but gone from them, and who screech like apes."

Though horrified by the Westernization of his homeland, Yokoi prospered as a lecturer on survival techniques. He unsuccessfully ran for Parliament on a platform that stressed simplicity and discipline and included measures such as enforced composting and converting golf courses into bean fields. On September 22, 1997, Yokoi died of heart failure at age 82.

The Battle of the Bands

A melancholy "Home Sweet Home" sustained both Northern and Southern armies before the Battle of Stones River.

✳ ✳ ✳ ✳

A BITTER WAR DOESN'T usually inspire its combatants to break into song, but that's exactly what happened on the night of December 30, 1862. With Union and Confederate forces bedding down yards from each other in Tennessee, military bands on both sides began battling in song. The North played "Yankee Doodle" and "Hail Columbia," and the South responded with "Dixie" and "The Bonnie Blue Flag." In a bittersweet moment, both sides drifted into "Home Sweet Home," as tens of thousands of soldiers sang along. For one brief period in the lonely holiday season, soldiers whose only goals were to kill each other and survive shared a song of empathy and comfort. But the chorus would die before dawn.

The Respite Over, the Fighting Begins

The next morning, the music was replaced by gunshots, screams, and groans as the Battle of Stones River began. One of the most brutal battles of the Civil War, it resulted in 24,000 casualties—almost one third of the battle's total forces.

Both sides were desperate for a win—particularly the Union. The South struck first, pushing Union forces back three miles. Neither army attacked on January 1, but Southern General Braxton Bragg took to the offensive again the next day. Northern artillery fought back, and a bloodied General William Rosecrans refused to yield. His forces thoroughly demolished advancing gray columns, and Bragg eventually retreated 25 miles south.

This win temporarily allowed President Abraham Lincoln to silence war critics. He told Rosecrans, "You gave us a hard-earned victory, which, had there been a defeat instead, the nation could hardly have lived over."

The Angel of the Battlefield

One of the Civil War's bravest, most influential figures was an ardent feminist who started off as a teacher and a patent clerk.

✳ ✳ ✳ ✳

WE NEVER KNOW the extent of what we can do until we're caught in a crisis, and sometimes that crisis brings out the best in us. Nowhere is this truer than in the life of Clara Barton. Barton was one of the greatest heroes of the Civil War. Her fearless work for the wounded earned her the nickname the "Angel of the Battlefield."

A Beginning in Medicine

Clara Barton quit a teaching career after she discovered that she couldn't earn the same pay that men did. She was the first woman ever hired by the U.S. Patent Office in Washington, D.C., but although they hired her as a clerk, they later dropped her title to "copyist" and paid her a mere 10 cents for every 100 words she copied. The Civil War allowed her true talents to come to the fore. In her free time, Barton began tending to wounded soldiers in the hospital. She quickly recognized the Army Medical Department's inability to care properly for so many casualties. At the time of the First Battle of Bull Run, the Union army had practically nothing in terms of a hospital corps. Taking on the daunting task herself, Barton placed an ad in a Massachusetts newspaper asking for donations of medical supplies. Ultimately, she founded an organization to collect and distribute provisions for wounded soldiers.

Cutting through official red tape, Barton won permission from the War Department to go into the field and help the wounded. She began personally delivering medical supplies to battlefields where they were most needed. Barton witnessed firsthand the chaos, inefficiency, and unsanitary nature of military medical care during combat. At the Second Battle of Bull Run, for example, she came upon hundreds of wounded Union soldiers who had been dumped on a hillside, where they'd lain in the summer sun without water for two days. The aftermath of the Battle of Cedar Mountain and the Union field hospital in Fredericksburg, Virginia, were even worse. These experiences strengthened Barton's resolve to help, which, in turn, led her to some of the grimmest, most deadly battlefields of the war.

Braving the Battle

The worst of these must have been at Antietam—the war's bloodiest single day of fighting. There, on September 17, 1862, 23,000 combatants were killed, wounded, or otherwise missing. The night before, Barton knew a major battle was near. Stuck at the end of the Army of the Potomac's long supply trail, she drove her wagons full of medical supplies all night to get to the battlefield. Arriving at midday, she was greeted by horrific slaughter. Dead and dying soldiers lay in heaps while frantic doctors were reduced to binding wounds with cornhusks. Like an answered prayer, Clara Barton arrived with an entire wagon-load of bandages. She dove into the fray, the only woman on the field, and indeed, the work was harrowing. At one point, a soldier begged her to cut a bullet from his cheek. She hesitated, as she had no surgical instruments and was not a doctor. But the young man pleaded with her, so she ended up cutting the bullet out with a pocketknife. As she cradled another wounded soldier in her arms to give him water, a bullet zipped past her, tearing the sleeve of her dress and killing the man she was helping. Rather than relent, she set upon creating a makeshift field hospital in a barn. When a surgeon she was assisting was suddenly killed, Barton took his place at the operating table. After the sun set, the doctors had no lanterns. Barton saved the day again by seeking out lanterns; then, she worked all night until she collapsed and was carried off the field.

Barton's own words describing the moment when the bullet that killed her patient nearly ended her own life best sums up her stoic and determined state of mind: "A ball had passed between my body and the right arm which supported him, cutting through the sleeve and passing through his chest from shoulder to shoulder. There was no more to be done for him and I left him to his rest. I have never mended that hole in my sleeve. I wonder if a soldier ever does mend a bullet hole in his coat?" For the duration of this long, tumultuous war, she paused in her labors only to recuperate from incapacitating illnesses, including typhoid fever.

Receiving Recognition

Clara Barton continued her service at Fredericksburg (where she worked beside Walt Whitman), Hilton Head, Petersburg, Richmond, and the siege of Fort Wagner. By this time, Union leadership had recognized her value, and Union General Benjamin Butler named her "Lady in Charge" of field hospitals for his Army of the James.

After the war, she helped identify some of the 13,000 unknown Union dead at the notorious Confederate prison camp at Andersonville. By establishing the Office of Correspondence, Barton helped families learn the fates of nearly 30,000 soldiers who were missing in action. To put this number in perspective, the U.S. military currently lists only 10,000 MIAs from Vietnam and Korea combined.

Barton remained busy as a lecturer and activist after the war, campaigning for women's suffrage. Four years after the war's end, she traveled to Europe for a much-needed and long-overdue sabbatical. Ironically, she arrived just in time for the Franco-Prussian War of 1870–1871. Signing on as a member of the International Red Cross, she was instrumental in getting the organization to broaden its focus from merely caring for the war-wounded to providing disaster relief. On her return home, Barton performed her most lasting act on behalf of suffering peoples everywhere—she founded the American Red Cross. In a sense, whenever we turn on the television today and see the Red Cross providing assistance after natural disasters, we are looking at Clara Barton's shining monument.

Enslaved Inside a Mountain: The Mittelwerk V-2 Rocket Facility

V-2 rockets rained random destruction upon England during the final months of World War II, but the production of the weapons cost far more human lives than the attacks ever claimed.

❋ ❋ ❋ ❋

THE ADVANCES IN rocketry during World War II would eventually lead to the Soviet Union's launch of *Sputnik* in 1957 and the United States reaching the moon 12 years later. The roots of the technology, however, began in Nazi Germany during the darkest days of the Second World War. Few—if any—of those celebrating the conquest of space in the decades that followed thought of the enormous suffering that made the Space Age a reality.

In the summer of 1943, the V-2 rocket facility at Peenemünde was suffering from increasingly effective Allied bombing raids. To protect their new weapon, the Nazis ordered thousands of Russian, Polish, and French concentration-camp prisoners to enlarge two gypsum-mine 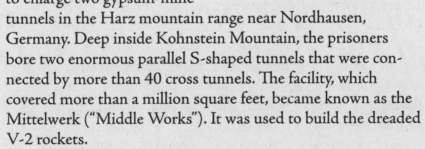 tunnels in the Harz mountain range near Nordhausen, Germany. Deep inside Kohnstein Mountain, the prisoners bore two enormous parallel S-shaped tunnels that were connected by more than 40 cross tunnels. The facility, which covered more than a million square feet, became known as the Mittelwerk ("Middle Works"). It was used to build the dreaded V-2 rockets.

Nearby, the Nazis opened the Mittelbau concentration camp to provide labor. The prisoners were literally worked to death in the harsh underground plant, where beatings, exhaustion, malnutrition, and illness were commonplace. Because the rockets were not designed for mass production, each V-2 had to be assembled and tested with custom parts. After several acts of sabotage, regulations required laborers to sign off on their work so that flaws could be traced back to whomever had assembled the rocket. Prisoners still engaged in minor acts of sabotage, such as cold-soldering connections and failing to completely weld components. However, if caught, inmates faced immediate death: They would be hung from cranes in the halls and left for a day or more as a warning to others.

Of the roughly 50,000 detainees working at the Mittelbau complex, approximately half did not survive. For every operational rocket that left the factory, six people had died making it. At least 11,000 of these prisoners were killed in April 1945, when the retreating SS guards obeyed orders to slaughter the prisoners in order to protect the secrets of the Mittelwerk plant. However, the American 104th Infantry Division and 3rd Armored Division discovered the facility soon after the SS retreat. An intelligence officer was quoted as saying that being in the tunnels was "like being in a magician's cave." Near the "magician's cave," the Americans had already discovered the Dora concentration camp, where many of the most horrific scenes of Nazi abuse were documented by film crews. Today, a memorial dedicated to the prisoners who died in the facility is located near the entrance to the tunnels.

---

* **Boise City, Oklahoma, was inadvertently bombed on July 5, 1943, when a B-17 pilot mistook the lights on the town square for his training target.**

* **In addition to a national ban on dancing, Mussolini banned beauty contests in Italy as immoral and banned the Marx Brothers' comedy *Duck Soup* (1933) because he perceived it as a direct attack on his character.**

No Picnic

When the Japanese attacked Pearl Harbor on December 7, 1941, 2,390 lives were lost and the United States was dragged into World War II. But it is a mistake to believe that no one was killed on the U.S. mainland as a result of the war.

* * * *

O N MAY 5, 1945, a Japanese balloon bomb killed six Americans who were picnicking near the town of Bly, Oregon. The Japanese launched some 9,000 balloon bombs against the United States during World War II. They attached incendiary and anti-personnel bombs to large hydrogen-filled balloons and released them into the jet-stream winds to float 5,000 miles across the Pacific Ocean and explode in the forested regions of Western states. Their hope was that this would cause widespread forest fires and divert U.S. manpower and resources away from the war in the Pacific.

Officially, only 285 of the balloon bombs reached North America, though experts estimate that approximately 1,000 made it across the Pacific Ocean. The U.S. military successfully orchestrated a media blackout of the bombings to deny the Japanese any publicity of success. The first balloon bomb landed near San Pedro, California, on November 4, 1944. The bombs traveled as far north as Canada, as far south as the Arizona–Mexico border, and as far east as suburban Detroit. The U.S. military shot some down, while others landed without exploding. None succeeded in creating the major forest fires Japan desired.

However, when a 13-year-old girl picnicking in the woods near Bly, Oregon, discovered one of the balloon bombs, she tried to pull it from a tree. It exploded, killing the girl, a minister's wife, and four other children. These six people were the only war casualties on the U.S. mainland during World War II. Had the Japanese perfected the balloon bombs and used biological weapons, that statistic might have been very different.

The Making of West Virginia

The Union wasn't the only entity to split during the Civil War:
The state of Virginia broke into two pieces, as well.

✳ ✳ ✳ ✳

WHEN THE WAR started, there was just one Virginia. In the hearts of many residents, however, the state was already divided. The eastern and southern parts were more aristocratic, with grand plantations in the countryside and wealthy cities near the Atlantic coast. The northwestern part was mountain country. As early as 1829, northwesterners complained that their counties didn't benefit from their taxes as much as eastern counties did. In 1830, a newspaper in the northwestern city of Wheeling said the area should separate from the rest of the state "peaceably if we can, forcibly if we must."

A Reason to Split

The Civil War sparked the separation. When Virginia joined the Confederacy, the residents of two dozen northwestern counties objected. They started their own governing body, which they called the Restored Government of Virginia.

The U.S. Constitution stipulates that states can't be formed from already-existing states without approval of that state's legislature. But Virginia had seceded from the Union, so the Restored Government of Virginia claimed to be the legitimate legislature and voted to separate from the rest of the state. Congress agreed. President Lincoln was reluctant to approve such a change during wartime, but in the end, he signed on. West Virginia officially became the 35th state on June 20, 1863.

The name West Virginia makes sense, but it wasn't a given when the northwestern leaders debated seceding from the state. Other proposed names were Allegheny, New Virginia, Augusta, and Kanawha. Imagine John Denver singing, "Kanawha, mountain mama, take me home...". It just doesn't have the same ring to it.

Night Witches on the Russian Front

Marina Raskova and her squadron of tenacious female pilots caused the Germans to lose sleep—and, ultimately, the war.

✳ ✳ ✳ ✳

IN NOVEMBER 1941, Stalingrad was under siege, the German Army was within 20 miles of Moscow, and more than three million Russians had been taken prisoner. If ever there was a time for Russian heroes, this was it. Into that breach stepped a true hero of the Soviet Union—Marina Raskova. The aviatrix with nerves of steel had won the accolades of the Russian people when she helped her three-woman team complete a flight across Siberia in 1938. (At the end of the grueling journey, she had bailed out into the frozen tundra when the plane's wings became heavy with ice and there was nothing left to throw out.)

Three years later, with the Germans howling at the gates, Raskova was granted permission by Stalin to organize three squadrons for defense of the homeland. These squadrons— the 586th Fighter Aviation Regiment, the 588th Night Bomber Regiment, and the 125th Guards Bomber Aviation Regiment—were composed entirely of women, from the pilots to the mechanics. Although all the squadrons performed laudatory feats, the 588th became famous as the *Nachthexen* (German for "night witches") for their daring and tireless night raids on German command posts and tactical targets.

Raskova's reputation notwithstanding, the female aviators encountered a great deal of prejudice and harassment during their service. For example, the 588th was provided with anti-quated, open-cockpit Po-2 biplanes, which had a top speed of 94 miles per hour (slower than most World War I–era planes) and could carry only two small bombs. Nevertheless, the intrepid women developed techniques that maximized the admittedly few advantages of their outmoded aircraft. They learned to fly low to the ground (sometimes just above the hedgerows) and to cut their engines before reaching their targets, which allowed them to glide over the German soldiers undetected until their ordnance had been delivered. They also learned that the planes' slow speed meant that they could out-turn the German Me-109s, which would stall if they tried to match the Po-2s' maneuvers. In fact, many German pilots simply gave up trying to shoot down Po-2s, though they were promised an Iron Cross if they could do so. To penetrate searchlight and antiaircraft emplacements, the Night Witches approached a target in groups of three. Once their planes were spotlighted, the two outside planes would break away and the center plane, left in the gap of darkness, would deliver its payload. The planes would then regroup and make the run again until each had dropped its ordnance.

The Night Witches endured countless hardships during the war. In some cases, the women flew as many as 18 missions a night. During the brutal winter of 1942–1943, the women would often have to lie on the wings of their planes to keep the gale-force winds and ice from blowing the light craft off the airfield. All told, the 588th completed more than 24,000 combat missions by the end of the war. Marina Raskova was killed in January 1943 while attempting to make an emergency landing, but the squadrons that she formed went on to fight in the skies above Berlin, bringing victory for the Soviet Union.

✳ Twenty-three of the pilots who served in the all-female 588th Night Bomber Regiment of the Soviet Air Force, known as the Night Witches, were decorated as Heroines of the Soviet Union, the USSR's equivalent of the Medal of Honor.

The Last Shot

The last shot of the American Civil War was fired in Alaska!

✳ ✳ ✳ ✳

I T ALL BEGAN with the *Sea King,* a steamer ship built as a British troop transport in the summer of 1863. A Confederate agent working in Britain noticed the vessel and went about purchasing it for the South. In October 1864, the *Sea King* went on a "trading voyage," ostensibly to India. However, it rendezvoused with another ship at Funchal, Madeira. Guns, officers, and other military items were brought aboard the *Sea King* to transform it into the Confederate warship *Shenandoah.* Lieutenant James I. Waddell was given the mission of finding and destroying the Union's waterborne shipping and commerce.

Waddell continued south, intending to prey on ships in the area between the Cape of Good Hope and Australia. He captured six enemy ships before arriving at Melbourne, Australia, on January 25, 1865, for repairs and fresh supplies.

By that time, the Confederacy was on life support. But Waddell was unaware, so he continued to terrorize American whalers in the North Pacific. On June 23, while in the Bering Sea, Waddell was shown a San Francisco newspaper that reported Lee's surrender at Appomattox. But elsewhere in the paper, Jefferson Davis stated that, "the war would be carried on with renewed vigor." Waddell took that as a sign that the war was still on, so he continued attacking whaling ships, taking 21 more prizes. Finally, on August 2, the *Shenandoah* encountered a British ship and Waddell learned that the war had indeed ended in April. The prizes he had taken in June were the last shots of the Civil War to be fired in anger.

After traveling approximately 44,000 miles, the *Shenandoah* was the only Confederate warship to circumnavigate the globe.

The Real "Man Who Never Was"

When a corpse washed ashore in Spain holding a briefcase full of plans to invade Sardinia and Greece, the Nazis thought that they'd made an astounding catch. They couldn't have been more wrong.

✳ ✳ ✳ ✳

THE ROUGH TIDES slapped against the southern Spanish coast in the spring of 1943, carrying with them the mangled corpse of a British major who appeared to have drowned after his plane crashed into the sea. The body, one of thousands of military men who had met their ends in the Mediterranean waters, floated atop a rubber life jacket as the current drifted toward Huelva, Spain. With a war raging in Tunisia across the sea, a drifting military corpse was not such an unusual sight.

But this body was different, and it drew the immediate attention of Spanish authorities who were sympathetic to German and Italian Fascists. Chained to the corpse was a briefcase filled with dispatches from London to Allied Headquarters in North Africa concerning the upcoming Allied invasions of Sardinia and western Greece. The information was passed on to the Nazis, who accepted their apparent stroke of good luck and anticipated an Allied strike on the "soft underbelly of Europe." Unfortunately for them, the whole affair was a risky, carefully contrived hoax.

Rigging the "Trojan Horse"

Operation Mincemeat was conceived by British intelligence agents to convince the Italians and Germans that the target of the next Allied landings would be somewhere other than Sicily, the true target. To throw the Fascists off the trail, British planners decided to find a suitable corpse—a middle-aged white male—put it in the uniform of a military courier, and float it and documents off the coast of Huelva, Spain, where a Nazi agent was known to be on good terms with local police.

The idea of planting forged documents on a dead body was not new to the Allies. In August 1942, British agents planted a corpse clutching a fake map of minefields in a blown-up scout car. The map was picked up by German troops and made its way to Rommel's headquarters. He obligingly routed his panzers away from the "minefield" and into a region of soft sand, where they quickly bogged down.

This deception, however, would be much grander. If the planted documents made their way up the intelligence chain, Hitler and Mussolini would expect an invasion far from the Sicilian coast that Generals Eisenhower, Patton, and Montgomery had targeted for invasion in July 1943.

The Making of a Major

Operation Mincemeat—spearheaded by Lieutenant Commander Ewen Montagu, a British naval intelligence officer, and Charles Cholmondeley of Britain's MI5 intelligence service—found its "host" in 1943 when a Welshman living in London committed suicide by taking rat poison; the substance produced a chemical pneumonia that could be mistaken for drowning. The two operatives gave the deceased man a new, documented identity: "Major William Martin" of the Royal Marines. They literally kept the "major" on ice while arrangements for his new mission were made. To keep Spanish authorities from conducting an autopsy—which would give away the body's protracted post-mortem condition—the agents decided to make "Major Martin" a Roman Catholic, giving him a St. Christopher medallion and a silver cross. They dressed the body, complete with a Royal Marine uniform and trench coat and gave him identity documents and personal letters (including a swimsuit photo of his "fiancée," an intelligence bureau secretary). With a chain used by bank couriers, they fixed the briefcase to his body.

Martin's documents were carefully prepared to show that Allied invasions were being planned for Sardinia and Greece (the latter bearing the code name "Operation Husky"). They also indicated that an Allied deception plan would try to convince Hitler that the invasion would take place in Sicily (the site of the real Operation Husky). With everything in order, the agents carefully placed the corpse in a sealed container—dry ice kept the body "fresh" for the ride out to sea.

The submarine HMS *Seraph* carried "Major Martin" on his final journey. On April 28, the *Seraph* left for the Andalusian coast, and two days later, the body of a Royal Marine officer washed ashore. Within days, photographs of the major's documents were on their way to Abwehr intelligence agents in Berlin.

Taking the Bait

Abwehr, Hitler, and the German High Command swallowed the story. After the war, British intelligence determined that Martin's documents had been opened and resealed before being returned by the Spanish. The German General Staff, believing the papers to be legit, had alerted units in the Mediterranean to be ready for invasions of Sardinia and Greece. They moved one panzer division and air and naval assets off the Peloponnese and disputed Italian fears of an impending invasion of Sicily.

The Allies captured Sicily in July and August 1943, and after the war, Commander Montagu wrote a best-selling account of Operation Mincemeat titled *The Man Who Never Was*. The book was made into a film a few years later.

Who was Major Martin? The original body appears to have been that of a 34-year-old Welsh alcoholic named Glyndwr Michael, and "Major Martin's" tombstone in Spain bears his name. Historians have debated the identity of "Major Martin," however, theorizing that a "fresher" corpse from a sunken aircraft carrier was substituted closer to the launch date.

Whoever the real "Major Martin" may have been, one thing is certain: He saved thousands of lives, and became a war hero and action movie star in the process—quite an accomplishment for a dead man!

Allan Pinkerton: Spying for the Union Cause

The exploits of Allan Pinkerton during the Civil War helped pave the way for the modern Secret Service.

✳ ✳ ✳ ✳

I N A LETTER to President Lincoln dated April 21, 1861, detective Allan Pinkerton offered his services and commented on one of the traits that would make him an icon of law enforcement for generations. "Secrecy is the great lever I propose to operate with," he wrote.

Establishing the Eye

Born in Scotland in 1819, Pinkerton came to the United States in 1842. He was a barrel builder by trade, but his keen observation and deduction skills led him to a career fighting crime. By age 30, he'd joined the sheriff's office of Cook County, Illinois, and had been appointed Chicago's first detective. He later joined attorney Edward Rucker to form the North-Western Police Agency, a forerunner of the Pinkerton Agency. As his corporate logo, Pinkerton chose an open eye, perhaps to demonstrate that his agents never slept. Clients began calling him "The Eye."

Pinkerton and his agents were hired to combat the growing number of train robberies, which became more of a problem as railroads expanded across the nation. George B. McClellan, president of the Ohio and Mississippi Railroad, took particular notice of Pinkerton's talents.

Wartime Duties

In 1861, Pinkerton's agency was hired to protect the Philadelphia, Wilmington, and Baltimore Railroad. In the course of their duties, the detectives learned of a preinaugural plot to kill President-elect Lincoln. They secretly took Lincoln into Washington before he was scheduled to arrive, thwarting the conspirators. Lincoln was sworn in without incident.

When the war began, Pinkerton was given the duty of protecting the president as a forerunner of today's Secret Service. He was also put in charge of gathering intelligence for the army, which was run by McClellan. Using surveillance and undercover work (both new concepts at the time), agents gathered vital information about the enemy. His operatives interviewed escaped slaves and tried to convince literate slaves to return to the South to spy. He used female spies, and he even infiltrated the Confederacy himself several times using the alias "Major E. J. Allen."

Uncertain Information

While much of this was invaluable, Pinkerton's work was tarnished by a seeming inability to identify enemy troop strengths. His reports on the enemy were detailed, including notes on morale, supplies, and movements. Yet the numbers of troops he provided were highly suspect.

In October 1861, as McClellan was preparing to fight, Pinkerton reported that Confederate General Joseph Johnston's troops in Virginia were "not less than 150,000 strong." In reality, they were fewer than 50,000. The next year, he reported the strength of Confederate General John Magruder's forces at Yorktown at about 120,000, when the true number was closer to 17,000.

After the true strength of these forces was discovered, Pinkerton was ridiculed. Some historians believe that he was unaware of the faulty information, but others insist that he intentionally provided inflated figures to support McClellan's conservative battle plans. The truth will likely never be known, as all of Pinkerton's records of the war were lost in the Great Chicago Fire of 1871.

Return to Civilian Life

After Lincoln relieved McClellan—one of Pinkerton's staunchest supporters—of his command, Pinkerton limited his spying duties and shifted his work back toward criminal cases, which included the pursuit of war profiteers. He ultimately returned to Chicago and his agency, working there until his death in 1884.

Defying Death:
The Rosenstrasse Protest

While most Germans cowered at the mention of the Gestapo, a group of housewives faced down the feared agents.

✳   ✳   ✳   ✳

UNTIL EARLY 1943, Jews with Gentile wives were exempt from deportation to labor and death camps. But during a "final roundup" in late February 1943, Heinrich Himmler's security forces arrested about 1,700 of these Jews and collected them at a welfare center on Rosenstrasse, in the heart of Berlin. From there, the Nazis planned to send them to the camps.

The men's outraged wives gathered in small groups at the welfare center and began to shout to the anonymous SS agents inside, "Give us back our husbands!"

Around 600 women showed up at first, and by the time the protest was over, 6,000 more women had joined the effort. Police scattered the women into the streets, threatening to shoot. SS guards even began setting up machine guns. As a Berlin housewife recalled, "Then ... we really hollered. Now, we couldn't care less ... Now they're going to shoot in any case ... We yelled, 'Murderer, murderer, murderer, murderer.'"

The Rosenstrasse Protest—and the publicity it attracted— quickly caught the attention of Joseph Goebbels, Hitler's propaganda minister and the man in charge of maintaining public unity and morale in the Reich's capital. Goebbels noted in his diary, "There have been unpleasant scenes ... The people gathered together in large throngs and even sided with the Jews." After a week of protests, Goebbels ordered the Gestapo to release the 1,700 Jews with non-Jewish German spouses. As his deputy recalled, Goebbels took this action to "eliminate the protest ... so that others didn't ... do the same." In the end, he never arrested the women who forced the dreaded Gestapo to back down.

The Watermelon War

This riot erupted due to some seedy behavior.

✳ ✳ ✳ ✳

To SAY THAT American-Panamanian relations were tense during the second half of the 19th century would be an understatement. After completing the trans-Panama railroad, the United States eliminated jobs that had been filled by native-born residents and put Americans in managerial roles. Race riots were commonplace, but one notable revolt began over a watermelon.

On April 15, 1856, the steamer *John L. Stephens* docked on Taboga Island en route to Panama City with 1,000 passengers. When they disembarked to visit a local market, a drunk American passenger named Jack Oliver grabbed a watermelon slice from a black vendor.

When Oliver refused to pay for it, the vendor pulled out a knife. While one of the men in Oliver's party tried to pay for the fruit, Oliver drew a gun. The gun went off accidentally and injured a bystander. A riot broke out as a mob of angry Panamanians rushed in and began beating the Americans, looting hotel rooms, and destroying property. When police tried to quell the disorder, one of them was shot, thus inciting them to join the melee. In the end, 15 Americans and 2 Panamanians were killed and dozens were wounded.

In an attempt to protect American economic interests in Panama, the U.S. military unlawfully occupied the isthmus and railroad station. Eventually, the occupation was abandoned, but the United States still pursued payment for the damage caused by the riot. The Republic of New Granada, which encompassed present-day Panama and Columbia, begrudgingly paid nearly $400,000 in damages. For years, relations remained sticky. The Watermelon War was a devastating conflict any way you slice it.

The Trials of a Ball-Turret Gunner

Depending on whom you talk to, the job of a ball-turret gunner (or belly gunner) was either the best or the worst job to have on a B-17 Flying Fortress flight crew during World War II. Chosen because of their small stature, ball-turret gunners had to withstand extremely cramped quarters, hours of subzero temperatures, extreme heights, and death—all without the benefit of wearing a parachute.

✳   ✳   ✳   ✳

All the Comforts of Home

During World War II, the B-17 Flying Fortresses were flown over enemy lines in thousands of combat missions. Each bomber depended on ten crew members; the smallest and most nimble were assigned to sit in the ball turret.

The ball turret was a small, bubblelike compartment made of glass and plastic that housed two .50-caliber machine guns. Mounted on the underside of the aircraft, the ball turret was so small that the gunner had to be lowered into the turret through an escape door after takeoff while the guns were pointing straight down. Once inside the turret, the gunner sat curled up in a fetal position with his feet secured in overhead stirrups, straddling a 13-inch armored-glass window that was his main vantage point during battle. The gunner looked through his knees and, with his left foot, operated a red sight that framed the target. The ends of the machine guns projected backward (as did the gunners), just inches from either side of the gunner's head. Handles on either side of the window allowed him to rotate the turret 90 degrees up and down and 360 degrees in either direction and fire the machine guns.

Bunny Suits to the Rescue

Unlike the other crew members' toasty, wool-lined flight suits, the ball-turret gunners were extremely limited by space, so they depended on thin fabric flight suits worn over electrically heated "blue bunny suits" to ward off the cold. After sliding into the turret, the gunners would plug into an electrical outlet to heat their blue bunny suits—that is, when the mechanism worked.

It was fairly common for ball-turret gunners to spend 10 to 12 consecutive hours suspended inside their turret—definitely in excess of one's average bladder-retention capability. "Relief tubes in the fuselage benefited the other crewmen but were of little use to a ball-turret gunner," said Harold "Diz" Kronenberg, a World War II ball-turret gunner. "At high altitudes, the relief tubes often froze up and were . . . of no value to anyone." Just in case, Kronenberg said, gunners always brought along an empty can.

The Good News and the Bad News

Many B-17 crew members considered the ball turret the worst assignment on the aircraft, but it was the safest. Because the turret could rotate up and down and in both directions, ball-turret gunners were the first to see incoming attackers. As a result, they experienced the fewest number of battle wounds.

Of course, it wasn't all high flying. In January 1943, ball-turret gunner Alan Magee was wounded during his seventh mission in a B-17 dubbed *Snap! Crackle! Pop!* During a daytime raid over Saint-Nazaire, France, a German fighter shot off part of the bomber's right wing, sending it into a tailspin toward the ground.

Magee managed to climb out of his ball turret but had to jump from the plane without a parachute, falling more than four miles before crashing through the glass roof of a train station. The roof helped break his fall, but Magee sustained 28 shrapnel wounds; several broken bones; damage to his nose, an eye, lung, and kidney; and he nearly severed an arm. He was held behind enemy lines until May 1945 when U.S. troops liberated him. For his efforts, he received the Air Medal for meritorious conduct and a Purple Heart.

Escape of the Journalists

Condemned for reporting for a "bastion of Northern propaganda," two New York Tribune reporters survived a harrowing escape from a Confederate prison camp.

✳ ✳ ✳ ✳

M OST OF THE POWs held within the walls of Civil War prison camps were soldiers captured on the battlefield. However, on May 3, 1863, reporters Albert Richardson and Junius Browne of the *New York Tribune* and Richard Colburn of the *New York World* became the first reporters to be captured during the conflict. A year and a half later, Richardson and Browne would stage one of the most thrilling and harrowing escapes of the war.

A Wrong Turn

The three men were attempting to join up with General Grant and his troops near Vicksburg, Mississippi, when they decided to take a shortcut to reach the front lines as quickly as possible. Unfortunately, it's hard to remain inconspicuous when you're riding a barge down the Mississippi River, and sure enough, they were spotted by the Confederates. The reporters' hopes of scooping their fellow correspondents on Grant's activities vanished when the Rebels took to the river after them. All three were captured and imprisoned in the local jail.

Northern Rag

Richardson and Browne were warned by other prisoners that the *Tribune* was not respected in the South. "Tell them you are correspondents of some less-obnoxious journal," they were advised. Indeed, a casual conversation with some Confederate officers confirmed that any reporter from the *Tribune* was likely to be hanged. Despite this, the two reporters decided to tell the truth. At first, all three reporters were promised release and given food and clothes by some of their counterparts from Southern newspapers. However, when Colburn was freed, the two *Tribune* reporters were told that because their newspaper was regarded as a bastion of Northern propaganda, they would be imprisoned for the remainder of the war. They were immediately transferred to Libby Prison in Richmond.

Southern Hospitality

Richardson and Browne would see the insides of seven prisons over the next 20 months, eventually ending up in Salisbury, North Carolina, where they met Confederate Lieutenant John Welborn, a member of the secretive fraternal organization Sons of America. This order believed that it was its duty to try to return as many Northern prisoners to their homes as possible, and Welborn risked his life to do just that. The Southern lieutenant helped the reporters secure assignments in the prison hospital, where they were able to obtain passes that allowed them to travel outside the walls without supervision. On December 18, 1864, Richardson and Browne told a guard that they were picking up some desperately needed medical supplies and walked out of the prison. This began an adventure that found the pair escaping from Southern patrols, living with local slaves, and joining up with Confederate deserters and federal sympathizers.

Free at Last

Nearly one month and 340 miles later, the ragged group of more than a dozen that had collected around the pair staggered into a Union camp near Knoxville, Tennessee. They had successfully run away from the dogs that were used to capture escaped prisoners and navigated the dangerous swamps and unforgiving mountains. They had come up against and escaped bushwhackers, the "vicious, passionate, bloodthirsty" men who captured escaped prisoners and slaves and would just as soon kill them as take them back alive. "I walked within the lines that divided freedom, enlightenment, loyalty from slavery, bigotry, treachery," Richardson wrote upon his arrival in the North. "Out of the jaws of death, out of the mouth of hell."

The Heroine of Auschwitz

Human courage did not always wear a soldier's uniform. It was sometimes found in unlikely places: One of the war's bravest fighters, for example, labored in the women's workshop of Hitler's most notorious concentration camp.

✳ ✳ ✳ ✳

ROZA ROBOTA WAS a Polish Jew whose family was gassed in the death chambers of Auschwitz II. Selected as "fit labor," 21-year-old Roza was put to work in the death camp's clothing-supply section. There she labored for two years, a well-liked, energetic girl who became a natural leader among the female inmates in her section.

In 1943, Roza was contacted by a member of the Jewish underground. Rumors had spread of orders to liquidate inmates at a faster rate before the advancing Soviet Army could uncover evidence of the Nazi Final Solution, and the underground wanted her help.

Roza accepted a dangerous role in the plot. She helped her contacts in the camp's munitions plant to smuggle out black powder in hidden pockets of workers' dresses. When the time was right, male inmates in the camp's underground would rise up, kill as many guards as possible, and blow up the camp's infrastructure while partisans liberated the prisoners.

For weeks, Roza and several other women removed small quantities of explosives in pouches in their dresses. One of the places that the stolen powder was hidden was with a Sonderkommando unit—a Jewish slave-labor detail charged with hauling bodies out of the gas chambers and operating the camp's crematoria. When the Sonderkommandos discovered that they were slated to be disposed of in the same ovens they operated, they staged a premature revolt. On October 7, 1944, one crematorium exploded, and the Sonderkommandos tried to overpower the guards. Four or five SS guards were killed, and some 600 inmates rioted before the revolt came to an end.

A Gestapo team was brought in to investigate, and by torturing inmates, they discovered the source of the powder that was used to destroy the crematorium. The investigators rounded up and brutally tortured Roza and three of her accomplices. Even after enduring weeks of torture, Roza revealed nothing. Knowing that she was about to be hanged, Roza quietly accepted her fate. She smuggled out a final note to her friends, which said, "Hazak v'amatz," meaning, "Be brave and strong."

On January 6, 1945, Roza Robota was taken before assembled munitions workers and hanged for her part in the plot to blow up the death machines. Two weeks later, as the Red Army approached, SS guards destroyed the remaining crematoria to keep evidence of their atrocities out of the hands of the Allies. In demolishing the death ovens, the heavily armed SS men committed an act of abject cowardice. By doing the same thing, a frail Jewish girl committed an act of supreme heroism.

---

* In 1938, *Time* magazine named Adolf Hitler its Man of the Year. Of course, *Time*'s selection was based on significance and influence—not on objective merit.

* During World War II, British pilot and double amputee Douglas Bader was forced to bail out of his aircraft, leaving one of his artificial legs behind. Under a special temporary truce agreed to by his German captors, the RAF delivered a replacement leg by parachute.

* Even after the Warsaw Ghetto Uprising was over, a few Jews somehow managed to survive in the ruins. Jewish prisoners clearing the rubble found a girl who was badly burned and near death. For reasons that remain unknown, the Gestapo gave her medical care until her health recovered—then they took her back to the ruins and shot her.

Confederates South of the Equator

They call them Confederados *in Brazil: Southerners who fled farther south, into the tropics, after losing the Civil War.*

✳   ✳   ✳   ✳

IN THE SOUTHERN Brazilian state of São Paulo, there survives a town named Americana where the great-great-grandchildren of Confederate rebels speak in a southern drawl with a Portuguese accent. They are descendants of the Confederados, a group of Southerners who settled in Brazil after the Civil War ended. Every year, a diminishing number of offspring reunite for the *Festa Confederada*, a celebration of Dixie culture. They serve deep-fried chicken, fly the Confederate flag, dress in antebellum fashion, and downplay the issue of slavery.

Fleeing the South

In the upheaval following the Civil War, a Confederate migration found tens of thousands of people escaping to Europe, Mexico, and beyond. While the 2,500 who settled in Mexico were ultimately forced to return to the United States, the 9,000 who continued "way, way down South" to Brazil found a dependable protector in its emperor, Dom Pedro II. He welcomed the Southerners—and their state-of-the-art cotton-farming know-how—to his country.

Pedro was gradually phasing out slavery in Brazil. In 1888, his daughter, Princess Isabel, would end the institution with the stroke of a pen by ushering in the Golden Law of abolition. But with a few exceptions, it appears that the Confederados didn't bring all the practices of their antebellum plantations to the tropics. A recent study by Brazilian researcher Alcides Gussi was able to find evidence of only 66 slaves owned by four families of Confederados in Brazil. Many of the immigrants were too poor to own slaves, and the rest relied largely on cheap local labor.

A New Home

By the end of the 1860s, Southerners were steaming for Brazil from the ports of New Orleans; Charleston; Newport News, Virginia; Baltimore; and Galveston, Texas, to settle in towns such as Americana and the nearby Santa Bárbara D'Oeste. "My grandfather came from Texas and built his house in the middle of a forest," 86-year-old Maria Weissinger told *The Atlanta Journal-Constitution* in 2003. The new land was fertile but rife with insects that carried deadly tropical diseases. Many refugees gave up and returned to the South, but about two in five stayed in their adopted homeland, intermarried with Brazilians, grew pecans and peaches, and built schools and universities.

Weakened Roots

After nearly 150 years, most of the Confederados have been absorbed into the populations of big cities. But in the industrial town of Americana, traces of the South survive, kept alive in part by a group called the Fraternity of American Descendents. Generations later, the lines between Confederados and Brazilians have become extremely blurred. Many of the Confederados in Americana no longer speak English, but most continue to travel to a cemetery in Santa Bárbara D'Oeste, where the graves tell the stories of their families' fading Confederate roots. In 1972, the cemetery was visited by one settler's great-niece: Rosalyn Carter, wife of future president Jimmy Carter, who was then governor of Georgia.

✳ Known as the Drummer Boy of Chickamauga, nine-year-old Johnny Clem was one of the youngest Union army soldiers. He ran away from home and tried to join an Ohio regiment, but they rejected him for being too young. He then tried to join the 22nd Michigan, which also rejected him, but he tagged along and served as a drummer and was allowed to enlist two years later. In 1864, he was discharged at the age of 13. In 1871, Clem rejoined the regular Army and remained in its service until 1915, when he retired as a brigadier general. He is buried at Arlington Cemetery.

Why Did the Nazis Keep a Record of the Holocaust?

The events of World War II were recorded to an extent far beyond that of preceding conflicts. Events were captured in print, photographs, and moving pictures. The most chilling of all was the exhaustive documentation of the Holocaust, much of it created by the very people who committed the crimes.

✳ ✳ ✳ ✳

KNOWLEDGE OF THE HOLOCAUST stems from many sources, the most compelling of which are the eyewitness accounts of victims. But there is another source that helps confirm the extermination's unthinkable scale, as well as the fates of individuals: the records kept by the Nazis themselves.

Seized by the liberating armies in the last days of the war, the documentation exists in various collections, but the bulk of the records have been under the care of the Red Cross for the last half-century. The files are extensive: millions upon millions of papers covering 16 miles of shelves. So why would a group of people intent on committing mass murder risk putting their activities in writing?

The answer may surprise you. In the opinion of Paul Shapiro, director of Holocaust studies at the United States Holocaust Memorial Museum, "They wanted to show they were getting the job done." Many accounts suggest that he may be correct.

Just a Job: "The Bureaucracy of the Devil"
A stereotypical but not entirely inaccurate image of the prewar German government is one of bureaucracy. Everything was documented, and paper authorizations were generated by the handful for the most mundane of tasks. This attitude extended into the war. The task of running an empire, even a despicable one, is complex, requiring extensive procedures and paper trails. Like many governments, Nazi Germany employed an array

of middle managers who wanted to prove their efficiency. The only way an official could show that he was performing up to par was to keep records.

Prisoners who were immediately executed required the least documentation, sometimes as little as a mere entry in the number of arrivals for the day. Those who stayed in the camps longer typically had more extensive records. Because of the sheer number of people involved—some 17 million in all—some startling documents survived, such as the original list of Jews transferred to safety in the factories of Oskar Schindler. Another file contains the records of Anne Frank.

Why Worry?

For most of the war, the Nazis showed little compunction about documenting their activities. In their minds, why should they? To whom would they be accountable? After all, many thought that the Third Reich would last a thousand years. In the closing months of the war, there was a slight reversal of this policy, and the commandants of some camps sought to destroy records and eliminate all witnesses as the Allied forces closed in. Fortunately, they were not able to erase the records of their own atrocities.

Private memoirs of the Holocaust also exist. Participants at all levels wrote letters about their experiences, and some SS guards took photographs of the camps and inmates with their personal cameras. Some Nazi leaders were also prone to recording daily activities; Joseph Goebbels kept a journal throughout the war, viewing it as a "substitute for the confessional."

Much like Goebbels's diary, the official records of the Holocaust have become an unintentional confession of a Nazi machine that had uncountable crimes for which to answer. The archive exists in Bad Arolsen, Germany, and was opened to the online public in late 2006. Survivors of the camps hope that its presence serves as a counterargument to those who inexplicably deny that the Holocaust ever happened and as a reminder that humankind must never allow it to happen again.

The Cost of the Civil War

The losses and the repercussions of the Civil War would be felt for generations to come.

✳ ✳ ✳ ✳

The Human Cost

THE CIVIL WAR was the bloodiest conflict in American history. If we combine all the deaths from all conflicts in U.S. history, nearly half would come from the Civil War. But because of fragmentary records, it can never be known precisely how many people served on both sides of the war, nor how many were killed, wounded, and captured.

Officially, the U.S. Army recorded 2,778,304 enlistments, but many soldiers enlisted more than once in different regiments. Of these, 178,975 were black and 3,530 were Native American. The U.S. Navy and Marine Corps enlisted 105,963.

Northern casualties can be tallied as follows:

Total deaths	359,528
Killed in battle	67,088
Mortally wounded	43,012
Died of disease	199,720
Died as prisoners of war	24,866
Killed by accident	4,114
Died from other causes	20,728
Wounded	275,175
Navy killed and mortally wounded	1,804
Navy died of disease and accidents	3,000
Navy wounded	2,226

Because many Confederate records were burned when Richmond fell in 1865 or were otherwise misplaced or destroyed during the war itself, exact numbers of Southern enlistments and casualties remains something of a mystery. It is estimated that Confederate enlistments fell somewhere between a low of 600,000 and a high of 1.4 million, with 1 million being a widely accepted estimate. The gray armies included more than 1,000 regiments, battalions, independent companies, and artillery batteries.

The best estimate of Southern deaths include 94,000 killed in battle or mortally wounded, and another 164,000 deaths from disease, for a total of 258,000. One incomplete summary of the Confederate wounded includes 194,026 names. The number of soldiers who died in Northern prison camps has been estimated to be somewhere between 26,000 and 31,000. However, figures for losses in specific battles change over the years, as historians discover new muster rolls or casualty lists and must revise the figures from earlier research.

Statisticians adding up the fighting in the war came up with 1,882 incidents in which at least one regiment was engaged. In 112 of these battles, one of the two sides had at least 500 combatants killed or wounded.

Monetary Costs

In addition to the human toll, the Civil War cost quite a bit of cash. Here are some estimates, figured in 1879 dollars and adjusted for inflation to 1999 dollars.

	1879 Estimate	1999 Estimate
Union	$3.2 Billion	$27.3 Billion
Confederacy	$2.0 Billion	$17.1 Billion

Female Espionage Agents: Working Undercover for the Allies

Women played vital roles in conducting acts of sabotage and gathering intelligence. Some paid for their bravery with their lives.

✳ ✳ ✳ ✳

IN EUROPE IN the 1940s, the "invisibility" of women often made them ideal operatives. They could eavesdrop in public and witness encounters with authorities unnoticed. Britain's Special Operations Executive (SOE) realized that female agents could be especially effective in the field.

"Hedgehog"

After France fell in 1940, Marie-Madeleine Fourcade helped establish a partisan resistance group called Alliance. Headquartered in Vichy, the group became known as "Noah's Ark" after Fourcade gave its members animal code names; her own code name was "Hedgehog." The group worked to obtain information about the German armed forces and passed the intelligence on to the SOE. The Alliance was among the first partisan groups organized with the help of the SOE, which supplied the French operatives with shortwave radios and millions of francs by parachute.

Fourcade was one of the Alliance's top agents, but she was caught four times by the Germans; she escaped or was released each time. Once she was smuggled out of the country in a mailbag. On another occasion, she escaped from prison by squeezing through the bars on the window of her prison cell. While Fourcade's luck held, other members of Noah's Ark were captured in 1944 during a partisan operation aiding the Allied advance in Alsace; they were later executed at the Natzweiler-Struthof concentration camp in France. Fourcade survived the war and wrote a book about her experiences, *Noah's Ark*, which was published in 1968. She died in Paris in 1989 at age 79.

"Louise"

Born to a French mother and an English
father, Violette Bushell Szabo joined SOE
after her husband, a Hungarian serving
in the Free French Army, was killed at the
Battle of El Alamein. She was given the
code name "Louise." Following intensive
espionage training, Szabo parachuted into
France near Cherbourg on April 5, 1944.
On her first mission, she studied the
effectiveness of resistance, and she subsequently reorganized a resistance network
that had been destroyed by the Nazis. She led
the group in sabotage raids and radioed reports
to the SOE specifying the locations of local factories that were
important to the German war effort. Szabo returned to France
on June 7 and immediately coordinated partisans to sabotage
communication lines. The Germans captured her three days
later, but she reportedly put up fierce resistance with her Sten
gun. Szabo was tortured by the SS and sent to Ravensbrück
concentration camp, where she was executed on February 5,
1945, at age 23. Three other female members of the SOE were
also executed at Ravensbrück: Denise Bloch, Cecily Lefort, and
Lilian Rolfe. Szabo became the second woman to be awarded
the George Cross (posthumously), and she was awarded the
Croix de Guerre in 1947.

Palmach Paratroops

Haviva Reik and Hannah Senesh were Eastern European
Jews who joined the SOE to help liberate their homelands.
Reik was born in Slovakia and grew up in Banska-Bystrica,
in the Carpathian Mountains. Senesh—a diarist, playwright,
and poet—was born in Budapest. The daughter of a well-
known playwright and journalist, she enjoyed a comfortable, secular life before discovering Judaism as a teenager.
Both women emigrated to Palestine in 1939 and joined the

Palmach, a paramilitary branch of the Zionist Haganah underground organization. Trained as parachutists, Reik and Senesh were two of more than 30 Palestinian Jews dropped behind German lines on secret SOE missions.

In March 1944, Senesh parachuted into Yugoslavia and, with the aid of local partisans, entered Hungary. She was almost immediately identified by an informer and arrested by the Gestapo. "Her behavior before members of the Gestapo and SS was quite remarkable," a comrade later wrote. "She constantly stood up to them, warning them plainly of the bitter fate they would suffer after their defeat." Though brutally tortured, she refused to give up her radio codes. On November 8, she was executed by a firing squad. "Continue the struggle till the end, until the day of liberty comes, the day of victory for our people," were her final written words.

In September 1944, Reik and four other agents parachuted into Slovakia to aid an uprising against the Fascist puppet government that had been installed by the Nazis and assist the Jews in the passage to Palestine. Back in her native Banska-Bystrica, she aided refugees, helped Jewish children escape to Palestine, and joined resistance groups in rescuing Allied POWs. In October, Nazis occupied the town. A few days later, Reik and her comrades were captured in their mountain hide-out by Ukrainian Waffen-SS troops. On November 20, they were executed. In 1952, the remains of Reik, Senesh, and five other SOE agents were buried in the Israeli National Military Cemetery on Mount Herzl in Jerusalem.

✳ Romania's contribution to distinguished female agents was Vera Atkins, who worked for the French section of the Special Operations Executive (SOE) as a recruiter, trainer, and facilitator of agents throughout Europe.

✳ Following the war, Atkins spent a year questioning former concentration-camp guards, who might have held clues to the fates of 118 agents who never returned to SOE.

Chapter 10

Happy Holidays

An *Eggs-cellent* Easter Tradition

Ancient confusion about animals in springtime led to a holiday icon.

✳ ✳ ✳ ✳

EASTER IS BOTH the most solemn and the most joyous event on the Christian calendar. But the history of this celebration is complicated, and modern Easter traditions include elements from a variety of spiritual practices. For example, the term *paschal* describes the Christian season but also refers to *Pesach*, the Jewish Passover.

Decorated eggs have long been a part of Easter observances. In the 13th century, the church didn't allow eggs to be eaten during Holy Week (the week prior to Easter Sunday). Eggs reappeared on the dinner table on Easter Day as part of the celebration. These eggs were often colored red to symbolize joy. Hunting for eggs and giving decorated eggs as gifts came later.

Why, though, does the Easter Bunny deliver the eggs? When did this tradition begin? The link between Easter and rabbits is ancient. The word *Easter* is likely derived from the name of the mother goddess Eostre, who was worshipped by the ancient Saxons of northern Europe. Her festival coincided with the lengthening days that marked the arrival of spring and the return of life after a barren winter. Eostre's emblem was the hare; rabbits and hares have often been regarded as symbols of fertility. In one legend, Eostre magically changed a beautiful bird into a hare that built a nest and laid eggs.

Myths and folk traditions frequently offer explanations for natural events. Hares raise their babies in "forms," a hollow in the ground of a field or meadow. Female hares, or does, often divide a litter among two or three forms for safety. Abandoned and empty forms attract plovers (a kind of wading bird), which occasionally move in and use the forms as nests for their own eggs. People saw hares in fields—seemingly hopping away from forms full of eggs—and concluded that the hares had laid the eggs.

As Christianity spread throughout the world, it absorbed pagan beliefs and practices, endowing them with Christian meaning. Church officials placed observances of the events surrounding the crucifixion of Jesus in the early spring. Eggs, the sources of new life, came to symbolize Christ's resurrection. The celebration of spring and the worship of the goddess Eostre became Christian Easter. In time, as distinctions between hares and rabbits were blurred, Eostre's hare became the Easter Rabbit.

The link between rabbits and Easter emerged most strongly in Protestant Europe during the 17th century, particularly in Germany. Boys and girls built "nests" with their caps and bonnets, and good children were rewarded with a "nest" full of colored eggs brought to them by the *Osterhas*, or Easter Rabbit. Variations of this practice came to America in the 18th century with German immigrants. Hunts for decorated Easter eggs left by the Easter Bunny became common in the 19th century.

In Your *Eostre* Bonnet

Easter, which celebrates Jesus Christ's resurrection from the dead, is thought by some to be nothing more than a pagan holiday. Despite all the bunnies and bonnets, though, its religious roots hold firm.

✳ ✳ ✳ ✳

SOME PEOPLE THINK that Easter has its origins in paganism because it falls roughly at the time of the spring equinox, when the earth comes back to life from the dead of winter.

It *is* a significant Christian holiday, true, but Easter is actually associated with Passover, the Jewish holiday that celebrates the Hebrews' release from bondage in Egypt. Jesus and his disciples were in Jerusalem to celebrate Passover when he was arrested, tried, and crucified. Following his resurrection, the disciples understood Jesus to be the sacrificial lamb that took away the sins of the world, a fulfillment of the Passover lambs that were sacrificed each year. Hence, the original name for Easter was *Pasch*, from the Hebrew word for Passover, *Pesach*.

The word "Eostre" derives from the Old English name for the month of April, and, as discussed in this book's previous article, the month of Eostre was named after a goddess. Much later, Jacob Grimm (of the Brothers Grimm) speculated that Eostre was named for the ancient German goddess Ostara. The reference could also come from the word *east*—the direction of the sunrise—or from the old Germanic word for "dawn," which makes sense, given that dawn comes earlier in the spring.

As for the date of Easter, which varies from year to year, it's calculated based on the lunar calendar and some complex ecclesiastical rules. These calculations include the spring equinox, which is when the sun is directly above Earth's equator. But whether the equinox and Easter fall close together is a matter of chance.

Breaking a Few Eggs for Art

With Eggshelland, the front yard of a little ranch house in Lyndhurst, Ohio, became a major tourist attraction.

✳   ✳   ✳   ✳

IN 1957, RON and Betty Manolio began blowing and painting Easter eggs to decorate their lawn. That first year, their display consisted of 750 eggshells in the shape of a cross. In recent years, their displays have averaged six or seven designs and 50,000 colored eggshells.

These displays use enameled eggshells to create colorful mosaics across the Manolios' lawn. Designed by Betty, they combine secular images, such as an Easter Bunny, with religious images. Each year's mosaic has a theme, such as a popular book, movie, or historical period.

Although the Manolios do reuse eggshells from year to year, they average breakage rates of 1,500 eggshells each year. In addition, northeast Ohio's unpredictable spring weather has broken more than 27,000 of the eggshells with hailstorms, snowstorms, and wind.

The Manolios collect eggs from a local restaurant for the following year's display. Ron is in charge of boring a hole in the eggshell. Betty empties the contents, then Ron smooths the jagged opening. They both paint them, using sign-painting enamel. Then, after transferring Betty's yard designs to graph paper, Ron pounds small pegs into the grass and places the eggs on them. Once Easter is over, the eggs are separated by color and stored for the next year.

Eggshelland has been featured on *The Today Show*, *Ripley's Believe It or Not*, and even in an 2009 an eponymous documentary. Anybody spending Easter on Cleveland's east side would do well to hop over to Lyndhurst to check out Eggshelland.

Should Auld Acquaintance Be Forgot?

Every December 31, as the clock strikes midnight, English-speaking people all over the world sing "Auld Lang Syne" to herald the new year. Although few people can claim to know all the words, or indeed what they mean, even fewer know the history of this New Year's Eve tradition.

❋　❋　❋　❋

THE SONG ITSELF dates as far back as the 17th century, but the custom of singing it at the start of a new year didn't begin until the 1930s. Scottish poet Robert Burns first published "Auld Lang Syne" in the mid-1790s, though the earliest mention of this traditional Scottish folk tune was made more than 100 years earlier. Translated from the Gaelic, *auld lang syne* literally means "old long since," but in this context it is better translated as "times gone by." The opening verse of the song asks if old friends and old times should be forgotten; the chorus answers no—we should take a drink of kindness and remember the times gone by.

The poignant sentiment and old-fashioned tune fit perfectly with the dawn of a new year, but it wasn't Burns who transformed the song into a New Year's Eve anthem. The musician Guy Lombardo first played "Auld Lang Syne" a few minutes before midnight at a New Year's Eve party in New York City in 1929. He and his orchestra were regulars on New Year's Eve radio—and then television—programs for the next 50 years. Lombardo's New Year's Eve show became so popular that the TV networks CBS and NBC competed over its broadcast rights. As a compromise, CBS broadcast the show until midnight and NBC took over afterward. This prompted Lombardo to play "Auld Lang Syne" at the stroke of midnight to signal the end of the old year and the start of the new.

New Year's Drops

New York City's ball-dropping tradition in Times Square has become synonymous with New Year's Eve. Here's how towns across America have gotten in on the act.

✳   ✳   ✳   ✳

✳ The highlight of the First Night event in Atlanta is the dropping of an 800-pound peach.

✳ Celebrators in Key West, Florida, ring in the New Year in three ways: by dropping a conch shell, a pirate wench from the mast of a tall ship, and a six-foot-tall red high-heel shoe carrying a drag queen.

✳ Known as the Walleye Capital of the World, Port Clinton, Ohio, drops a 20-foot, 600-pound walleye each year at midnight. During the celebration, the town serves walleye chowder, walleye sandwiches, walleye cinnamon chips, and walleye popcorn. A local winery has even created a "walleye white."

✳ In 2000, the fishing community of Point Pleasant, New Jersey, dropped a ten-foot wooden replica of an Atlantic baitfish. The fish, dubbed Mo the Millennium Mossbunker, was covered with 1,500 Mylar scales.

✳ A sardine is dropped in Eastport, Maine. The 22-foot structure is made of lumber and chicken wire and is decorated with silver lamé. Between appearances, it is stored at a restaurant called The Pickled Herring.

✳ Pennsylvania boasts an unusual lineup of New Year's Eve celebrations. Residents of Lebanon witness the dropping of a 100-pound, 16-foot stick of bologna; in Dillsberg, it's an eight-foot-tall papier-mâché pickle. The folks in Falmouth lower a stuffed goat to honor the town's goat races, and in Wilkes-Barre, a chunk of coal "transforms" into a diamond as it descends.

Misremembering Memorial Day

On Memorial Day, people in the United States remember the nation's war dead. But how well do they remember the origins of the holiday?

❋   ❋   ❋   ❋

ASK THE AVERAGE person when Memorial Day originated and you'll probably get an answer as vague as "after World War I or II." But the origins of Memorial Day stretch back to an earlier century and a conflict that divided the country before serving to unite it.

In the years following the Civil War, people in the North and South often decorated the graves of loved ones lost in battle. These informal remembrances evolved into a celebration called Decoration Day in dozens of cities both north and south of the Mason-Dixon Line.

The first national Decoration Day took place on May 30, 1868 (when flowers were likely to be in bloom across the nation), by order of the commander of the Grand Army of the Republic, General John Logan. In his General Order No. 11, Logan decreed that May 30 be set aside for "decorating the graves of comrades who died in defense of their country during the late rebellion."

Of course, language like that didn't make the new holiday very popular with the Southern states: They refused to acknowledge it until after World War I, when it was expanded to include those who had died in *all* American wars. By that time, many people had begun calling it Memorial Day, but it didn't become an official federal holiday until 1971. At that time, the holiday was moved to the last Monday in May—to the chagrin of many who felt that the solemnity of the celebration was not enhanced by making it part of a "long holiday weekend."

Happy Juneteenth!

Taking root in Galveston, Texas, an obscure holiday has become anything but a trivial affair.

* * * *

I F ASKED TO name an important holiday that occurs during the month of June, some might suggest Father's Day. But Texas has a significant holiday called Juneteenth (a combined term for June 19), which originated in Galveston in 1865. It predates Father's Day (1910), but it is not as widely known. If other states begin to follow the Lone Star State's lead, however, this holiday could end up as a major day of remembrance.

Roots

On June 19, 1865, two months after the Confederacy surrendered in the Civil War, an interesting incident occurred in Galveston. Without telephones to deliver news from the various fronts or a timely communications network to relay news from Washington, important information often arrived slowly. When a contingent of Union soldiers rode into the city with news that President Abraham Lincoln had issued an Emancipation Proclamation freeing American slaves, it is not clear how much they were spreading news and how much they were enforcing a law that had been kept secret from the slave population. Were slaves surprised by the development, or had they become impatient with what had already been declared their due? Either way, never before had an American president gone firmly on record with such a proclamation, and never before had slaves been promised a legitimate end to their suffering.

First Celebrations

As the oldest nationally celebrated commemoration of America's end to slavery, it's surprising that the holiday is still obscure. Originating as Emancipation Day on that pivotal date of June 19, 1865, the holiday is currently celebrated by various states in a rather loose fashion, with different dates set aside for its observance. Texas became the first state to proclaim Juneteenth an official state holiday, on June 3, 1979. Since that time, the holiday has become firmly entrenched in Texas culture, and the celebration of slavery's end is gaining ground in other states as well. Washington, D.C., holds its observance on April 16—the very day that President Lincoln signed the Compensated Emancipation Act of 1862. More than half of the 50 states have established Juneteenth as a holiday or a special day of recognition.

Current Celebrations

Mimicking the joy that slaves felt when they first learned of their freedom, Juneteenth revelers cut loose and have a good time. In Texas, many families host elaborate family reunions to coincide with the holiday, and the mid-June date generally assures good weather for outdoor cookouts. From there, the sky's the limit. The day has been used to play, visit friends, unwind, and relax, but most importantly, it's a time to reflect: "Juneteenth is arguably one of the most important days in our country's history, as well as African American history," said Florida Congresswoman Corrine Brown in praise of the holiday.

---

* Dred Scott was a slave who sued to gain his freedom before the Supreme Court. The case began in 1846 but wasn't decided until 1857. After the Court's ruling against Scott, his former master's son purchased Scott and his wife and set them free.

* Almost 39 percent of the Confederacy's population were slaves.

Giving Thanks in a Time of War

We may think of Thanksgiving as a celebration of pilgrims, parades, football, shopping sales, and eating until we're fit to burst. But Thanksgiving wasn't a national holiday until Lincoln made it one.

✳ ✳ ✳ ✳

Harvest festivals have been held in North America since the 17th century to celebrate the year's bounty. After the founding of the nation, George Washington and other presidents occasionally declared periods of national "thanksgiving." The holidays were more popular in some states than others—in 1858, the governors of 25 states and two territories proclaimed various days of thanksgiving.

After the war's onset, Sarah Josepha Hale, the editor of the influential *Godey's Lady's Book* monthly magazine, appealed to President Lincoln to declare a national holiday of thanksgiving. Lincoln, seeing a chance to boost war morale, complied and on October 3, 1863, he declared the final Thursday of November a national Thanksgiving Day.

In his speech, he reminded Americans of the country's industriousness and abundant bounties and asked them to praise God accordingly. "In the midst of a civil war of unequalled magnitude and severity... peace has been preserved with all nations, order has been maintained," Lincoln said in his proclamation. "I do therefore invite my fellow citizens in every part of the United States... to set apart and observe... a day of Thanksgiving and Praise to our beneficent Father who dwelleth in the Heavens."

And since that proclamation, Thanksgiving has been celebrated annually in the United States. The official date of Thanksgiving has shifted over the course of time, though never by much. In 1941, Congress decided that Thanksgiving would be celebrated as it is today—on the fourth Thursday in November.

The Cola Claus

Although the Coca-Cola Company helped to popularize jolly ol' Santa Claus, it cannot take credit for creating the ubiquitous Christmas image.

✳ ✳ ✳ ✳

NOTHING SAYS "CHRISTMAS" like the image of a white-whiskered fat man in a red suit squeezing down a chimney with a sack full of toys. But Santa Claus didn't look that way until the 1930s, when the Coca-Cola Company used the red-robed figure to promote its soft drinks.

Santa Claus evolved from two religious figures: St. Nicholas and Christkindlein. St. Nicholas was a real person—a monk who became a bishop in the early fourth century and was renowned as a generous gift-giver. The folkloric Christkindlein (meaning "Christ child") was assisted by elfin helpers and would leave gifts for children while they slept.

Until the early 20th century, Santa Claus was portrayed in many different ways. He could be tall and clad in long robes like St. Nicholas or small with whiskers like the elves who helped Christkindlein.

In 1881, Thomas Nast, a caricaturist for *Harper's Weekly*, first drew Santa as a merry figure in red with flowing whiskers— an image that's close to the one we know today. In 1931, the Coca-Cola Company first employed Haddon Sundblom to illustrate its annual advertisements, choosing a Santa dressed in red and white to match its corporate colors. By then, however, this was already the most popular image of Santa Claus, one that was described in detail in a *New York Times* article in 1927. If Coca-Cola had really invented Santa Claus, children would likely be leaving him soda and cookies on Christmas Eve.

New Year's Traditions

New Year's celebrations worldwide focus on hope for prosperity and health in the future. Whether you throw dishes at your friends or free your tortured goldfish, may your tradition bring you good luck.

✳   ✳   ✳   ✳

✳ In Cambodia, people roam the streets with squirt guns full of tinted water (red, pink, or yellow) in search of friends to anoint with a "colorful" future.

✳ In Wales, young men soak evergreen branches (symbols of good luck) in water and use them to sprinkle their friends' and relatives' homes.

✳ In Burma, you stand a good chance of being doused with buckets of water for good luck.

✳ In Puerto Rico, celebrants toss pans full of water out their windows in the belief that they're also tossing out evil spirits.

✳ South American celebrants hang dummies stuffed with newspaper and firecrackers off the fronts of their homes. At midnight, someone strikes a match and . . . *kaboom!*

✳ In an interesting gesture of good will, Danish citizens throw old dishes at their friends' front doors. A large pile of porcelain pieces indicates a wide circle of friends.

✳ In Vietnam, New Year's Eve is celebrated with the planting of a tree in the family garden. The tree is decorated with red streamers and lots of bells in an effort to ward off evil spirits.

✳ For days prior to the new year in China, the family home must be meticulously cleaned. On New Year's Day, all cleaning activities must cease in the belief that any accumulated good fortune might be swept away and disposed of.

* The New Year's Eve dinner is the biggest celebration of the Chinese New Year and typically features vast quantities of food. In northern China, a central feature of the dinner is dumplings, which are thought to symbolize the promise of wealth because they have a shape that's similar to that of a Chinese gold nugget.

* In the southern United States, Hoppin' John (black-eyed peas and rice) is prepared on New Year's Day for luck and wealth. Some families also include collard greens—the peas represent coins, and the greens represent paper money.

* People in Cuba and Mexico eat 12 grapes at midnight to ensure 12 happy months.

* In the Philippines, tables are piled high with food at midnight to ensure abundance for the coming year. Some also have a centerpiece that consists of seven round fruits. The round shape symbolizes money, and seven is a lucky number.

* Tradition in Anglo-Saxon countries has been that the first visitor after midnight brings the household luck for the coming year. Dark-haired men, recognized as the luckiest, often offer their neighbors token gifts of money, bread, or coal.

* On New Year's Day in Germany, people tell fortunes by dropping molten lead into cold water. The lead takes a shape, and each shape represents something different: A heart or circle indicates an impending wedding, a ship means a journey, and a pig signifies plentiful food.

* In Japan, many New Year's parties are festooned with paper lobster decorations. These are symbols of longevity because the crustacean's curved back resembles a hunched elderly person.

Groundhog Reckoning Day

In the midst of our winter woes, we celebrate the groundhog—
but is it a meteorological marvel or a marketing gimmick?

✳ ✳ ✳ ✳

A Shadowy Assertion

THE WORLD'S GREATEST weather forecaster, according to common belief, is essentially a large squirrel. As the story goes, if the groundhog sees its shadow on February 2, there will be six more weeks of winter. No shadow means an early spring. How did this myth get started?

Lacking the Weather Channel, our ancestors relied on numerous methods of predicting the weather. One had to do with animals—such as the groundhog, or "woodchuck"—that hibernate in winter. If hibernating animals were out and about in the early part of the year, spring must be near.

The idea makes some sense from a meteorological standpoint as well. To see shadows, you need clear skies, and in winter, that means cold weather. An overcast sky provides the earth with a blanket of warmth that could mean the end of winter. And since Groundhog Day falls on February 2—roughly halfway between the winter solstice and spring equinox—it has long been seen as a natural turning point for the weather.

How did the lowly woodchuck become the bearer of winter weather reports? The notion seems to have begun with German settlers in Punxsutawney, Pennsylvania, where groundhogs are plentiful. In the late 1800s, a group of friends formed the Punxsutawney Groundhog Club to conduct an annual search for groundhogs. In 1887, the group introduced Punxsutawney Phil and proclaimed him to be the only "official" weather-predicting woodchuck. That title didn't last long: Many other cities and towns have since named their own meteorological rodents, but none have caught on like Punxsutawney Phil.

You're *Way* Off, Columbus

Columbus Day is a holiday that honors the man who discovered America. But according to historians—and the man himself—he missed the mark.

✳ ✳ ✳ ✳

CHRISTOPHER COLUMBUS CROSSED the Atlantic in 1492, convinced he could find a route to the Far East (or "the Indies") that wouldn't require him to sail all the way around the Horn of Africa. Today, we celebrate Columbus Day to honor the fact that he discovered America.

But did he? The truth is that Columbus landed in the Bahamas and from there went to Cuba (he thought it was China) and Hispaniola (present-day Haiti and the Dominican Republic, which he thought was Japan). On his second trip, Columbus returned to Hispaniola, and it wasn't until his third voyage that he finally landed in America—South America, that is, in what is now Venezuela.

Columbus made one last voyage across the Atlantic in 1502, hoping for definitive proof that he'd found a western route to the Indies. Instead, he discovered St. Lucia, Honduras, Costa Rica, and the Isthmus of Panama. By that time, another Italian mariner, Amerigo Vespucci, had sailed along the coast of South America and proposed that it was not Asia at all but an entirely new continent.

Columbus was nothing if not resolute. He continued to insist that he had discovered a new route to the Indies until the day he died in 1506. A year later, a German mapmaker included the newfound lands on a world map and called them America, in honor of Vespucci. It was the first time that the name had been used. Columbus wasn't around to complain, and the name stuck.

For the Love of Mom

Who doesn't love Mother's Day? As it turns out, the woman credited with organizing the first "official" Mother's Day wasn't a fan.

✳ ✳ ✳ ✳

THE ANCIENT GREEKS worshiped the concept of motherhood. Rituals that honored Rhea, the mother of the gods, took place in spring and were said to be wild ecstatic parties full of drumming, dancing, and drinking. The Roman version of this festival—Matronalia, which was dedicated to Juno Lucina, the goddess of childbirth—was not as wild, but a Roman mother could at least expect some gifts. With the advent of Christianity, such festivals disappeared, but eventually the church also provided a way to honor mothers.

In the 16th century, the fourth Sunday of Lent was set aside as "Mothering Sunday," a time when people were expected to visit the church where they were baptized. This day ultimately became a family reunion of sorts. Back then, many children were sent away from their parents to work as apprentices or servants. Their visits home for Lent were known as "going a-mothering." The children would bring their mothers wildflowers or small presents. "Mothering Sunday" is still celebrated in Ireland and parts of the United Kingdom.

From Mothering Sunday to Mother's Day

Today, if you are anything like 96 percent of the American public, you participate in Mother's Day by spending money. Mother's Day is one of the biggest gift-giving holidays of the year. In fact, we spend more only during the Christmas season. Interestingly, the holiday got its start from a group of early feminist activists.

During the American Civil War, a woman by the name of Ann Reeves Jarvis formed "Mother's Day Work Clubs" to help improve living conditions (among other things) for soldiers on both sides of the conflict. Her efforts inspired the suffragist Julia

Ward Howe (who was famous for writing "The Battle Hymn of the Republic") to call for a "Mother's Day for Peace." Howe was convinced that if women had been running things, the horrors of the Civil War could have been prevented. In 1870, she wrote a "Mother's Day Proclamation" that urged women to unite against all war. But neither woman's efforts amounted to any official recognition of Mother's Day.

The cause gained new life in 1905, after Ann Reeves Jarvis's death. Her daughter, Anna M. Jarvis, took up the "Mother's Day" mission—and it eventually took over her life.

This Wasn't What I Had in Mind...

It was Anna M. Jarvis's grief over her mother's death that most likely prompted her, on May 10, 1908, to stage what is considered to be the first "official" Mother's Day. She gave white carnations to each of the mothers at her church and encouraged her fellow parishioners to set the day aside as one of commemoration and gratitude. The idea spread, and in the years that followed, Jarvis lobbied business leaders and politicians for a national mother's holiday. Finally, in 1914, Woodrow Wilson issued a proclamation that set aside the second Sunday in May to honor mothers. The country quickly embraced this idea in a distinctly American way—by shopping.

The commercialization of Mother's Day enraged its founder. Jarvis saw what she'd envisioned as a holy day become another excuse to sell merchandise. In fact, when she saw "Mother's Day" carnations being sold to raise money for veterans, she tried to stop the sale and was arrested for disturbing the peace. She even filed a lawsuit to stop Mother's Day altogether. She died, childless, at the age of 84, regretting Mother's Day's existence. She never knew it, but her final medical bills were partly paid for by florists.

Flubbed Headlines

* "Death Causes Loneliness, Feeling of Isolation"
* "'Save the Whales' Trip Cut Short After Boat Rams Whale"
* "Absentee Votes Can Be Made in Person"
* "Milk Drinkers Are Turning to Powder"
* "Child's Stool Great for Use in Garden"
* "N.J. Judge to Rule on Nude Beach"
* "Babies Are What the Mother Eats"
* "Joint Committee Investigates Marijuana Use"
* "Deaf College Opens Doors to Hearing"
* "Stud Tires Out"
* "Survivor of Siamese Twins Joins Parents"
* "Safety Experts Say School Bus Passengers Should Be Belted"
* "Meridian Woman Training for 20004 Games"
* "Plane Too Close to Ground, Crash Probe Told"
* "Man Found Dead in Cemetery"
* "Miners Refuse to Work After Death"
* "Stolen Painting Found by Tree"
* "Cold Wave Linked to Temperatures"
* "Some Pieces of Rock Hudson Sold at Auction"
* "Man Minus Ear Waives Hearing"
* "Include Your Children When Baking Cookies"
* "Harrisburg Postal Employees Gun Club Members Meet"
* "Have You Driven a Fjord Lately?"
* "Woman Improving After Fatal Crash"
* "Gators to Face Seminoles With Peters Out"
* "Missippi's Literacy Program Shows Improvement"
* "Fried Chicken Cooked In Microwave Wins Trip"

Chapter 11

Pop Culture and Games

The Good Old Days of Soda Fountains

In an age when drive-through coffee shops serve up iced mochachinos, it's easy to forget that chrome-topped soda fountains once held a place of distinction in American culture.

✳   ✳   ✳   ✳

The Golden Age

IN 1819, THE first soda-fountain patent was granted to Samuel Fahnestock. His nifty invention combined syrup and water with carbon dioxide to make fizzy drinks—and they caught on instantly.

The first soda fountains were installed in drugstores, which were sterile storefronts originally intended only to dispense medicines. To attract more business, pharmacists started to sell a variety of goods, including soda drinks and light lunch fare. That way, customers could come in to shop, take time out for some refreshment, and possibly shop a bit more before they left.

Typical soda fountains (the name for both the invention and the shops where the fountains could be found) featured long countertops, swivel stools, gooseneck spigots, and a mirrored back bar, all of which helped attract the attention of young and old alike. Soda fountains were also installed in candy shops and ice cream parlors. Before long, freestanding soda fountains were being built across the country.

Two of the world's most popular beverages got their starts at soda fountains. In 1886, Coca-Cola was first sold to the public at the soda fountain in a pharmacy in Georgia. Pepsi's creator, Caleb Bradham, was a pharmacist who started to sell his beverage in his own drugstore in 1898.

Soda-fountain drinks had to be made to order, and this task was typically performed by male clerks in crisp white coats. Affectionately referred to as "soda jerks" (for the jerking motion that was required to draw soda from the spigots), these popular mixologists added an entertaining element to soda fountains. Think of a modern-day bartender juggling bottles of liquor to make a drink: Soda jerks performed roughly the same feats, except that they used ice cream and soda.

Birth of the Brooklyn Egg Cream

In Brooklyn, New York, candy shop owner Louis Auster created the egg cream, a fountain drink concoction that actually contained neither eggs nor cream.

You make an egg cream any way you like it, but a basic recipe for one combines a good pour of chocolate syrup, twice as much whole milk, and enough seltzer water to fill the glass. (In New York, an egg cream isn't considered authentic unless it's made with Fox's "U-Bet" chocolate syrup.)

The foam that rises to the top of the glass resembles egg whites, which may be how the drink got its name. Some claim that the original chocolate syrup contained eggs and cream; others say that the name "egg cream" comes from the Yiddish phrase *ekt keem*, meaning "pure sweetness"; and still others believe that when kids ordered "a cream" at the counter, it sounded like "egg cream." Whatever the etymology, the drink is legendary among soda-fountain aficionados. Auster claimed that he often sold more than 3,000 egg creams a day. With limited seating, this meant that most customers had to stand to drink them, prompting the traditional belief that if you want to really enjoy an egg cream, you have to consume it while standing up.

Several beverage companies approached Auster to purchase the rights to the drink and bottle it for mass distribution, but trying to bottle an egg cream was harder than they thought: The milk spoiled quickly, and preservatives ruined the taste. Thus, the egg cream remained a soda-fountain exclusive.

Sip and Socialize

Prohibition and the temperance movement gave soda fountains a boost of popularity during the 1920s, as they served as a stand-in for pubs. Booze became legal again in 1933, but by that time, fountains had become such a part of Americana that few closed shop. During the '50s, soda fountains became the hangouts of choice for teenagers everywhere.

It wasn't until the '60s that the soda fountain's popularity began to wane. People were more interested in war protests and puka beads than Brown Cows and lemon-lime-flavored Green Rivers. As more beverages became available in cans and bottles and life became increasingly fast-paced, people no longer had time for the leisurely pace of the soda shop.

Some fountains survive and still serve frothy egg creams to customers on swivel stools, and many of these establishments attempt to appeal to a wide audience by recreating that old-fashioned atmosphere.

Without a Clue

There's nothing puzzling about the origin of the crossword puzzle.

✳   ✳   ✳   ✳

THE FIRST PUBLISHED crossword puzzle was created and constructed in 1913 by an editor and journalist named Arthur Wynne, who worked for the daily newspaper the *New York World*. Wynne was asked by his editors to create a new puzzle for the paper that would challenge, entertain, and educate. Wynne—who was originally from Liverpool, England—designed a format similar to Magic Squares, a popular word game he played as a child.

Wynne's puzzle, which was originally dubbed "Word-Cross," first appeared in the Sunday, December 21, 1913, edition of *The World*. It was diamond-shaped, contained no internal black squares, and provided one free solution (the word *fun*) to get the semantic search started. The clues were not separated into across and down divisions; instead, a numbering system was used to guide the riddle researcher. The first clue to intrigue potential puzzlers was "what bargain hunters enjoy," which, of course, is "sales." Wynne also compiled the first book of crossword puzzles, which hit the bookshelves in 1924.

In February 1922, the first British crossword was printed in *Pearson's Magazine*, a monthly publication that specialized in essays on the arts and politics and helped spawn the careers of such notable authors as H. G. Wells and George Bernard Shaw. The first crossword to appear in *The Times*, England's national newspaper, was published on February 1, 1930.

The invention of cryptic crosswords, in which each clue is a puzzle in itself, is usually credited to Edward Powys Mathers, a brilliant English scholar, translator, poet, and linguist, who devised more than 650 puzzles for the *Observer* newspaper under the pseudonym Torquemada.

When Fads Go Bad

The leisure suit was created with the best of intentions.

✳   ✳   ✳   ✳

THE 1970S ARE often referred to as the "Me Decade"—and with good reason. Coming off the freewheeling, free-loving '60s, folks embraced social decadence with an enthusiasm not seen since the 1920s. Swinging, cocaine, pornography—whatever your vice, it was available for the taking.

But in the eyes of many, the true nadir of the Me Decade was the leisure suit, a ghastly fashion trend that tried to meld the respectability of the business suit with the comfort of "casual Friday," with predictably disastrous results.

As if the Polyester Weren't Bad Enough...

Almost always made of polyester, leisure suits had buttoned jackets with wide lapels and large pockets, as well as pants that typically flared into bell-bottoms. Adding to their garish appearance was large, decorative stitching that sometimes contrasted with the suit's color.

And what colors! Unlike the muted tones of the traditional business suit, leisure suits were manufactured in a rainbow of hues, including burnt orange, cinnamon, crimson, pink, powder blue, saffron, tangerine, and umber.

Because leisure suits were designed for leisure, ties were a no-no. As a result, most men complemented it with a silk or polyester shirt worn with the top buttons undone; the collars were worn outside the jacket. This allowed for easy accessorizing with some '70s "bling."

Wardrobe for the Wealthy

Here's the really interesting thing about the leisure suit: While it is most closely associated with the '70s, it was actually introduced shortly after World War II as casual vacation-wear for the rich.

The first true leisure suits were produced by Louis Roth Clothes and were made of wool gabardine. They had belted jackets with an inverted pleat in back; came in bright, nontraditional colors; and sold for a whopping $100, which was nearly four times what stores such as Sears charged for dress suits.

Leisure suits remained a fashion novelty until the 1970s, when they hit the scene like a brightly colored postmodern sledgehammer. Manufacturers produced them in innumerable styles, and they quickly became the polyester plumage of the hip and groovy. Leisure suits even appeared in television shows such as *The Six Million Dollar Man* and futuristic movies such as the 1975 James Caan hit *Rollerball*.

The timing was perfect: The leisure suit offered a compromise between the sloppy hedonism of '60s hippie-wear and the stodgy formality of the traditional suit. At their height, they were sold in almost every department store and men's clothier, and they were ubiquitous on the business and social scenes. Regardless of class or status, everyone was equal in a leisure suit.

Feeling Groovy?

Then came the inevitable backlash. Among the first to rail against the leisure suit were respected fashion experts such as John Molloy, the author of *Dress for Success*, who declared leisure suits completely inappropriate for the office. (In a 1977 article, Molloy labeled such trends "fads for fools.")

But even more critical to the leisure suit's demise were dissatisfied wearers, who quickly realized that the outfits weren't as groovy as they had been led to believe. Because they were made of polyester, leisure suits were uncomfortably hot, especially in the summer. They also had a tendency to stretch, bag, snag, and shrink. As a result, just a few short years after its second introduction, the leisure suit was consigned to the back of the closet, never to be worn again—except as gag wear for '70s-themed costume parties.

LEGO Love

❋ British inventor Hilary Fisher Page actually patented the interlocking brick years before LEGO came into the picture. Called the "Kiddicraft Self-Locking Building Brick," the idea was copied, more or less, by LEGO founder Ole Kirk Christiansen. The LEGO brick arrived in 1958.

❋ The term "LEGO" comes from the Danish phrase *leg godt*, which means "play well." *Lego* also means "I put together" in Latin.

❋ There are 62 LEGO bricks for every one of the world's 7 billion inhabitants.

❋ Little LEGO people are called "minifigures," or "minifigs" for short. They came on the LEGO scene in 1974 and, at that time, had no arms.

❋ LEGO also makes small tires for building cars, trucks, and moon buggies. The LEGO factory manufactures around 306 million tiny rubber tires every year—more than any other tire manufacturer in the world.

❋ Six LEGO bricks can be arranged in 915,103,765 different ways.

❋ Worldwide, seven LEGO sets are sold every second.

❋ In 2003, the longest LEGO chain ever was built by 700 children in Switzerland. It was 1,854 feet long, had 2,211 chain links, and was made from 424,512 LEGO bricks.

❋ More than 400 billion LEGO bricks have been produced since the company's inception. Stacked on top of each other, this is enough to connect the Earth to the Moon ten times over.

Down in the Dumps

Sorry, E.T. can't phone home—he's buried in a landfill in New Mexico.

✳ ✳ ✳ ✳

A Lovable Alien

IN 1982, THE world was introduced to a lovable alien named E.T.: The Extra-Terrestrial in an exceedingly successful movie of the same name. The film went on to win four Oscars and and spawned a flurry of international movie merchandising. One success story was found in Hershey's ability to boost the sales of its struggling brand Reese's Pieces, the candy that, in the movie, entices E.T. out of hiding and into Elliot's bedroom. Not all merchandising is created equal, however; thus begins the sad tale of Atari and the first video game based on a movie.

At the time, video-game arcades were extremely popular in the United States. In 1981, Americans spent $5 billion playing more than 75,000 hours worth of arcade games, out-grossing both the film and recording industries. Atari's VCS (Video Computer System), more commonly known as the Atari 2600, was a new but successful player in the home video-game market. Atari's goal was to bring the arcade into consumers' homes with cartridge versions of their favorite games.

Dumped

What Atari didn't take into consideration was that the consumer wasn't necessarily excited to play a game about a movie involving a boy and his alien friend. It didn't help that the game was so rushed into production that it proved dull and difficult to play. Stores returned millions of E.T. game cartridges, and many copies were simply never shipped out from Atari's plant in Texas. In the end, 14 truckloads of the unsold games were driven to a landfill in Alamogordo, New Mexico, where they were dumped, crushed, and covered in a layer of concrete to discourage scavengers.

The Dumb Blonde Stereotype

How did a woman with blonde hair become synonymous with "airhead"?

✳ ✳ ✳ ✳

I F YOU WERE to do an informal survey of Caucasian American women, you would find that one out of three has some shade of blonde hair. Of these, only a small percentage are natural blondes; the rest lighten their locks with the help of chemicals, partaking in a ritual that goes back thousands of years. It turns out that in ancient Greece, blonde hair was all the rage.

Dyeing to Be Blonde

The original blonde bombshell was Aphrodite, the Greek goddess of love. Aphrodite embodied the erotic. She was said to have risen out of the ocean in a wave of sea foam, which explains her white skin and flowing blonde tresses. Aphrodite's paleness set her apart from her dark-haired Mediterranean worshippers. What better way to revere than to imitate? Hellenic women (and men) used saffron, oils, lye, and even mud to yellow their hair. The demand for blonde hair was so great that the Romans kidnapped fair-haired barbarians just to keep their wigmakers well-supplied. Thus began one of the longest-lasting fashion trends in Western culture.

Since then, the only time that blonde has gone out of style was during the Middle Ages, when it was considered sinful for a woman to show her hair in public, much less treat it with bleaches. The Catholic Church tried to make blondness a symbol of chastity and innocence: The Virgin Mary was depicted with golden hair (as was Jesus at times). However, despite sermons and writings denouncing the evil, carnal power of colored hair, it did not lose its allure. By the time of Queen Elizabeth I, the use of hair dyes was in vogue again.

Playing the Part

In the 1700s, a Parisian courtesan by the name of Rosalie Duthe became the first famously dumb blonde. It is said that she never spoke but could seduce any man simply by looking at him and entrancing him with her elaborately styled golden hair. She was parodied in intellectual circles and was even featured in a comic play in which she became the laughingstock of Paris. As women of "ill repute" increasingly colored their hair to boost business, the association between colored hair and cheap, wanton behavior deepened.

Hooray for Hollywood!

It took Hollywood to combine innocence and sexuality. Starlets such as Jean Harlow and Marlene Dietrich paved the way for modern-day goddess—and ultimate blonde icon—Marilyn Monroe. Even today, decades after her death, she remains as alluring and enduring as Aphrodite herself. A succession of Tinseltown deities—including Brigitte Bardot, Farrah Fawcett, and Paris Hilton—all fit the "dumb blonde" stereotype that was placed on Monroe. There is, of course, no connection between a woman's hair color and her intelligence. Marilyn Monroe was, in fact, a woman of calculating intelligence. (For example, she was known to relax between shoots by reading weighty authors such as Thomas Paine and Heinrich Heine.)

There is, however, a kernel of truth to the "dumb blonde" stereotype. A group of French academics conducted a study that suggests blonde hair does influence intelligence—men's intelligence. It turns out that the sight of a blonde woman makes men dumber! Men were shown a series of photographs of women with various hair colors and then given a series of basic knowledge tests; those who were shown blonde women scored the lowest. In the end, the joke seems to be on men!

"It takes a smart brunette to play a dumb blonde."

—Marilyn Monroe

Freaky Facts: Fashion

✳ Napoleon regularly wore black silk handkerchiefs as part of his wardrobe, and he steadily won battle after battle. But in 1815, he decided to vary his attire and donned a white handkerchief before heading into battle at Waterloo in present-day Belgium. He was defeated, and it led to the end of his rule as emperor.

✳ The shoe has historically been a symbol of fertility. In some Eskimo cultures, women who can't have children wear shoes around their necks in the hope of changing their childbearing luck.

✳ In the Middle Ages, pointy-toed shoes were all the rage. The fad was so popular that King Edward III outlawed points that extended longer than two inches. The public didn't listen, and eventually the longest points were 18 inches or more!

✳ Catherine de Medici popularized high-heel shoes for women when she wore them for her 1533 wedding to Henri II of France, who later became king. However, several sources say that men had been wearing heels long before that to keep their feet from slipping off stirrups while riding on horseback. A century later, when King Louis XIV of France wore high heels to boost his short stature, the trend became popular with the nobility.

✳ When Joan of Arc was burned at the stake, she was condemned for two crimes: witchcraft and wearing men's clothing.

✳ In the 16th century, men wore codpieces for numerous reasons. The frontal protrusions held money, documents, and whatever else they needed to carry.

Girl Scout Beginnings: Southern Dames and Warfare Games

Youth movements and bloody wars generally don't go hand-in-hand, but the origin of the scouting movement for both girls and boys is inextricably linked with patriotism and war.

✳ ✳ ✳ ✳

THE FOUNDER OF Scouting, Robert Stephenson Smyth Baden-Powell, doesn't have much in common with the founder of the Girl Scouts in the United States, Juliette Gordon Low—at least not on paper. For one, Baden-Powell, an Englishman, started his career as a military officer. He fought in many wars in the late 1800s, and as a military leader, he often incorporated wilderness-survival skills when he trained his men. When his troops were besieged in the Second Boer War, a group of local youths helped out by carrying messages and supplies. This got Baden-Powell thinking about hands-on youth education. If young people were taught survival and leadership skills, basic military maneuvers, and community service, they would also learn about teamwork and civic responsibility.

And thus the Boy Scouts were born. In 1908, Baden-Powell published *Scouting for Boys*, which is considered the first edition of the *Boy Scout Handbook*. From there, troops formed throughout the United Kingdom and around the world. The Scouting movement was international in focus from the beginning, since Baden-Powell's philosophy included multicultural acceptance and the benefits of physical and mental growth for all children. So it's not surprising that soon after the formation of the Boy Scouts, girls were enthusiastically welcomed into the movement as well.

The first Girl Scouts appeared in the form of Girl Guides, a group that was formed in 1910 by Agnes Baden-Powell, Robert's sister. And then, in 1912, Juliette Gordon Low founded the Girl Scouts in the United States. By the time Low learned about the Scouting movement, she had spent a lifetime in community service. When she met Agnes Baden-Powell in 1911, her plan had been to emigrate to France. But after learning about Agnes's new youth movement, she moved to Scotland to lead a Girl Guides troop. She soon realized, however, that she wanted to bring what she had learned back to the United States.

Low wasted no time. Once back in Savannah, she called a friend and is reported to have exclaimed "I've got something for the girls of Savannah, and all of America, and all the world, and we're going to start it tonight!" On March 12, 1912, Low convened the first meeting of the American Girl Guides. The name was changed to "Girl Scouts" the following year.

The movement spread with dizzying speed. There was an excitement in encouraging girls to take part in activities that were so different from the mechanic drudgery of school lessons. From outdoor adventure activities to bake sales to volunteer work at hospitals, girls were expected to do it all. Education was transformed from writing and copying to doing. The organization's philosophy was learning through experience and making the community a better place in the process. The Girl and Boy Scout movements actively recruited minority members as well as those with physical and mental disabilities. An effort was also made to create groups in poorer areas.

World War I broke out soon after the creation of the Girl Scouts, and the movement's root in helping one's country during war was immediately put to practical use. Girl Scout troops volunteered in hospitals, sold war bonds, and learned how to conserve resources. Since that time, the Girl Scouts have become the largest worldwide educational organization for girls.

"Ours is a circle of friendships united by ideals."

—JULIETTE GORDON LOW

Popular Hairstyles Through the Ages

Here are some of the more extreme, time-consuming, gravity-defying looks that have graced our scalps through the ages.

✳ ✳ ✳ ✳

Powdered Wigs

IN THE 18TH century, hygiene was hardly what it is today. Head lice was a rampant problem, so wigs were worn to cover shaved heads. But powdered wigs themselves were hotbeds for lice, roaches, and other critters. Still, wigs were worn until the end of the 18th century, when an expensive tax on white powder caused the trend to die down.

Pompadour

In a pompadour, the hair is gelled back—giving it a sleek, wet look—and the front is pushed forward. The look is often associated with the 1950s, when it was worn by the likes of Elvis Presley and Johnny Cash. But Madame de Pompadour, a mistress of King Louis XV and a fashionista in her time, originally donned the look in the 1700s. The style is still worn by retro rockabilly types, including Brian Setzer.

The Flip

This spunky, youthful style was mega-popular among women throughout the 1960s. Shoulder-length hair was back-combed or teased slightly at the top, and then the ends were curled up in a "flip" with rollers or a curling iron. Depending on the age of the woman and her willingness to push the envelope, the flip was combined with the bigger, puffier bouffant. Mary Tyler Moore sported the classic flip on *The Dick Van Dyke Show*, and Jackie Kennedy had her own more conservative version. Later, the style became so ubiquitous that it was nicknamed "beauty pageant hair" or "Miss America hair," because for years nearly every contestant in these competitions sported flip after perfect flip.

The Mop Top

The influence of The Beatles on popular culture was unlike anything the world had ever seen. Girls and boys alike mimicked the boyish charm of these Liverpool lads, especially when it came to this long and floppy hairstyle.

Afro

Birthed by the black-pride movement, the afro represents a time when African Americans stopped applying harsh chemicals to their hair in order to achieve something close to Caucasian texture and style. In the 1960s, the afro was a political statement, but today it is worn as fashion. Anyone with extremely curly hair can adopt the style.

Mohawk

Originally worn in Native American cultures, the mohawk reappeared sometime in the 1970s as part of the punk rock movement. At its most extreme, the mohawk, called the *mohican* in the United Kingdom, featured a stripe of hair sticking straight up and running down the middle of the head, with the sides shaved. It made a statement about not giving in to social standards. However, thanks to the advent of the "fauxhawk" (small spike, not shaved on sides), the mohawk of today is much more watered-down.

Mullet

Also appearing sometime in the 1970s, the mullet soared to popularity in the '80s among rockers and rednecks. But the origins of the short-on-the-top-and-sides, long-and-free-in-the-back look are a bit mysterious. Some say that hockey players originated the 'do, and others think that it's an offspring of rock 'n' roll (David Bowie was an early adopter)—but most agree that the style should be retired.

Greed Is Good

For decades, the origins of the popular board game Monopoly were hidden behind a myth that was propagated by the toy company Parker Brothers. But the real story reflects a dark truth about American business.

✳   ✳   ✳   ✳

THE GOAL OF Monopoly is for players to get rich through the ruthless acquisition of property at the expense of the other players. Any player can acquire property and win—no matter their station in life—as long as they play shrewdly with cut-throat tactics. It's no small wonder that Monopoly rose to prominence during the Great Depression and has remained extremely popular—it encapsulates the American Dream.

Darrow's Our Hero

As the Parker Brothers version goes, Charles Darrow was an unemployed salesman in Germantown, Pennsylvania, who was struggling to make ends meet during the Depression. Thinking back to better days when he and his family spent summers in Atlantic City, Darrow began drawing the streets of the East Coast fun capital on a kitchen tablecloth. He added or built tiny hotels and houses, eventually turning the set-up into a miniature Atlantic City.

Darrow began making and selling the games for a few dollars each, offering them to department stores in Philadelphia. He was so encouraged that he attempted to sell his brainchild to Parker Brothers—one of America's premier game manufacturers. But they turned him down, claiming that the game had "52 fundamental gaming errors," including everything from overly complicated rules and a dull central concept to an unusually long playing time.

Undaunted, Darrow produced 5,000 copies and began selling them to FAO Schwartz and other toy stores. Eventually, one fell into the hands of Sally Barton, the daughter of George Parker, the founder of Parker Brothers. Parker Brothers bought the game from Darrow and gave him royalties on all games sold, and Darrow retired as a millionaire. The end.

Go Straight to Jail; Do Not Collect $200

The origins of the game go all the way back to 1904, when a Quaker woman named Elizabeth Magie-Phillips developed and patented The Landlord's Game to illustrate political economist Henry George's tax principles. George called for a single tax based on land ownership because he felt it would encourage equal economic opportunities and discourage land speculation. Despite any similarities between Monopoly and The Landlord's Game, the goal of the latter was to criticize land acquisition—not celebrate it—by illustrating that landowners have all the financial advantages in society.

As years passed, the game's popularity spread. The rules evolved and the name changed from The Landlord's Game to such variations as Finance, Auction Monopoly, and finally Monopoly. At least eight different people or groups were associated with a Monopoly-type game, including a hotel manager named Charles Todd, who may have introduced it to Darrow.

Truth and Fables

The Parker Brothers version is more of an all-American, pull-yourself-up-by-the-bootstraps story—but it's not the whole truth. Darrow may have added the chance cards and the railroads, but he did not invent the original game concept. In fact, after Parker Brothers bought Darrow's version of the game, they quietly settled with the inventors of the other versions to protect their investment. Parker Brothers began to spread their revised origin story, which was so widely accepted that in 1970, Atlantic City erected a plaque near a corner of Park Place in Darrow's honor.

There She Is ...

The early days of the Miss America beauty pageant were anything but tasteful.

* * * *

The Fall Frolic

IN 1920, ATLANTIC City hotel owner H. Conrad Eckholm had an idea for a way to attract folks to town after Labor Day—typically the slowest time of the year for tourism. The "fall frolic" was moderately successful, and it was decided that the event should be repeated the following year. However, one important addition was made to the frolic: a beauty contest.

But this was Atlantic City, a city that reveled in spectacle. Its celebrated boardwalk contained, among other curiosities, a grandiose Moorish estate where the owner fished from his bedroom window. No mere bevy of bathing-suit beauties would compete at this contest! To flesh out the theme and heighten the theatrics, the event coordinators hired 80-year-old Hudson Maxim to preside over the Fall Frolic as Father Neptune. Trident in hand and white beard flowing, Father Neptune floated onto the beach in a seashell barge along with his retinue of beauty queens on September 8, 1921. The Fall Frolic beauty pageant was officially underway; its winner would be crowned "Miss America."

Neptune Reigns Supreme

Clearly the star of the show, Father Neptune was dubbed "His Bosship of the Briny" and "Blue Blood of the Breakers" by the local press. The 1921 Fall Frolic was a wild success. It featured lifeguard boat races, fireworks, and vaudeville shows. It seemed as if everyone was in bathing suits, including the police. A bawdy, Mardi Gras-type affair known as "Neptune's Frolique" was held, in which women showed off their "nude limbs." Judging of the pageant itself was based on the crowd's applause for which girl they liked best, as well as a panel of "experts." The first Miss America was 16-year-old Margaret Gorman of Washington, D.C.

Substance Over Form

The 1923 pageant featured 57 contestants, including Miss Alaska. Although her arduous trek from the snowy hinterlands to the sandy beaches was described as having forced her to use "dog-sled, aeroplane, train, and boat," in reality she was married and lived in dogsled-free New York City. Exposed as a fraud, she was barred from the competition by red-faced pageant officials. She, in turn, filed a $150,000 lawsuit.

Other problems arose that year when several contestants appeared in the Bather's Revue (now known as the "swimsuit competition") in revealing (for the time) one-piece bathing suits. The other women, who had worn more traditional and modest bathing attire, immediately staged a "beauty strike." They didn't know, they protested, that "form was to be considered in selecting America's most beautiful bathing girl." Problems continued as Miss Philadelphia was knocked unconscious when the chairs on which the contestants were standing collapsed in a heap.

Shedding Its Image

More disasters followed over the next few years. Against regulations that restricted the contest to single women, there were several instances when married women (including one with a baby) competed. Rumors that the final results were fixed didn't help matters. Eventually—and somewhat incredibly—the pageant was seen as hurting rather than helping Atlantic City's image. It was canceled in March 1928.

The pageant was resurrected briefly in 1933; this time, it featured Miss New York's collapse on stage due to an abscessed tooth, Miss Oklahoma's emergency appendectomy, another married contestant, and three others who were disqualified for residing in states other than those claimed.

When the Miss America pageant returned once more in 1935, it had been restructured and regulated to more closely resemble today's glossy version. No one, it seemed, wanted to recall the "good old days."

'Toon Truths

Read on for fascinating facts about a few of your favorite cartoons of yesteryear—and some contemporary ones too!

✳ ✳ ✳ ✳

✳ During World War II, the Nazi propaganda newspaper *Das Schwarz* tried to discredit Superman by labeling the Man of Steel a Jew.

✳ *South Park* originated as a joke Christmas-card video that was shared by Hollywood insiders. Among them was actor George Clooney, who distributed multiple copies to friends.

✳ The full names of the eponymous characters from the show *Ren & Stimpy* are Ren Höek and Stimpson J. Cat. By the way, Ren is a certified Asthma-Hound Chihuahua.

✳ *The Flintstones* was the first animated cartoon series with recurring characters to air in prime time, premiering on September 30, 1960.

✳ The Reynolds Tobacco Company was an early sponsor of *The Flintstones*, so first-season episodes featured commercials in which Fred and Barney enjoyed a smoke—something you would never see on a children's program today.

✳ Over the years, *The Flintstones* featured several real-life celebrities, including singer Ann-Margret (Ann-Margrock), songwriter Hoagy Carmichael (Stoney Carmichael), and actor Tony Curtis (Stoney Curtis).

✳ According to *Esquire* magazine, 58.2 percent of men surveyed found Betty Rubble more desirable than Wilma Flintstone.

✳ Jean Vander Pyl, who voiced Wilma, gave birth on February 22, 1963—the same day that Wilma gave birth to Pebbles on the show.

* It took *The Flintstones* producers two days of intensive meetings to decide whether the Flintstones would have a boy or a girl.

* *The Flintstones* proved so popular that a spin-off movie, a spy parody titled *The Man Called Flintstone*, was released in 1966.

* On *The Jetsons*, George and Jane Jetson had an odd sleeping arrangement. In some episodes, their beds are apart. In others, their beds are pushed together. And in a few, they sleep in the same bed.

* On *Jonny Quest*, the character of Jonny's pal Hadji was based on the Indian actor Sabu.

* A young Tim Matheson, who later played Eric Stratton in *National Lampoon's Animal House*, provided the voice of Jonny Quest.

* Tweety Bird, the frequent foil of Sylvester the Cat, was originally pink. However, studio censors thought that he looked a little too naked, so they demanded he be yellow instead.

* Actress Mae Questel, who voiced Olive Oyl in the early *Popeye* cartoons, also voiced Popeye in a handful of shorts while Popeye's regular voice actor, Jack Mercer, was overseas during World War II.

* The Rio de Janeiro Tourist Board threatened to take legal action after an episode of *The Simpsons* portrayed the city as rife with crime and overrun by slums. (Executive producer James L. Brooks later apologized.) However, it wasn't the first time a municipality whined about its portrayal on the show: New Orleans city officials complained following the airing of a season-four episode in which a musical production of *A Streetcar Named Desire* portrayed the city as "home of pirates, drunks and whores."

Forget Getting In: Studio 54

*If you want to understand disco, then you first need to understand
the nightclub called "Studio."*

✳ ✳ ✳ ✳

Pre-function

IT BEGAN WITH an odd couple: Steve Rubell and Ian Schrager.
Rubell was loud and outgoing, Schrager quiet and businesslike.
They ran steak houses in the NYC metro area, but they wanted
to operate at the top of the Big Apple's food chain: Manhattan.
They teamed with young Peruvian party promoter Carmen
d'Alessio, who would handle PR.

At 254 W 54th St. in Manhattan stood an old CBS TV studio.
Rubell and Schrager leased it and started costly, rococo renova-
tions, complete with a huge man-in-the-moon that had a hanging spoon rising up to its nose. One could hardly find a more
obvious way to say "Inhale Cocaine Here." From the address and
the building's history came the logical—if unimaginative—name:
Studio 54. It was a big place, larger than a football field. It would
need to be.

Party On

Opening Night was April 26, 1977. In those days, New York
normally reacted to new discos like husbands react to their wives'
new shoes: *That's nice, hon; is there any pizza left?* But hundreds
could smell the hedonism in the Manhattan air.

A young employee, Joanne Horowitz, convinced Rubell that
she could lure some glitterati. The early opening hours were
slow, but soon a platoon of celebrities appeared: Margaux
Hemingway, Mick and Bianca Jagger, Salvador Dali, Donald and
Ivana Trump, Brooke Shields, Cher, and more. Frank Sinatra,
Henry Winkler, and Warren Beatty didn't get in—nor did most
of the throng waiting outside. The doorman, 19-year-old Marc
Benecke, was the most envied college kid in town.

Studio became *the* place—to which you could only hope to gain entrance. Unless you were a major celebrity, or lucky, you didn't make it inside. Behind the doors, nearly anything went. Studio served alcohol, naturally, but others inside dished cocaine, Quaaludes (Rubell's favorite appetizer), and similarly illicit party favors. Nudity and open sexual activity—straight, bi, or gay— were common.

Hungover and Throwing Up

Incredibly, Rubell was operating the club on one-day caterers' permits rather than a liquor license. Three weeks after opening, the liquor authorities simply denied the daily permit and sent a police raid. (*They* got in.) Studio spent five months as a disco juice bar until the license affair was straightened out, but the club didn't lose its mojo.

But Rubell should have left the PR to d'Alessio. In late 1978, he bragged to reporters that Studio had made about $7 million the previous year, asserting that only the Mafia made more. Might as well rename Studio "Audit 54." On December 14, 1978, the IRS showed up. The very walls of the club were stuffed with cash, and the safe yielded detailed incriminating records. As for drugs, any investigator incapable of finding those in Studio would have been hopeless.

After a lengthy investigation, Rubell and Schrager were charged with tax evasion, skimming, and more in June 1979. Studio threw one last wingding on February 4, 1980, before their jail time began.

After being released after 13 months in the lockup, Schrager and Rubell went into the hotel business together. Ian Schrager remains a very successful hotelier, though as of April 2009, he was facing financial troubles at at least one of his establishments.

In 1985, Rubell tested HIV positive but didn't take care of himself. He died on July 25, 1989, at age 45; his official cause of death was listed as hepatitis.

The Secret Life of Big Tex

Every year since 1952, Big Tex—a 52-foot, 6,000-pound cowboy standing at the entrance to the "Million Dollar Midway"— greets attendees of the State Fair of Texas. But Big Tex has a big secret: He started out as Santa Claus in 1948 in the small town of Kerens, Texas.

✳   ✳   ✳   ✳

HOWELL BRISTER, SECRETARY of the Kerens Chamber of Commerce, was concerned that the citizens of Kerens were traveling the 70 miles to Dallas to shop for Christmas. He needed a gimmick to encourage locals to spend their money closer to home, and he finally decided on a 49-foot Santa Claus made of iron-pipe drill casing and papier-mâché. The statue was outfitted with a Santa suit and hat, as well as a beard that was created from 7-foot lengths of unraveled rope.

The promotion proved a huge success during the 1948 holiday shopping season, but the novelty quickly wore off. In 1951, the State Fair of Texas purchased the giant Santa for $750,000 and transported him to Dallas. Santa was transformed into a cowboy and dubbed "Big Tex" in time to make his debut at the 1952 State Fair.

Santa's Makeover

To achieve the transformation, his nose was straightened and his eyes, which appeared to be winking lasciviously, were corrected. The H. D. Lee Company donated a plaid shirt and denim jeans, and a pair of size 70 boots and a 75-gallon hat completed the ensemble.

In 1953, Big Tex was installed with a device that allowed his mouth to move automatically, and over the years, seven different people have performed the booming "H-O-W-D-Y" that is his signature greeting. In 2009, the seventh voice of Big Tex was

Bill Bragg, who took over the job on Big Tex's 50th birthday in 2002. Bragg broadcasts from a small booth 60 times a day for the 24 consecutive days that the fair runs.

Improvements over the Years

Additional updates have made Big Tex stronger and given him more character. His original papier-mâché head was ultimately replaced with one made of fiberglass. In 1997, his body was rebuilt with a cagelike skeleton of 4,200 feet of steel rods. In 2000, a mechanical arm was installed so that Big Tex could wave at visitors, and animatronics were added to allow him to move his head.

Big Tex has to be a snappy dresser, and his clothing has also undergone changes over the years. In 1965, he donned a 300-pound, 15-foot-by-60-foot Mexican serape. In 1975, Big Tex greeted visitors in a candy-striped shirt for the Yankee Doodle Dandy-themed fair. And in 1982, he showed his school spirit by sporting an orange University of Texas T-shirt.

The current cowboy outfit sported by Big Tex was specially designed by the Williamson–Dickie Company of Fort Worth. It took a team of eight employees to assemble it. The custom shirt covers Big Tex's 30-foot chest and 181-inch arms with room for his 100-inch neck. The buttons are three and a half inches in diameter, and the shirt is made of 70 yards of blue denim and 80 yards of awning material. Tex's matching denim pants are size 284W with a 200-inch inseam, a 56-inch fly, and rivets that are three and a half inches in diameter. These hefty jeans weigh 65 pounds. The outfit is completed with a 75-gallon hat measuring five feet and size 70 boots measuring seven feet and seven inches.

So when fall arrives and the fair kicks into gear at its 50-acre park in Dallas, listen for the booming voice of Big Tex—and don't forget to wave.

Men's Societies

A fraternal organization is a group of men who bond through rituals, handshakes, and sometimes uniforms. They usually have overlapping missions—whether they emphasize fellowship, patriotism, religion, or philanthropy—and most are active in their communities. Here are some of the most recognizable, along with notable members past and present.

✳ ✳ ✳ ✳

Moose International, Inc. Founded in 1913, the Family Fraternity, often called the Loyal Order of Moose (and Women of the Moose), is a nonsectarian and nonpolitical organization. Moose International headquarters in Mooseheart, Illinois, oversees 2,000 lodges, 1,600 chapters, and approximately 1.5 million members throughout the United States, Canada, Great Britain, and Bermuda. According to the group's mission statement, the moose was selected as the society's namesake animal because "it is a large, powerful animal, but one which is a protector, not a predator." Moose members are active in their communities, contributing nearly $90 million worth of service every year to charities and social causes in their hometowns.

Famous Moose members: presidents Franklin D. Roosevelt and Harry S. Truman, actor Jimmy Stewart, athletes Arnold Palmer and Cal Ripken Sr., and U.S. Supreme Court Chief Justice Earl Warren.

The Benevolent and Protective Order of Elks of the United States of America was founded in 1868, making it one of the oldest fraternal organizations in the country. With headquarters in Chicago, the order includes more than a million members. The Elks' mission is to promote the principles of charity, justice, brotherly love, and fidelity; encourage belief in God; bolster patriotism; cultivate fellowship; and actively support community charities and activities. A major component of the Elks' mission is working with and mentoring youngsters.

Famous Elks: presidents John F. Kennedy and Gerald Ford, actor Clint Eastwood, football coach Vince Lombardi, and baseball greats Casey Stengel and Mickey Mantle.

Masons, also known as Freemasons, belong to the oldest fraternal organization in the world. Today, there are more than 2 million Freemasons in North America. Freemasonry, or Masonry, is dedicated to the "Brotherhood of Man under the Fatherhood of God." Masonry's principal purpose is "to make good men better." Since the origins of Masonry have been lost, no one knows exactly how old the movement is, but many historians believe that it arose from the powerful guilds of stonemasons of the Middle Ages. In 1717, Masonry became a formal organization when four lodges in London formed England's first Grand Lodge. The oldest Masonic jurisdiction on the European continent is the Grand Orient de France, which was founded in 1728.

Famous Masons: presidents George Washington and James Monroe, composer Wolfgang A. Mozart, astronaut John Glenn, actor John Wayne, and escape artist Harry Houdini.

Lions Clubs International remains the world's largest service organization, with 45,000 clubs and 1.3 million members in 200 countries around the world. Its international headquarters is in Oak Brook, Illinois. The organization was founded in 1917 in the United States and became international in 1920 when the first Canadian club was established in Windsor, Ontario. All funds raised by Lions Clubs from the general public are used for charitable purposes, and members pay all administrative costs. Since the Lions Clubs International Foundation began in 1968, it has awarded nearly 8,000 grants (totaling $566 million) to assist victims of natural disasters, fight physical and mental disabilities, and serve youth causes.

Famous Lions: President Jimmy Carter, race-car driver Johnny Rutherford, explorer Admiral Richard Byrd, and basketball star Larry Bird.

Hot Pants

The 1970s saw more than its share of bad fashion trends, but few did more to set back the burgeoning women's liberation movement than hot pants.

✳ ✳ ✳ ✳

BASICALLY JUST REALLY short shorts, hot pants were typically made of satin or some other clingy fabric. It took a certain physical build to look good in hot pants, but they were amazingly popular nonetheless.

Hot pants were introduced in Paris and London in the late 1960s and early 1970s in response to the midi-skirt—a calf-length dress that fashion designers were promoting as a more mature answer to the miniskirt. But young women didn't want to show less leg— they wanted to show more.

It took a while for hot pants to go mainstream. Fashion magazines were reluctant to promote them, so at first they were worn primarily by women who felt empowered by their outrageous sensibility. Only later did European magazines finally admit to their popularity.

The story was completely different in the United States, where the skimpy shorts were an immediate hit. In fact, Bloomingdale's was literally mobbed when the store first started selling them in January 1971. It was *Women's Wear Daily* that first coined the term "hot pants," and it didn't take long for the short shorts to become high fashion. Celebrities such as Marlo Thomas and former First Lady Jackie Onassis were photographed wearing the daring duds, and designers were soon producing them in every style, from mink to sequins.

But like most fashion fads, hot pants quickly lost their appeal and—as with the leisure suit—were banished to the back of the closet.

Nylon Evolution

Back in the day, women put on their underwear and hosiery—and then a garter belt or another support device to hold up the hosiery. Sometimes the garter belt would snap or the hose would bag, resulting in oh-so-attractive "elephant knees." What was a woman to do?

❋　❋　❋　❋

THE SOLUTION CAME with the development of panty hose, a convenient one-piece garment that combined panties and stockings. Actress-dancer Ann Miller, who starred in such films as *Easter Parade* (1948) and *Kiss Me Kate* (1953), claims to have worn the first panty hose in the early 1940s as a time-saver when filming dance numbers. At that time, stockings were commonly sewn directly to costumes because garter belts were impractical for dancers. But when the stockings tore, a new pair had to be resewn. So Miller asked a hosiery manufacturer to sew the stockings to a pair of briefs. The first attempt was too short for Miller's long gams, but the second pair fit just right.

In 1959, Allen Gant Sr. of North Carolina-based Glen Raven Mills introduced panty hose to the American public, revolutionizing the foundation-garment industry. In 1965, Glen Raven Mills introduced the first seam-free panty hose, which conveniently coincided with the advent of the miniskirt. Over the years, numerous other styles have been produced, most notably "Control Top" pantyhose, which contain a reinforced panty to make the wearer appear more slender.

Panty hose were a huge success from the moment they were introduced because women found them convenient and comfortable. Today, the hosiery industry produces an estimated two billion pairs of panty hose each year. And while the majority of wearers are women, men have also been known to slip into a pair. Football players and other athletes occasionally wear panty hose to help them stay warm, to reduce chafing, and to improve circulation.

The ABCs of SCRABBLE

In the late 1930, Alfred Mosher Butts of Poughkeepsie, New York, lost his job. With all that free time, in 1938, he created his own board game, which eventually became known as SCRABBLE. Today, millions of SCRABBLE sets are sold each year in North America. How much do you know about SCRABBLE?

✳   ✳   ✳   ✳

✳ When Butts first invented SCRABBLE, he called it LEXIKO. Later, the game was renamed CRISS CROSS WORDS. Obviously, neither name took, and the game was christened SCRABBLE in 1948.

✳ According to legend, SCRABBLE struggled to find an audience in its first two years. It wasn't until the early 1950s that sales started to boom, thanks to an assist from the president of Macy's department store in New York City. The story goes that the business tycoon discovered SCRABBLE when he was on vacation and was so taken with it that he ordered several sets for his store. Suddenly, everyone had to have a SCRABBLE game. It has been extremely popular ever since.

✳ How popular is SCRABBLE? More than 150 million sets have been sold worldwide.

✳ Businessman James Brunot bought the rights to Butts's game in 1948. Brunot renamed the game SCRABBLE, a word that means "to grope frantically."

✳ Determining how many Ws, Ts, Ps, and other letters to include in the game was no easy task. Butts pored over the front pages of *The New York Times* to figure out how often each letter of the alphabet showed up. This determined how many tiles each letter would get. There was one exception: Butts only included four S tiles; he didn't want players to simply make words plural.

✳ The secret to SCRABBLE success is to rely on unusual words to get out of jams. For example, there are 121 SCRABBLE words that contain no vowels. Unfortunately, if you don't have a Y, you'll only be able to use 20 of them. Some of these vowel-free words include *crypt, cysts, dry, fly, fry, nymph,* and *myth.* Some of the 20 Y-free no-vowel words include *tsk, brr, psst,* and *crwth.* A crwth, by the way, is an archaic stringed musical instrument. It's usually associated with Welsh music.

✳ You may also need to use a word that has a Q but no U. These are rare. In fact, *The Official SCRABBLE Players Dictionary* only features 24 of these words. Some of them include *qadi,* which is an Islamic judge; *faqir,* a Muslim or Hindu monk; *sheqel,* an ancient unit of weight; and *qindar,* an Arabian currency.

✳ Two-letter words are vitally important in championship-level play. These words don't generally give you a lot of points, but they do help players out of tough spots. *The Official SCRABBLE Players Dictionary* lists 102 two-letter words. Some are common, such as *am, an, be,* and *ox.* Others, though, are quite unusual. Do you know what a *za* is? It's slang for "pizza," as in, "Let's order a za tonight." How about *qi?* That's the ancient Chinese term for the vital energy that supposedly flows through our bodies.

✳ *The Official SCRABBLE Players Dictionary* is usually updated every five years. For the fourth edition, which was released in 2005, more than 3,300 new words were added. Thanks to the words *za* and *qi,* the dictionary now, for the first time ever, contains two-letter words that use the Z and Q tiles. This is important for tournament players, because these two tiles are worth ten points each— the highest number of points for any letters in the game.

Chapter 12

The Supernatural

Circle Marks the Spot: The Mystery of Crop Circles

The result of cyclonic winds? Attempted alien communication? Evidence of hungry cows with serious OCD? There are many theories regarding how crop circles, or grain stalks flattened in recognizable patterns, have come to exist. Most people dismiss them as pranks, but there are more than a few who believe that there's something otherworldly going on.

✳   ✳   ✳   ✳

Ye Olde Crop Circle

SOME EXPERTS BELIEVE that the first crop circles date back to the late 1600s, but there isn't much evidence to support them. Other experts cite evidence of more than 400 simple circles 6 to 20 feet in diameter that appeared worldwide hundreds of years ago. The kinds of circles they refer to are still being found today, usually after huge cyclonic thunderstorms pass over a large expanse of agricultural land. These circles are much smaller and not nearly as precise as the geometric, mathematically complex circles that started appearing in the second half of the 20th century. Still, drawings of and writings about these smaller circles lend weight to the claims of believers that the crop circle phenomenon isn't a new thing.

The International Crop Circle Database reports stories of "UFO nests" in British papers during the 1960s. About a decade or so later, crop circles fully captured the attention (and the imagination) of the masses.

No, Virginia, There Aren't Any Aliens

In 1991, two men from Southampton, England, came forward with a confession. Doug Bower and Dave Chorley admitted that they were responsible for the majority of the crop circles found in England during the preceding two decades.

Inspired by stories of "UFO nests" in the 1960s, the two decided to add a little excitement to their sleepy town. With boards, string, and a few simple navigational tools, the men worked through the night to create complex patterns in fields that could be seen from the road. It worked, and before long, much of the Western world was caught up in crop circle fever. Some claimed that the crop circles were irrefutable proof that UFOs were landing on Earth. Others said that God was trying to communicate with humans "through the language of mathematics." For believers, there was no doubt that supernatural or extraterrestrial forces were at work. But skeptics were thrilled to hear the confession from Bower and Chorley, since they never believed the circles to be anything but a prank in the first place.

Before the men came forward, hundreds of crop circles appeared throughout the 1980s and '90s, many of them not made by Bower and Chorley. Circles "mysteriously" occurred in Australia, Canada, the United States, Argentina, India, and even Afghanistan. In 1995, more than 200 crop circles were reported worldwide. In 2001, a formation that appeared in Wiltshire, England, contained 409 circles and covered more than 12 acres.

Many were baffled that anyone could believe that these large and admittedly rather intricate motifs were anything but man-made. Plus, the more media coverage crop circles garnered, the more new crop circles appeared. Other people came forward, admitting that they were the "strange and unexplained power" behind the circles. Even then, die-hard believers dismissed the hoaxers, vehemently suggesting that they were either players in a government cover-up, captives of aliens who were forced to throw everyone off track, or just average Joes looking for 15 minutes of fame by claiming to have made something that was clearly the work of nonhumans.

Scientists were deployed to ascertain the facts. In 1999, a well-funded team of experts was assembled to examine numerous crop circles in the UK. The verdict? At least 80 percent of the circles were, beyond a shadow of a doubt, created by humans. Footprints, abandoned tools, and video of a group of hoaxers caught in the act all debunked the theory that these crop circles were created by aliens.

But Still...

So if crop circles are nothing more than hoaxers having fun or artists playing with a unique medium, why are we still so intrigued by them? Movies such as *Signs* (2002) capitalize on the public's fascination with the phenomenon, and crop circles still capture headlines. Skeptics will scoff, but from time to time, there is a circle that doesn't quite fit the profile of a manmade prank.

There have been claims that fully functional cell phones cease to work when the caller steps inside certain crop circles. Could it be caused by some funky ion-scramble emitted by an extra-terrestrial force? Some researchers have tried to re-create the circles and succeeded, but only with the use of high-tech tools and equipment that would not be available to the average prankster. If all of these circles were made by humans, why are so few people busted for trespassing in the middle of the night? And why don't they leave footprints behind?

Eyewitness accounts of UFOs rising from fields can hardly be considered irrefutable evidence, but there are several reports from folks who swear that they saw spaceships, lights, and movement in the sky just before crop circles were discovered.

Stranger than Fiction: Doppelgängers

A perplexing number of people have reportedly encountered their doppelgängers. Read on for some of the most famous examples.

✳ ✳ ✳ ✳

Haven't We Met Before?

ACCORDING TO MANY sources, *doppelgänger* is a German word meaning "double goer" or "double walker." Essentially, a doppelgänger is defined as a person's twin, although not in a Doublemint Gum sense. Rather, a doppelgänger is often described as a very pale, almost bloodless version of the person. Its appearance usually means impending danger or even death for its human counterpart, although there have been instances in which the doppelgänger foretold the future or simply showed up and didn't cause any harm.

Interestingly, the doppelgänger is such a well-known phenomenon that Sigmund Freud tackled it in a paper titled "The Uncanny." In it, he theorized that the doppelgänger is a denial of mortality by humans. After they leave that denial behind, the double remains as "the ghastly harbinger of death."

Deathly Doppelgängers

Many famous people have reported seeing their doppelgängers, and it usually wasn't a good thing. One of the most famous people to be visited by such a creature was President Abraham Lincoln, who reported seeing his doppelgänger in a mirror in 1860, just after his election. As he later described it to his friend Noah Brooks, the double was "five shades" paler than himself. Lincoln's wife interpreted this as an omen that he would be elected to a second term but would not live through it. She was eerily on the mark.

Near the end of her long reign, England's Queen Elizabeth I reportedly saw a pale and wizened double of herself laid out on her bed. She died soon afterward in 1603. Renowned poet Percy Bysshe Shelley supposedly encountered his doppelgänger in Italy; the figure pointed toward a body of water. Shortly after, on July 8, 1822, Shelley drowned while sailing.

In 1612, English poet John Donne was traveling abroad in Paris when the doppelgänger of his pregnant wife appeared before him. Although the double was holding a newborn baby, she looked incredibly sad. "I have seen my dear wife pass twice by me through this room, with her hair hanging about her shoulders, and a dead child in her arms," Donne told a friend. He later found out that at the precise moment the apparition appeared to him, his wife had given birth to a stillborn child.

Sometimes a person won't see his or her doppelgänger, but somebody else will, often with the same unfortunate result. That was the case with Pope Alexander VI, who was a man given to murder, incest, and other manner of foul deeds. According to some stories, Alexander plotted to kill a church cardinal for his money. He brought poisoned wine to a dinner with the cardinal but forgot to bring an amulet he owned that he believed made him invulnerable to poison. According to lore, Alexander sent a church official back to get it. When the official entered Alexander's room, he saw a perfect image of the pope lying atop a funeral bier in the middle of the room. That night at dinner, Alexander drank his own poison by mistake. He died a few days later on August 18, 1503.

Hello, It's . . . Me?

Not all doppelgängers sound Death's clarion call. In 1905, a severe influenza outbreak prevented a member of the British Parliament named Sir Frederick Carne Rasch from attending a session. However, during the session, a friend—Sir Gilbert

Parker—looked over and saw Rasch. Another member also reported briefly seeing Rasch, who, as it turned out, had never left his home. (When Rasch finally returned to Parliament, he became annoyed whenever someone poked him in the ribs to make certain it was really him.)

In 1771, while traveling, Wolfgang von Goethe encountered himself wearing unfamiliar clothes and heading in the opposite direction. Eight years later, Goethe found himself traveling the same road, heading in the same direction as his double had—and wearing the same clothes that before had seemed unfamiliar.

Nineteenth-century French writer Guy de Maupassant once watched as his double sat down near him and dictated what he was writing. Maupassant's doppelgänger experiences became so common that he wrote about them in his short story "Lui."

Twin Teachers

One of the most celebrated cases in doppelgänger lore occurred in 1845 in Latvia. Emilie Sagée was a popular French teacher at a school for upper-class girls. The students often talked about how Sagée seemed to be in two places at once: One student would report seeing her in a hallway, but another would say she had just seen Sagée in a classroom.

One day, while Sagée was writing on the blackboard, her double appeared right beside her, moving its hand in unison with the teacher. Another time, as Sagée helped a young girl dress for a party, the girl looked in the mirror to see two Sagées moving in perfect harmony, working on the dress.

While reports like these and similar incidents involving Sagée didn't seem to distress the students, it freaked out their parents, who began pulling their children out of the school. According to some stories, the headmaster decided that he would have to let Sagée go. When told of this, the teacher reportedly lamented that she had lost nearly two dozen teaching positions throughout her career for the same reason.

Three Sides to Every Story

Few geographical locations on Earth have been discussed and debated more than the three-sided chunk of ocean between the Atlantic coast of Florida and the islands of Bermuda and Puerto Rico known as "the Bermuda Triangle."

* * * *

OVER THE CENTURIES, hundreds of ships and dozens of airplanes have mysteriously disappeared while floating in or flying through the region commonly called "the Bermuda Triangle." Myth-mongers suggest that alien forces are responsible for these dissipations. Because little or no wreckage from the vanished vessels has ever been recovered, paranormal pirating has also been cited as the culprit. Other theorists suggest that leftover technology from the lost continent of Atlantis—mainly an underwater rock formation known as "the Bimini Road" (which is situated just off the island of Bimini in the Bahamas)— exerts a supernatural power that grabs unsuspecting intruders and drags them to the depths.

A Deadly Adjective

Although the theory of the Triangle had been mentioned in publications as early as 1950, it wasn't until the '60s that the region was anointed with its three-sided appellation. In the February 1964 edition of *Argosy* magazine, columnist Vincent Gaddis wrote an article that discussed the various unexplained disappearances that had occurred there over the years and designated the area where myth and mystery mixed as the "Deadly Bermuda Triangle." The use of the word *deadly* suggested the possibility that UFOs, alien anarchists, supernatural beings, and metaphysical monsters reigned over the region. The mystery of Flight 19, which involved the disappearance of five planes in 1945, was first noted in newspaper articles in 1950. But its fame was secured when it was fictitiously featured in Steven Spielberg's 1977 alien opus, *Close Encounters of the Third*

Kind. In Hollywood's view, the pilots and their planes were plucked from the sky by friendly aliens and later returned safely to terra firma by their abductors.

In 1975, historian, pilot, and researcher Lawrence David Kusche published one of the first definitive studies that dismissed many of the Triangle theories. In his book *The Bermuda Triangle Mystery—Solved,* he concluded that the Triangle was a "manufactured mystery," the result of bad research, lazy reporting, and, occasionally, deliberately falsified facts. Before weighing anchor on Kusche's conclusions, however, consider that one of his next major publications was a tome about exotic popcorn recipes.

Explaining Odd Occurrences

Other pragmatists have insisted that a combination of natural forces—a double whammy of waves and rain that created the perfect storm, for example—was most likely the cause of these maritime misfortunes. Other possible "answers" to the mysteries include rogue waves (such as the one that capsized the *Ocean Ranger* oil rig off the coast of Newfoundland in 1982), underwater earthquakes, hurricanes, and human error. Every day, the Coast Guard receives almost 20 distress calls from amateur sailors attempting to navigate the slippery sides of the Triangle. Modern-day piracy—usually among those involved in drug smuggling—has been mentioned as a probable cause for odd occurrences, as have unusual magnetic anomalies that screw up compass readings. Other possible explanations include the Gulf Stream's uncertain current, the high volume of sea and air traffic in the region, and even methane hydrates (gas bubbles) that produce "mud volcanoes" capable of sucking a ship into the depths.

Other dramatic and disastrous disappearances amid the Bermuda Triangle include the USS *Cyclops,* which descended to its watery repository without a whisper in March 1918 with 309 people aboard. Myth suggests supernatural subterfuge, but the reality is that violent storms or enemy action were the likely culprits. The same deductions were discussed and similar conclusions reached in 1812 after the sea schooner *Patriot*, a commercial vessel, was swept up by the sea with the daughter of former vice president Aaron Burr on board.

Flight 19

The incident that cemented the Triangle's notoriety as a map point of the macabre was the disappearance of Flight 19. On December 5, 1945, five TBM Avenger torpedo bombers seemingly dropped off radar screens while on an authorized overwater training flight from the Naval Air Station in Fort Lauderdale, Florida. Lieutenant Charles Carroll Taylor was supervising the flight but was not out front in the lead position. At some point, and for reasons that are unclear, Taylor assumed the lead, only to become confused. Instead of guiding the bombers back to Fort Lauderdale, he ended up flying as far as 200 miles out to sea, east of the Florida peninsula. In his last transmitted message, Taylor said, "We'll have to ditch . . . we all go down together." The five planes and 14 crew members were lost without a trace, and despite numerous missions to recover remnants of the planes throughout the years, no missing links to the incident have ever been found.

Although a number of other aircraft had met similar fates before Flight 19—including four U.S. Navy Lockheed PV-1 Venturas in 1943—the 1945 occurrence caught the attention of conspiracy theorists. Hollywood cited alien abduction as the answer, but scientists and military officials suggest a far simpler solution: Experts believe that Taylor became disorientated during the flight and ran out of fuel while trying to find his way home. Like lemmings led to the ledge, his squadron flew as he flew, eventually joining him in the brine below.

Further sensationalizing the incident was the disappearance of one of the PBM Mariner rescue seaplanes that were sent out to search for Flight 19. As the Mariner scoured the area for traces of the lost flight, the tanker SS *Gaines Mills* reported that the plane exploded in midair. All 13 crew members died.

Mystery Spot

Speculation about Area 51's purpose runs the gamut from a top-secret test range to an alien research center. One thing is certain: The truth is out there somewhere.

✳ ✳ ✳ ✳

LOCATED NEAR THE southern shore of the dry lakebed known as "Groom Lake" is a large military airfield—one of the most secretive places in the country. It is fairly isolated from the outside world, and little official information has ever been published about it. The area is not included on any maps, yet nearby Nevada State Route 375 is listed as "The Extraterrestrial Highway." Although referred to by a variety of names—including "Dreamland," "Paradise Ranch," and "Watertown Strip"—this tract of mysterious land in southern Nevada is most commonly known as "Area 51."

Conspiracy theorists and UFO aficionados speculate that Area 51 is everything from the storage location of the rumored crashed Roswell, New Mexico, spacecraft to a secret lab where experiments are conducted on matter transportation and time travel.

The truth is probably far less fantastic and far more scientific. Used as a bomb range during World War II, the site was abandoned as a military base at the end of the war. So much for being home to that alien spacecraft from Roswell (it "crashed" in 1947). The land wasn't used again until 1955, when it became a test range for the Lockheed U-2 spy plane and, later, the USAF SR-71 Blackbird.

Whether Area 51 was ever used to house UFOs isn't known for certain, but experts believe that it was probably a test and study center for captured Soviet aircraft during the Cold War. In 2003, the federal government actually admitted that the facility exists as an Air Force "operating location," but no further information was released.

Famous UFO Sightings

Unidentified flying objects, foo fighters, ghost rockets—whatever you call them, strange and unclassified objects in the sky remain one of the world's truly mysterious phenomena. Here are some of the most famous UFO sightings.

✳ ✳ ✳ ✳

The Battle of Los Angeles

On FEBRUARY 25, 1942, just weeks after Japan's attack on Pearl Harbor and America's entry into World War II, late-night air-raid sirens sounded a blackout order throughout Los Angeles County in California. A silvery object (or objects) was spotted in the sky, prompting an all-out assault from ground troops. For a solid hour, antiaircraft fire bombarded the unidentified craft with some 1,400 shells, as numerous high-powered searchlights followed its slow movement across the sky. Several witnesses reported direct hits on the invader, though it was never downed. After the "all clear" was sounded, the object vanished, and it has never been identified.

The Washington Flap

In two separate incidents that occurred just days apart in 1952, numerous objects were detected high above Washington, D.C., moving erratically at speeds as fast as 7,000 miles per hour. At one point, separate military radar stations detected the same objects simultaneously. Several eyewitnesses viewed the objects from the ground and from air-control towers, and three pilots spotted them at close range, saying that they looked like the lit end of a cigarette or like falling stars without tails. The official Air Force explanation was "temperature inversion," and the sightings were labeled "unexplained."

The Marfa Lights

The town of Marfa, which is located in western Texas, is home to what many believe is the best concentration of "ghost lights" in the nation. Almost nightly, witnesses along Highway 67 can peer across the flatland north of the Chinati Mountains and spot glowing orbs of various colors and sizes, bobbing and floating among the brush. It's an event that's reportedly been witnessed since the 1880s. Though several scientists have conducted studies on the strange lights, no one has been able to determine their origin. Nevertheless, local officials have capitalized on the phenomenon and constructed an official roadside viewing area.

The Hill Abduction, aka the Zeta Reticuli Incident

By the 1960s, a number of people had reportedly seen UFOs but hadn't actually encountered aliens personally. But on September 19, 1961, Barney and Betty Hill found themselves being chased by a spacecraft along Route 3 in New Hampshire. The object eventually descended upon their vehicle, at which time Barney witnessed several humanoid creatures through the craft's windows. The couple tried to escape, but their car began shaking violently and they were forced off the road. Suffering lapses in memory from that moment on, the Hills later recalled being taken aboard the ship, examined, and questioned by figures with very large eyes. The incident was known only to locals and the UFO community until the 1966 publication of *The Interrupted Journey* by John Fuller.

Fire in the Sky

After completing a job along Arizona's Mogollon Rim on November 5, 1979, Travis Walton and six fellow loggers spotted a large spacecraft hovering near the dark forest road leading home. Walton approached the craft on foot and was knocked to the ground by a beam of light. Then he and the craft disappeared. Five days later, Walton mysteriously reappeared just outside of town. He said that during his time aboard the spacecraft, he had struggled to escape from the short, large-headed creatures that performed experiments on his body. Neither Walton nor any of his coworkers has strayed from the details of their story since it occurred.

The *Apollo 11* Transmission

When American astronauts made that giant leap onto the surface of the moon on July 20, 1969, they apparently weren't alone. Although the incident has been repeatedly denied, believers point to a transmission from the lunar surface that was censored by NASA but was reportedly picked up by private ham-radio operators: "These babies are huge, sir! Enormous! . . . You wouldn't believe it. I'm telling you there are other spacecraft out there, lined up on the far side of the crater edge. They're on the moon watching us!"

The Rendlesham Forest Incident

In late December 1980, several soldiers at the Royal Air Force base in Woodbridge, Suffolk, England, saw a number of strange lights among the trees just outside the east gate. When they went to investigate, they spotted a conical or disk-shaped object hovering above a clearing. The object seemed aware of their presence and moved away from them, but the men eventually gave chase. No hard evidence has been provided by the military, but the event is often considered the most significant UFO event in Britain. The country's Forestry Commission has since created a "UFO Trail" for hikers near the RAF base.

JAL 1628

On November 17, 1986, as Japan Airlines Flight 1628 passed over Alaska, military radar detected an object on its tail. When the blip caught up with the cargo jet, the pilot reported seeing three large craft shaped like shelled walnuts, one of which was twice the size of an aircraft carrier. The objects matched the airplane's speed and tracked it for nearly an hour. At one point, the two smaller craft came so close to the JAL jet that the pilot said he could feel their heat. The incident prompted an official FAA investigation and made worldwide headlines.

The Phoenix Lights

In March 1997, hundreds—if not thousands—of witnesses throughout Phoenix, Arizona, and the surrounding area caught sight of what was to become the most controversial UFO sighting in decades. For at least two hours, Arizona residents watched an array of lights move across the sky, and many reportedly saw a dark, triangular object between them. The lights, which varied in color, were even caught on videotape. Nearby military personnel tried to reproduce the event by dropping flares from the sky, but most witnesses weren't satisfied with what was deemed a diversion from the truth.

Roswell

Undoubtedly the most famous UFO-related location, the name Roswell immediately brings to mind flying-saucer debris, men in black, secret military programs, alien autopsies, weather balloons, and government cover-ups. The incident that started it all occurred during the first week of July 1947, just before Roswell Army Air Field spokespeople claimed that they had recovered parts of a wrecked "flying disc" from a nearby ranch. The report was quickly corrected to involve a weather balloon instead, which many insist was part of a cover-up. In subsequent years, people claiming to have been involved in the recovery effort began to reveal insider information, insisting that not only was the wreckage of extraterrestrial origin but also that autopsies had been performed on alien bodies recovered from the site. Ever since, the name of this small New Mexico town has been synonymous with UFOlogy, making Roswell a popular stop for anyone interested in all things alien.

The Great Texas Airship Mystery

Roswell may be the most famous alleged UFO crash site, but did Texas experience a similar event in the 19th century?

✳ ✳ ✳ ✳

ONE APRIL MORNING in 1897—six years before the Wright Brothers' first flight and 50 years before Roswell—a huge, cigar-shaped UFO was seen in the skies above Texas. It was first spotted on November 17, 1896, about a thousand feet above rooftops in Sacramento, California. From there, the craft traveled to San Francisco, where hundreds of people saw it.

A National Tour

Next, the object crossed the country, where it was observed by thousands. Near Omaha, Nebraska, a farmer reported that the ship was on the ground. When it returned to the skies, it headed toward Chicago, where it was photographed on April 11, 1897, making this the earliest UFO photo on record. On April 15, near Kalamazoo, Michigan, residents reported loud noises coming from the craft. Two days later, the UFO attempted a landing in Aurora, Texas, which was nearly deserted, so its broad, empty fields would have been an ideal landing strip.

No Smooth Sailing

However, at about 6 A.M. on April 17, the huge airship "sailed over the public square and ... collided with ... Judge Proctor's windmill and went to pieces with a terrific explosion, scattering debris over several acres of ground, wrecking the windmill and water tank and destroying the judge's flower garden."

That's how Aurora resident S. E. Haydon described the events for *The Dallas Morning News*. The remains of the ship seemed to be strips and shards of a silver-colored metal. Just one body was recovered. The newspaper reported, "while his remains are badly disfigured, enough of the original has been picked up to show that he was not an inhabitant of this world."

On April 18, reportedly, that body was given a Christian burial in the Aurora cemetery, where it may remain to this day. A 1973 effort to exhume the body and examine it was blocked by the Aurora Cemetery Association.

A Firsthand Account

Although many people have claimed that the Aurora incident was a hoax, an elderly woman who was interviewed in 1973 clearly recalled the crash from her childhood. She said that her parents wouldn't let her near the debris from the craft. However, she described the alien as "a small man."

Aurora continues to attract people who are interested in UFOs. Nearby Fort Worth may be home to U.S. government experts in alien technology. Immediately after the Roswell UFO crash in 1947, debris from that spaceship was sent to Fort Worth for analysis.

Is There Any Trace Left?

The Aurora Encounter, a 1986 movie, documents the events that began when people saw the craft attempt a landing on the Proctor farm. Metal debris was collected from the site in the 1970s and studied at North Texas State University. One fragment appeared to be iron but wasn't magnetic; it was shiny and malleable rather than brittle, as iron should be.

As recently as January 2008, UFOs have appeared in the north central Texas skies. In Stephenville, a pilot described seeing a low-flying object "a mile long and half a mile wide." Others who saw the ship said that its lights changed configuration, so it wasn't an airplane. The government declined to comment.

Today, a plaque at the Aurora cemetery mentions the spaceship, but the alien's tombstone—if it ever actually existed—was stolen many years ago.

The Philadelphia Experiment

In 1943, the Navy destroyer USS Eldridge *reportedly vanished, teleported from a dock in Pennsylvania to one in Virginia, and then rematerialized—all as part of a top-secret military experiment. Is there any fact to this legend?*

✳ ✳ ✳ ✳

The Genesis of a Myth

THE STORY OF the Philadelphia Experiment began with the scribbled annotations of a crazed genius—Carlos Allende, who in 1956 read *The Case for the UFO* by science enthusiast Morris K. Jessup. Allende wrote chaotic annotations in his copy of the book, claiming, among other things, to know the answers to all the scientific and mathematical questions that Jessup's book raises. Jessup's interests included the possible military applications of electromagnetism, antigravity, and Einstein's Unified Field Theory.

Allende wrote two letters to Jessup, warning him that the government had already put Einstein's ideas to dangerous use. According to Allende, at some unspecified date in October 1943, he was serving aboard a merchant ship when he witnessed a disturbing naval experiment: The USS *Eldridge* disappeared, teleported from Philadelphia, Pennsylvania, to Norfolk, Virginia, and then reappeared in a matter of minutes. The men onboard the ship allegedly phased in and out of visibility or lost their minds and jumped overboard, and a few of them disappeared forever. This strange activity was part of an apparently successful military experiment to render ships invisible.

The Navy Gets Involved

Allende could not provide Jessup with any evidence for these claims, so Jessup stopped the correspondence. But in 1956, Jessup was summoned to Washington, D.C., by the Office of Naval Research, which had received Allende's annotated copy of Jessup's book and wanted to know about Allende's claims and his written comments. Shortly thereafter, Varo Corporation—a private group that does research for the military—published the annotated book, along with the letters Allende had sent to Jessup. The Navy has consistently denied Allende's claims about teleporting ships, and Varo's impetus for publishing Allende's annotations is unclear. Jessup committed suicide in 1959, leading some conspiracy theorists to claim that the government had him killed for knowing too much about the experiments.

The Facts Within the Fiction

It is not certain when Allende's story was deemed the "Philadelphia Experiment," but over time, sensationalist books and movies have touted it as such. The date of the ship's disappearance is usually cited as October 28, though Allende himself could not verify the date. However, the inspiration behind Allende's claims is not a complete mystery.

In 1943, the Navy was indeed conducting experiments, some of which were surely top secret, and sometimes they involved research into the applications of some of Einstein's theories. The Navy had no idea how to make ships invisible, but it *did* want to make ships undetectable to enemy magnetic torpedoes. Experiments such as these involved wrapping large cables around Navy vessels and pumping them with electricity in order to scramble their magnetic signatures.

Questionable Witness

According to Edward Dudgeon, a crew member on the USS *Engstrom*, invisibility experiments of the questionable kind took place in Philadelphia in August 1943 on both the USS *Eldridge* and the *Engstrom*. Dudgeon, claims that after the ships had vanished and rematerialized, some of the sailors themselves mysteriously disappeared from a local bar. However, further investigation revealed that the missing crew members had merely slipped out the back door of the bar to avoid punishment for underage drinking. And the *Eldridge* went to Norfolk that night to pick up ammunition but was back in Philly by morning.

The *Really* Secret Service

Backroom swamis may have given fortune-telling a bad name, but powerful people still seek ways to consult with dead relatives and discover future happenings.

✳ ✳ ✳ ✳

Spiritual Speed Dating

JUST WHEN YOU think you know somebody: William Lyon Mackenzie King was Canada's prime minister for 22 years, but it wasn't until after his death in 1950 that his interest in the supernatural was revealed. With the help of mediums, King allegedly was able to contact the spirits of Leonardo da Vinci, Florence Nightingale, Robert Louis Stevenson, Anne Boleyn, Queen Victoria, his favorite dog, and, of course, his late mother. King was hooked. After starting with direct-voice mediums, he moved on to table-tapping, automatic writing, numerology, and tea-leaf reading. King eventually gave up on mediums because he came to believe that he was a psychic himself.

The Doubting Dutch

Since the beginning of time, humans have been searching for ways to foretell what that tomorrow will bring. But as long as there have been seers, there have also been skeptics and scoffers who are quick to jump on the miscues and ambitions of people with possible extrasensory powers. When you're a high-profile head of state, discretion is key; that's where Queen Juliana of the Netherlands erred.

In 1956, skeptics were delighted when the queen was forced to sever ties with her faith healer, Greet Hofmans. The fear was that Hofmans was meddling in Dutch foreign policy and had far too much influence on the queen.

Stars over the Rose Garden

American presidents have also sought the wisdom of psychics and astrologers, though it must be said that their First Ladies were often the ones to spearhead this interest. Among the believers were Mary Todd Lincoln, Florence Harding, and Nancy Reagan.

Nancy's involvement with astrologers such as Joan Quigley and Jeane Dixon are well documented in tell-all books, and Ronald admitted that he was superstitious and regularly read newspaper astrology columns. In fact, as California's governor, he signed legislation that removed astrologers from the state criminal code, thus making them more legitimate. These beliefs made the Reagans the butt of many jokes. In 1976, psychic Jeane Dixon missed a Reagan prediction, so Nancy fired her, showing that psychics, like musicians, are only as good as their last gig. Dixon was replaced by Quigley, who would boldly brag that she'd helped end the Cold War. Maybe yes, maybe no, but she certainly sold a lot of books.

While Calvin Coolidge, Franklin D. Roosevelt, and Richard Nixon were also rumored to be pro-astrology, Teddy Roosevelt let it all hang out when he signed on as a founding member of the American Society for Psychic Research.

Corporate Due Diligence

The ancient Romans would not go into battle without approval from their College of Augurs. The Greeks relied on the Oracle of Delphi. In our hip, sophisticated age, it seems that some high-flying moguls won't cut their mega-million dollar corporate deals without a thumbs-up from the "intuitionists" or "mentalists" that the companies keep on retainer.

One of these intuitionists is Laura Day, a New York City mother who averages five clients monthly—typically corporate businesses, law firms, and entertainment honchos. Her going rate is $10,000 a month per client. Granted, Day has to be on call around the clock, and she isn't paid overtime, but every job has a downside. It seems that the business of fortune-telling has come a long way from sideshow crystal ball readings.

Alleged Celebrity UFO Sightings

It's not just moonshine-swilling farmers in rural areas who claim to have seen UFOs hovering in the night sky. Plenty of celebrities have also reportedly witnessed unidentified flying objects—and have been happy to talk about their experiences afterward.

✳ ✳ ✳ ✳

Jimmy Carter

Not even presidents are immune to UFO sightings. During Jimmy Carter's presidential campaign of 1976, he told reporters that in 1969, before he was governor of Georgia, he saw what could have been a UFO. "It was the darndest thing I've ever seen," he said of the incident. He claimed that the object that he and a group of others watched for ten minutes was as bright as the moon. Because he filed a report on the matter, Carter was often referred to as "the UFO president" after being elected.

Jackie Gleason

Actor and comedian Jackie Gleason was best known for his work on *The Honeymooners*. He was also supposedly a paranormal enthusiast who claimed to have witnessed several unidentified objects flying in the sky. Gleason's second wife, Beverly, claimed that in 1974 President Nixon took Gleason to the Homestead Air Force Base in Florida, where he saw the wreckage of a crashed spaceship and the bodies of dead aliens. The incident had such a profound affect on him that he gave up drinking alcohol, at least for a while. Gleason was so inspired by the visit that he later built a house in upstate New York that was designed to look like a spaceship and was called "The Mother Ship."

John Lennon

In 1974, former Beatle John Lennon claimed that he witnessed a flying saucer from the balcony of his apartment in New York City. Lennon and his girlfriend May Pang later described the craft as circular with white lights around its rim; they said that it hovered in the sky above their window. Lennon talked about the event frequently and even referenced it in his songs "Out of the Blue" and "Nobody Told Me," which contains the lyric, "There's UFOs over New York and I ain't too surprised ..."

Ronald Reagan

Former actor and U.S. president Ronald Reagan witnessed UFOs on two occasions. Once during his term as California governor (1967–1975), Reagan and his wife Nancy arrived late to a party because they witnessed a UFO while driving along the Pacific Coast Highway and had stopped to watch the event.

Reagan also confessed to a *Wall Street Journal* reporter that in 1974, when the gubernatorial jet was preparing to land in Bakersfield, California, he noticed a strange bright light in the sky. The pilot followed the light for a short time before it suddenly shot up vertically at a high rate of speed and disappeared from sight. Reagan stopped short of labeling the light a UFO, of course. As actress Lucille Ball said in reference to Reagan's first alleged UFO sighting, "After he was elected president, I kept thinking about that event and wondered if he still would have won if he told everyone that he saw a flying saucer."

William Shatner

For decades, the man who played Captain Kirk in the original *Star Trek* series claimed that an alien had saved his life. When the actor and a group of friends were riding their motorbikes through the desert in the late 1960s, Shatner was inadvertently left behind when his bike wouldn't restart after he drove it into a giant pothole. Shatner said that he spotted an alien in a silver suit standing on a ridge and that it led him to a gas station and safety. Shatner later stated in his autobiography, *Up Till Now*, that he made up the part about the alien during a television interview.

Ohio's Mysterious Hangar 18

An otherwordly legend makes its way from New Mexico to Ohio when the wreckage from Roswell ends up in the Midwest.

✳ ✳ ✳ ✳

EVEN THOSE WHO aren't UFO buffs have probably heard about the infamous Roswell Incident. (See page 357.) But what most people don't know is that allegedly, the mysterious aircraft was recovered (along with some alien bodies) and stored just outside of Dayton, Ohio.

✳ ✳ ✳ ✳

Something Crashed in the Desert

While the exact date is unclear, sometime during the first week of July 1947, a local Roswell rancher named Mac Brazel went to check his property for fallen trees and other damage after a night of heavy storms and lightning. Brazel allegedly came across an area of his property that was littered with strange debris unlike anything he had ever seen before. Some of the debris even had strange writing on it.

Brazel showed the debris to a few neighbors and then took it to the office of Roswell sheriff George Wilcox, who called authorities at Roswell Army Air Field. After speaking with Wilcox, intelligence officer Major Jesse Marcel drove out to the Brazel ranch and collected as much debris as he could. He then returned to the airfield and showed the debris to his commanding officer, Colonel William Blanchard. Upon seeing the debris, Blanchard dispatched military vehicles and personnel to the Brazel ranch to see if they could recover anything else.

"Flying Saucer Captured!"

On July 8, 1947, Colonel Blanchard issued a press release stating that the wreckage of a "crashed disk" had been recovered. The bold headline of the July 8 edition of the *Roswell Daily Record* read: "RAAF Captures Flying Saucer on Ranch in Roswell

Region." Newspapers around the world ran similar headlines. But then, just hours after the press release was issued, General Roger M. Ramey—commander of the Eighth Air Force in Fort Worth, Texas—retracted Blanchard's release for him and issued another statement saying that there was no UFO and that Blanchard's men had simply recovered a fallen weather balloon.

Before long, the headlines that had earlier touted the capture of a UFO read, "It's a Weather Balloon" and "'Flying Disc' Turns Up as Just Hot Air." Later editions even ran a staged photo of Major Marcel, who was first sent to investigate the incident, kneeling in front of weather-balloon debris. Most of the general public seemed content with the explanation, but there were skeptics.

Whisked Away to Hangar 18?

Those who believe that aliens crash-landed near Roswell claim that large portions of the alien spacecraft were brought out to the Roswell Air Field under cover of darkness and loaded onto aircrafts. Those planes were then supposedly flown to Wright-Patterson Air Force Base, which is located just outside of Dayton. After the planes landed, they were taken to Hangar 18 and unloaded. And according to legend, their cargo is still there.

There are some problems with the story, though. For one, none of the hangars at Wright-Patterson Air Force Base are officially known as "Hangar 18," and there are no buildings designated with the number 18. Rather, the hangars are labeled 1A, 1B, 1C, and so on. There's also the fact that none of the hangars seem large enough to house an alien spacecraft. But just because there's nothing listed as Hangar 18 on a Wright-Patterson map doesn't mean it's not there. Conspiracy theorists believe that hangars 4A, 4B, and 4C might be the infamous Hangar 18. As for the size of the hangars, it's believed that most of the wreckage has been stored in giant underground tunnels and chambers beneath the hangar, both to protect the debris and to keep it safe from prying eyes. It is said that Wright-Patterson is currently conducting experiments on the wreckage to see if scientists can reverse-engineer the technology.

So What's the Deal?

The story of Hangar 18 has only got stranger as the years have gone by, thanks in part to the government's Project Blue Book, a program designed to investigate reported UFO sightings across the United States. Between 1947 and 1969, Project Blue Book investigated more than 12,000 UFO sightings before being disbanded. And where was Project Blue Book headquartered? Wright-Patterson Air Force Base, naturally.

Then in the early 1960s, Arizona senator Barry Goldwater—himself a retired major general in the U.S. Army Air Corps (and a friend of Colonel Blanchard)—became interested in what, if anything, had crashed in Roswell. When Goldwater heard about Hangar 18, he wrote directly to Wright-Patterson and asked for permission to tour the facility but was quickly denied. He then approached another friend, General Curtis LeMay, and asked if he could see the "Green Room" where the secret UFO was being held. Goldwater claimed that LeMay gave him "holy hell" and screamed at him, "Not only can't you get into it, but don't you ever mention it to me again."

In 1982, retired pilot Oliver "Pappy" Henderson attended an Air Force reunion and announced that he was one of the men who had flown alien bodies out of Roswell in a C-54 cargo plane. His destination? Hangar 18 at Wright-Patterson. Although no one is close to a definitive answer about the place, it seems that the legend of Hangar 18 will never die.

"I certainly believe in aliens in space. They may not look like us, but I have very strong feelings that they have advanced beyond our mental capabilities.... I think some highly secret government UFO investigations are going on that we don't know about—and probably never will unless the Air Force discloses them."

—BARRY GOLDWATER, FORMER U.S. SENATOR

Chapter 13

History

Ms. President

When Ohio native Victoria Woodhull ran for president in 1872, some called her a witch, while others said she was a prostitute. In fact, the very idea of a woman even casting a vote for president was considered scandalous—which may explain why Woodhull spent election night in jail.

✳　✳　✳　✳

KNOWN FOR HER passionate speeches and fearless attitude, Victoria Woodhull was a trailblazer for women's rights. Woodhull advocated revolutionary ideas, including gender equality and women's right to vote. "Women are the equals of men before the law and are equal in all their rights," she said. America, however, wasn't ready to accept such radical ideas.

Early Independence

She was born Victoria Claflin in 1838 in Homer, Ohio. Young Victoria only attended school sporadically and was primarily self-educated.

When she was 15 years old, a 28-year-old doctor named Canning Woodhull asked for her hand in marriage. But the marriage was no paradise, and she soon realized that her husband was an alcoholic. She experienced more heartbreak when her son, Byron, was born with a mental disability. Victoria spent the next few years touring as a clairvoyant with her sister Tennessee; she finally succeeded in divorcing her husband in 1864. Two years later, she married Colonel James Blood, a Civil War veteran who believed in free love.

In 1866, Woodhull and Blood moved to New York City. Spiritualism was in vogue, and Woodhull and her sister worked as clairvoyants and discussed social and political hypocrisies with their clientele. Among their first customers was Cornelius Vanderbilt, the wealthiest man in America. He advised them on business matters and gave them stock tips. When the stock market crashed in September 1869, Woodhull made a bundle by buying instead of selling during the ensuing panic. That winter, she and her sister opened their own brokerage business. They were the first female stockbrokers in American history, and they did so well that two years after arriving in New York, Woodhull told a newspaper she had made $700,000.

Bigger Ambitions

Woodhull had more far-reaching ambitions, however. On April 2, 1870, she announced that she was running for president. In conjunction with her presidential bid, Woodhull and her sister started a newspaper, *Woodhull & Claflin's Weekly*, which highlighted women's issues, including voting and labor rights. It was another breakthrough for the two—they were the first women to publish a weekly newspaper.

On January 11, 1871, Woodhull also became the first woman to speak before a congressional committee. As she spoke before the House Judiciary Committee, she asked that Congress change its stance on women's suffrage. Woodhull's reasoning was elegant in its simplicity: She was not advocating a new constitutional amendment granting woman the right to vote; instead, she reasoned, women already had that right. The Fourteenth Amendment says, "All persons born or naturalized in the United States . . . are citizens of the Unites States." Since voting is part of the definition of being a citizen, Woodhull argued, women, in fact, already possessed the right to vote.

Woodhull, a persuasive speaker, actually swayed some congressmen to her point of view, but the committee chairman remained hostile to the idea of women's rights and made sure that the issue never came to a floor vote.

Her Opponents Could Dish It Out, But . . .

Woodhull had better luck with the suffragists. In May 1872, before 668 delegates, Woodhull was chosen as the presidential candidate of the Equal Rights Party; she was the first woman ever chosen by a political party to run for president. But her presidential bid soon foundered. Woodhull was on record as being an advocate of free love, which opponents argued was an attack on the institution of marriage. (For Woodhull, it had more to do with the right to have a relationship with anyone she wanted.) Rather than debate her publicly, her opponents made personal attacks.

That year, Woodhull caused an uproar when her newspaper ran an exposé about the infidelities of one of her detractors, Henry Ward Beecher. Woodhull and her sister were thrown in jail and accused of publishing libel and promoting obscenity. They would spend election night of 1872 behind bars as incumbent Ulysses Grant defeated Horace Greeley for the presidency.

One Last Milestone

Woodhull was eventually cleared of the charges against her (the claims against Beecher were proven true), but hefty legal bills and a downturn in the stock market left her embittered and impoverished. She moved to England in 1877, shortly after divorcing Colonel Blood. By the turn of the century, she had become wealthy once more, this time through marriage to a British banker. Fascinated by technology, she joined the Ladies Automobile Club, where her passion for automobiles led Woodhull to one last milestone: In her 60s, she and her daughter Zula became the first women to drive through the English countryside.

The *Hindenburg*

*On May 6, 1937, German zeppelin LZ-129—immortalized in
history as the* Hindenburg—*caught fire at Lakehurst Naval Air
Station, New Jersey. More than 30 people died in the incident.*

✳   ✳   ✳   ✳

TODAY, A HYDROGEN gasbag sounds like a scary way to fly. In
the 1930s, however, it was expensive but swift, part of the
great novelty of passenger air travel. In 1936, the *Hindenburg*
made ten round-trip flights to New York and seven to Brazil. It
could cross the Atlantic in two days for $400 one-way (which
would be roughly $6,000 today). The fastest ocean liners took
approximately four days to make the same trip.

The paying customer enjoyed opulent comfort, fine dining, and
even a cocktail lounge. A 200-foot promenade deck featured
breathtaking aerial views. Barf bags were unnecessary; no one got
airsick because they could feel very little movement.

The *Hindenburg* was the largest airship ever made—an immense
symbol in a political culture where physical size was vitally
important. It flew over the 1936 Berlin Olympics, Nazi swasti-
kas on its fins, trailing the Olympic flag from its gondola. The
Hindenburg served as a transatlantic billboard for the resurgent
Germanic pride embodied in Nazism. In those tense prewar
years, the airship's Lakehurst arrival was a major media event.

Werner Franz was a 14-year-old cabin boy working on the
Hindenburg. A water tank above him ruptured at just the right
time to soak him, thus helping him resist the fire. Passenger
Werner Doehner was just eight when
the great airship caught fire. His
mother (also a survivor) threw him out
the window. He later learned his father
and sister had perished. He rarely spoke
about the disaster for most of his life.

Fast Facts

✱ "Some Assembly Required": It took an estimated 30 million hours for Stone Age builders to complete the rock structure at Stonehenge, England.

✱ Of the Seven Wonders of the Ancient World, only one has survived the test of time: the two-million-block Great Pyramid of Giza.

✱ The Great Wall of China, at 3,977 miles, is the world's longest artificial structure. If you broke it up for materials, you could build 120 Great Pyramids.

✱ The Greeks had a working steam engine in the first century B.C.—about 1,700 years before an English engineer patented one. To the ancients, the steam engine was just an amusing toy.

✱ In 1931, the immense Boulder Dam on America's Colorado River was renamed the "Hoover Dam" after the incumbent Republican president, Herbert Hoover. When Democrat Franklin Roosevelt defeated Hoover the next year, Congress changed the name back to "Boulder Dam." After Roosevelt died, Congress changed the name back to "Hoover Dam."

✱ Constructing a tunnel under the English Channel has been discussed since the time of Napoleon. A brief attempt was made in the 1870s, but the British government refused to allow a tunnel that could give French soldiers a route to invade England. The "chunnel" finally opened in 1994.

✱ The first calendar was invented by the Egyptians more than 5,000 years ago to let everybody know when the Nile River was scheduled to flood its banks.

✱ The first bomb dropped by the Allies on the city of Berlin during World War II claimed an unusual casualty—the only elephant at the Berlin Zoo.

A*head* of His Time

It's long been claimed that Dr. Joseph-Ignace Guillotin, the presumed creator of the guillotine, was put to death during the French Revolution by the decapitating contraption that bears his name. It would be the ultimate irony—if only the story were true.

✳ ✳ ✳ ✳

B EFORE WE TAKE a closer look at this long-lived myth, we should probably clear up a larger misconception: Joseph Guillotin did not invent the guillotine. Mechanical beheading devices had long been used in Germany, Italy, Scotland, and elsewhere, though it was the French who made them (in-)famous.

The Good Doctor

Guillotin, a respected physician and member of the French National Assembly, opposed the death penalty. However, he realized that public executions weren't about to go out of style anytime soon, so he sought a more "humane" alternative to being drawn and quartered, which was the usual way that impoverished criminals were put to death.

A quick beheading, Guillotin argued, was far more merciful than being hacked apart by a dull ax. And it had the added benefit of making executions socially equal, since beheading had been, until then, a method of execution reserved for aristocratic convicts who could buy themselves quicker, kinder deaths.

Guillotin hooked up with German engineer and harpsichord maker Tobias Schmidt, who built a prototype of the guillotine as we know it today. For a smoother cut, Schmidt suggested a diagonal blade instead of the traditional round blade.

Heads Will Roll

The guillotine's heyday was during the French Revolution in 1789 and afterward. After King Louis XVI was imprisoned, the new civilian assembly rewrote the penal code to make beheading by guillotine the official method of execution for all convicted criminals—and there were a lot of them.

The first person to lose his head was Nicolas Jacques Pelletie, who was guillotined at Place de Greve on April 25, 1792. King Louis XVI felt the blade a year later, and thousands more followed. The last person to be publicly guillotined was convicted murderer Hamida Djandoubi, who died on September 10, 1977, in Marseilles.

Joseph Guillotin survived the French Revolution with his head attached, though he was forever stigmatized by his connection with the notorious killing machine. He died in 1814 from an infected carbuncle on his shoulder, and his children, not wanting to be associated with their father's grisly past, later petitioned the French government for the right to change their last name.

A common belief often associated with the guillotine is that people who are beheaded remain conscious for several agonizing seconds—and even respond to stimuli. Whether or not this is true remains open to debate. Many scientists believe that death by beheading is almost instantaneous, while others cite anecdotal evidence that suggests the deceased are well aware of what has happened to them.

Indeed, stories abound of "experiments" during the height of the guillotine boom in which doctors and others made agreements with condemned prisoners to determine once and for all if the head lived on for moments after being severed.

Most doctors agree that a brain may remain active for as long as 15 seconds after a beheading. Whether the individual is actually aware of what has transpired remains a medical mystery that likely will never be solved.

The Triangle Shirtwaist Fire

A horrific fire brought attention to the lives of sweatshop workers in America—and resulted in labor reform.

✳   ✳   ✳   ✳

O**N A** SATURDAY afternoon in March 1911, workers at the Triangle Shirtwaist Factory in New York City were getting ready to go home after a long day. They were tidying up their work spaces and brushing fabric scraps off the tables and into large bins. Someone on the eighth floor carelessly threw a match or cigarette butt into one of those bins, and within minutes, flames overtook the factory floors.

Panicked workers—most of them female immigrants—rushed to evacuate, and many on the ninth floor became trapped. There were two exit doors on that floor, but one was blocked by fire and the other was locked—a precaution that owners deemed necessary to prevent thefts by workers. The terrified laborers were faced with two choices: wait for rescue (and likely die in the fire) or jump from the windows. Many chose to jump. Overall, 146 workers died in the tragedy.

Austrian immigrant Rose Rosenfeld survived by figuring out how the executives were handling the situation. Seventeen-year-old Rosenfeld hopped a freight elevator to the roof, where she was rescued by firefighters. When her bosses tried to bribe her to testify that the doors hadn't been locked, Rosenfeld refused. The tragedy brought about an investigation of the welfare and safety of sweatshop workers, which resulted in new labor laws.

Rosenfeld's anger at the needless death lasted a very long lifetime. She promoted workplace safety reform by retelling her story. She died on February 15, 2001, at age 107—the last survivor of the Triangle Shirtwaist fire.

A Fair to Remember

When the gates of the World's Columbian Exposition opened in 1893, the fate of Chicago was forever changed.

✳ ✳ ✳ ✳

Long before television and the Internet, expositions provided a way to galvanize resources and people—and make money, of course. When it was announced that the House of Representatives was accepting petitions from cities to hold America's world's fair to commemorate the 400th anniversary of Columbus' voyage to the New World, competition was fierce.

New York, St. Louis, and Washington, D.C., all clamored for the honor of hosting the fair, but none fought harder than Chicago, which at that time was a city of immigrants and class struggle, known for manufacturing and not much else. City leaders believed that if they could nab the fair, they could change the way America and the world saw the city they loved.

On February 24, 1890, after months of deliberation, the House of Representatives gave Chicago its wish—as long as it raised another $5 million. The city's elite forked over the cash, and the fair was Chicago's.

Boss Burnham & Crew

After Chicago was given the green light, there was a lot to do. A governing body was put in place to oversee what would be officially called the World's Columbian Exposition. The fair would be held on 600 acres of land, and heading up the architectural plans was Burnham & Root, a firm that had helped rebuild Chicago after a devastating fire less than 20 years prior. Daniel Burnham was not only an architect, he was a crackerjack businessman whose input in the planning stages of the fair shaped the entire event. Burnham wisely chose Frederick Law Olmsted to help create the fairground environment; Olmsted was the best landscape architect money could buy. Artist Francis Millet supervised all painting, and aspects of the design process related to sculpture were handled by Augustus Saint-Gaudens.

Burnham was a fan of the Beaux Arts style of architecture and so the fair, he decided, would have a decidedly classical look and feel. Initially, the buildings were to be built near Michigan Avenue, smack-dab on the most primo lakefront area of the city, but it was vetoed by the millionaires there. The fair was moved south to Jackson Park.

Within a year, 40,000 skilled laborers were hard at work. The layout was simple: The Court of Honor would make up the center of the fairgrounds and would include such structures as the Electricity Building and the Palace of Fine Arts. Olmsted used the natural geography of the area to influence his plans: Waterways threaded through the park and emptied into Lake Michigan. Sculpture abounded, including a gilded statue of the Republic, located in the Grand Basin at the center of the Court of Honor.

At a time when the world was dingy with factory soot, the grandeur of these buildings and waterways defied comprehension. Burnham's enormous structures were covered with gleaming white stucco, which gave them a regal, heavenly quality. Photos showing the construction of these buildings made their way to the newspapers of the world, and anticipation of what was to become the world's most extraordinary exposition intensified.

If You Build It . . . Everyone Will Come

The gates to the world's fair opened in May 1893, and in walked more than 100,000 people on the first day alone. When fairgoers got to Chicago, they were presented with 65,000 exhibits, amusement rides, and parks and boulevards. An elevated railway system shuttled people around, and electric boats ferried folks across the many waterways. Hundreds of "Columbian Guards" and undercover detectives patrolled the grounds, while the cleanup crew kept the fair neat and tidy. Merchants numbered in the hundreds, selling such newfangled treats as hamburgers, bubbly drinks (commonly known as "soda" these days), and Cracker Jack. Scott Joplin performed a new kind of music called "ragtime," and a group of hula dancers introduced their style of expression. Visitors who rode the first Ferris wheel marveled at the size and scope of the fair.

Those who attended the world's fair took home stories of invention and innovation. Word spread that Chicago was "the White City," named for its gleaming buildings and the illumination of gas and electric lights. Industry in Chicago soared, and many saw the city as a thriving center of commerce.

Closing Time

The fair closed its gates in October 1893. What became of the buildings and exhibits? Many of the artifacts filtered out of the city. The hugely popular Ferris wheel spent some time in a North Side neighborhood, but it was eventually moved to St. Louis for the 1904 world's fair.

As for the buildings, most of them were razed. The majority of the structures weren't built to last, after all; only the Palace of Fine Arts remains. After the fair ended, it housed the Field Columbian Museum (known today as the Field Museum of Natural History) until 1920, and then in 1933, the building reopened as the Museum of Science and Industry.

Even though the majority of the buildings were destroyed, the layout of the fair influenced public works projects in America throughout the decades to come. The National Mall and countless college campuses took cues from the architecture and landscaping exhibited at the Columbian Exposition.

Indeed, Chicago's daily hustle and bustle proves that the legacy of the world's fair lives on. It put the city on the map and brought a world in transition into the 20th century with style, innovation, and hard work. Now *that's* the Chicago way.

Captain Bligh and the Loss of the *Bounty*

In 1787, the HMS Bounty *was a ship on a simple mission: to collect samples of the Tahitian breadfruit tree for use as a food source on the slave plantations of the West Indies. Instead, it would become infamous for an act of mutiny.*

✳   ✳   ✳   ✳

A Benevolent Bligh?

WILLIAM BLIGH WAS a mariner of no small experience. Having gone to sea at age 15, he was later praised by the legendary James Cook for his work as sailing master of the *Resolution* on Cook's third voyage. Bligh, in turn, learned a great deal from Cook and put a number of the famed explorer's techniques into practice when he became captain of the *Bounty*. He set about readying his ship for the voyage: the mundane task of loading supplies and overseeing the construction of a greenhouse that was designed to safely transport the botanical cargo. He also demanded that the ship's lifeboats be replaced with more suitable vessels—a request that would later save his life. Along with the physical preparations, Bligh instituted many shipboard practices that were perhaps avant-garde for the time but were clearly designed for the direct benefit of the crew. He changed the ship's routine to give the men a full eight hours of rest each night, rather than the traditional four. He took particular care to provide for cleanliness and prevent disease among his men through rigorous enforcement of bathing and dietary regulations.

A Disgruntled Sailor

Of course, Bligh was not without his faults. He was witheringly sarcastic and not above publicly humiliating his men in front of their shipmates if they did something incorrectly. In fact, he was described as having a "wonderful capacity for breeding rebellion." Unfortunately, the *Bounty* was a small ship, about 90 feet in length by a scant 25 feet wide, and there was no escaping Bligh's invective once a sailor attracted his wrath. One man in particular, Fletcher Christian, came to bear the brunt of Bligh's assaults. Christian was popular with his shipmates and island ladies alike but was known to suffer sudden bouts of depression, during which he skulked around the ship in a black silence. He also had a melodramatic streak, describing his existence as "hell." He planned to throw himself overboard if his call to mutiny failed.

The life-threatening voyage to Tahiti, followed by months of easy living on the island, created a powder keg when the sailors resumed their normal shipboard routine on the return trip. In the meantime, the personalities of Bligh and Christian combined to produce the explosive spark. Bligh's temper was on edge; he had discovered that his crew had allowed sails to rot and had neglected to wind the ship's chronometer—cardinal sins on any sailing ship. Shortly after leaving Tahiti, Bligh came to believe that someone was pilfering coconuts from the ship's supply and publicly accused Christian of being the thief. The altercation proved to be the breaking point for Christian. On April 28, 1789, he led the mutineers in seizing Bligh as he slept and setting him and 18 loyalists adrift in a small boat to face almost certain death.

Aftermath and Legacy

Bligh's seamanship proved equal to the task. He led his men to safety on an incredible 41-day journey of survival in a boat that would doubtless be judged unfit for passengers by modern Coast Guard standards. British naval regulations of the time demanded a court martial on the loss of any vessel, but the proceeding found Bligh blameless. His actions were reinforced by the presence of the loyal sailors who had risked their lives to join him.

Bligh was promoted and given a new command; he went on to attain the rank of admiral. In 1793, Christian was murdered in his island paradise as his mutineers clashed with natives.

Send Me to Timbuktu?

The city of Timbuktu lives in modern English vocabulary as a somewhat mythical place that's remembered for its unique, lyrical name. If someone were actually to travel to Timbuktu, what (if anything) would they find there?

✳   ✳   ✳   ✳

THE REAL TIMBUKTU is a small city in northwestern Africa, in central Mali, about 500 miles from the Atlantic coast on the Niger River. Its roughly 30,000 inhabitants are mostly of Tuareg, Songhai, Fulani, and Moorish heritage. Most of Timbuktu's residents are Sunni Muslims.

The Name

The Tuaregs, a nomadic Berber people of the Sahara region, founded Timbuktu sometime around A.D. 1000. The story goes that a well-respected elderly lady named Buktu lived near a well ("tin" in Tuareg). Nomads who needed to leave things behind for safekeeping entrusted them to Buktu and said that they had left their possessions at "Tin Buktu." Buktu is long gone, but her name has endured through Timbuktu. Even today, there is a well in Timbuktu that is said to be that of Buktu herself.

The Place

Timbuktu began as an encampment and grew into a town, becoming an important stop on the trans-Saharan trade route. Salt mined from the Sahara went south and west; slaves and gold went north toward the Mediterranean. Even though Timbuktu changed hands among African empires, it developed into a prestigious Islamic cultural and religious center.

In its peak era, beginning in about 1330, Timbuktu had 100,000 residents, including an astonishing 25,000 students. The prized turban of a Timbuktu scholar proclaimed its wearer to be a devout Muslim who was steeped in Islamic learning. In order to receive the lowest of four degrees conferred in Timbuktu, the student had to memorize the entire Koran. Learned scholars coming to Timbuktu from afar required extra teaching to bring their knowledge up to local standards. In terms of prestige, it might be fair to call Timbuktu the Oxford or Harvard of the medieval Islamic world.

This golden era ended in 1591, when Moroccans armed with the latest gunpowder weapons conquered the Songhai Empire. The Moroccan conquest didn't kill Timbuktu, but the city was mortally wounded, as trade routes shifted after the year 1600. Carrying goods across the sea became safer and faster than hauling everything across the Sahara. The city became something of a backwater, yet it remained an important destination for dedicated students seeking immersion in Islam.

By the 1800s, Timbuktu was only known as a legend to Europeans. A French exploration society offered a handsome bounty to anyone who visited the city and returned to describe it. At least one explorer perished in the attempt to locate Timbuktu, but an intrepid Frenchman named René-Auguste Caillé finally returned with an account of the contemporary city. His report would have made a lousy tourist brochure, as he found only some mud huts threatened by the rising sands of the Sahara. The only remarkable aspects of Timbuktu, Caillé said, were its centers of Islamic learning.

The French captured Timbuktu in 1893, incorporating it into their immense West African domain. In 1960, the Republic of Mali achieved independence. At that time, Timbuktu hadn't changed a lot since Caillé's visit. However, the city's prominence has risen in subsequent years. Today, Timbuktu is sometimes called "The Mecca of Africa" for the prestigious Islamic study courses that are offered at the city's Sankore Mosque. Additionally, hundreds of thousands of priceless historical documents can still be found in Timbuktu. Refusing to be forgotten again, the city has even successfully managed to keep the Sahara's drifting sands at bay.

Riding with the Pony Express

Neither rain, nor sleet, nor dark of night could keep the Pony Express from running mail across the Wild West.

✳ ✳ ✳ ✳

PRIOR TO THE Pony Express, it took weeks to deliver mail in the West, and that was done either by boat or stagecoach. Three men—William H. Russell, Alexander Majors, and William B. Waddell—believed that with the Pony Express they could reduce the delivery time down to a mere ten days.

Planning the Route

In early 1860, the men began by picking locations that would serve as "hubs" for the Pony Express: Sacramento, California, and St. Joseph, Missouri. Mail would arrive at each of these locations either by boat (Sacramento) or train (St. Joseph), and then it would be handed off to the Pony Express riders to deliver. All in all, the total one-way mail route would run approximately 2,000 miles.

While researching the route, it was determined that a pony at full gallop could travel approximately ten miles a day. Stations were constructed every ten miles or so along the route between the two hubs. An estimated 119 to 153 stations were built (sources vary); at these locations, riders could get fresh ponies. Now all that remained was to hire riders willing to make the trek.

Finding the Right Men

In the spring of 1860, posters went up in Sacramento and St. Joseph that read:

Wanted: Young, skinny, wiry fellows not over 18.

Must be expert riders, willing to risk death daily.

Orphans preferred.

Wages $25 per week

In order to keep the load as light as possible, applicants couldn't weigh more than 125 pounds. Most of the weight would come from the mail, which was carried in the pockets of a *mochila* (Spanish for "pouch"), which were placed over the horse's saddle. Riders would be expected to travel alone, day and night, regardless of the weather. They could stop at one of the Pony Express stations to switch ponies and get water, but only after traveling 75 to 100 miles could the rider switch out with another rider and rest.

Since the riders would be traveling alone, they were subject to the elements and attacks from wild animals, Native Americans, and bandits. Originally, the riders were armed with guns, but eventually most of the weapons were taken away to lighten the load.

The First Trip

On the opening day of the Pony Express—April 3, 1860— two riders left their stations; one traveled west from St. Joseph and one traveled east from Sacramento. It was approximately midnight on April 13 when the first Eastern mail arrived in Sacramento. The following day, the Western mail arrived in St. Joseph. All told, the first westbound trip was made in 9 days and 23 hours; the eastbound journey took 11 days and 12 hours.

The End of the Pony Express

Incredibly, despite the success of the first run, the Pony Express never really caught on. In fact, stagecoaches were still used to carry the bulk of the mail throughout the East Coast and central United States. The Civil War broke out in 1861, which also put a damper on things.

On October 24, 1861, the transcontinental telegraph connected Sacramento with Omaha, Nebraska, thereby joining the Midwest with the West. Two days later, the Pony Express officially went out of business. During its 18 months in existence, the Pony Express earned approximately $90,000. Their net losses, however, were more than double that.

The Donner Party

Farmers George and Jacob Donner and businessman James Reed succumbed to the land fever that was sweeping the country. Their disasterous trek became infamous.

✳   ✳   ✳   ✳

O N APRIL 16, 1846, the Donner party, comprised of nine wagons carrying 87 people, left from Springfield, Illinois. Their journey would take them 2,500 miles across plains, deserts, the Great Basin, and three mountain ranges. It was necessary to arrive at Sutter's Fort in California ahead of the snows that would blanket the Sierra Nevada Mountains.

On July 20, the group came upon the Little Sandy River in present-day Wyoming. There they made the fatal mistake of attempting a shortcut listed in Lansford Hastings's *The Emigrant's Guide to Oregon and California*. Although Hastings described this route in his publication, no one had ever tested it.

The first obstacle the party encountered was getting their wagons through dense brush. Next, they faced a maze of canyons that took them a month to navigate. Finally, they came to the Great Salt Lake Desert. The group struggled through five blistering days and freezing nights to make the 80-mile trek that they read would be only 40 miles.

By the time the group made it out of the desert, it was early September. They soon noticed snow flurries—the first sign of the coming winter—but they trudged on. In early November, they became trapped in the snowy Sierra Nevada Mountains, 150 miles from their destination. All that they could do was build rough cabins and wait until the worst of the winter was over. Starving, the group was trapped for four months. Before they were rescued, 40 had died, and some of the forlorn survivors had resorted to cannibalism. The last living member of the Donner party was Margaret Breen, who died in 1935.

A Brief History of Underwear

From fig leaves to bloomers to thongs, people have covered themselves a little or a lot. Here is a brief history of the undergarment.

✳ ✳ ✳ ✳

✳ The earliest and most simple undergarment was the loin-cloth—a long strip of material worn between the legs and around the waist. King Tutankhamun was buried with 145 of them, but the style didn't go out with the Egyptians: Loincloths are still worn in many Asian and African cultures.

✳ Men in the Middle Ages wore loose, trouserlike undergarments called *braies*, which one stepped into and tied around the waist and legs about mid-calf. To facilitate urination, braies were fitted with a codpiece—a flap that buttoned or tied closed.

✳ Medieval women wore a close-fitting undergarment called a *chemise*, and corsets began to appear in the 18th century. Early versions of the corset were designed to flatten a woman's bustline, but by the late 1800s, they were altered to give women an exaggerated hourglass shape.

✳ In the late 1800s and early 1900s, chastity was a big concern for married or committed couples. During that time, many inventors received patents for "security underwear" for men. These devices were meant to assure "masculine chastity." They ensured that men refrained from sexual relations with anyone other than the person with the key to open their particular device.

✳ The thong made its first public U.S. appearance at the 1939 World's Fair, when New York Mayor Fiorello LaGuardia required nude dancers to cover themselves, if only barely.

Worth a Thousand Words

Civil War photographers blazed a new trail to provide a visual record of the conflict and paved the way for modern photojournalism.

✳ ✳ ✳ ✳

IN TODAY'S DIGITAL world, photographers have a relatively easy time pursuing their craft. It would be quite a task for most modern shutterbugs to endure the extreme demands placed upon legendary photographer Mathew Brady and his peers a century and a half ago. The technical ability of photography during the Civil War was in no way up to the task of documenting the actual battles—photographers could really only hope to record the aftermath, with the dead lying where they fell or lined in rows for hasty burials. Despite that, the images with which we are familiar today are still extremely powerful and continue to have an immediacy that no advanced technology can surpass. Given the conditions in which these photographers worked, their accomplishments are little short of miraculous.

Road Blocks

One of the primary obstacles photographers had to face was transporting themselves and their equipment. The "film" was glass plate negatives, some measuring a mammoth 11 by 17 inches. Simply moving these from one locale to another was arduous, given the coarse terrain and the less-than-smooth ride afforded by horse-drawn wagons. A day spent capturing images could be lost forever simply by taking a spill. Moreover, taking photos was slow business. The time necessary for an exposure could run anywhere from 5 to 20 seconds, which tended to preclude asking combatants to hold still.

Timing Is Everything

A typical "shoot" would go something like this: A photography team (two people minimum) would rush to a battlefield after word arrived of a major engagement. Speed was of the essence, as everyone wanted to scoop the competition. Arrival at the battlefield would have to be timed just so. If they were too late, the armies may have already cleared the battlefield of the dead and, therefore, any compelling subject matter—all that would be left was an empty field. If they arrived too soon, however, they risked life and limb by being caught up in the battle itself, not to mention the chance of being captured by the combatants. Indeed, one of photographer Alexander Gardner's associates was briefly detained by lingering Confederate troops after the Battle of Gettysburg. Gaining permission and cooperation was necessary not only to access the site but also to get a handle on what had happened and where. Ghoulish tourists and souvenir hunters were another hazard, as they frequently picked the landscapes, and the casualties, clean of anything of value.

The photographers would typically scout for compelling subject matter, each with their own signature style. Brady was partial to including landmarks and vistas in his work, while his former employee Gardner seemed to focus exclusively on lifeless combatants. It must be remembered that adding to the challenge of working smoothly in the field was the unavoidable stench of decaying bodies, both human and equine. Burial parties would naturally concentrate on retrieving fallen comrades, so dead horses and mules were typically given last priority in refuse removal. That the photographers could stand to be in this environment for as long as was necessary to do their work speaks volumes about their dedication.

Strike a Pose

Though most of the striking battlefield images that are known to us today record soldiers at their place of death, studies of the photos more than a century later have revealed that not every photographer was intent on simply documenting what he found. One photo historian has revealed demonstrable evidence that Gardner and his assistants were not above "improving" upon warfare's aesthetic qualities by physically moving and "posing" corpses into more inspiring arrangements, sometimes transporting the decaying remains great distances. There, a dropped weapon might be added for poignancy.

After a commercially viable scene was selected, the team was ready to begin its work. Wet-plate photography, which was universally employed throughout the Civil War, meant the arduous task of applying a syrupy concoction (known as collodion and made from an egg-white derivative) to a spotless piece of glass on location. Next, chemicals were applied in complete darkness to render the sticky coating photosensitive. At this point, the tacky plate was considered "film" and had to be placed into a film holder and rushed out into a camera, which presumably had already been set into position by a team member. After an image was recorded, an assistant would hustle the film container to the darkroom wagon for development, mindful of the need to begin processing before the plate dried. With enough team members to assure a smooth, practiced routine, the entire process could be completed in as little as ten minutes per exposure.

✳ Battlefield photographers were, above all, entrepreneurs. They labored over generating the most salable product. This meant providing images in the preferred format of the day—"stereo-view." To achieve this effect, three-dimensional images were captured with dual-lens cameras. Direct precursors to Viewmasters of more recent vintage, handheld viewers provided the illusion of depth to the beholder. Many households of the time owned one, thereby affording gawkers the chance to experience the war in the comfort of their own homes. Updated reissues of these stereo-view images became widely available in the 1960s—the war's centennial decade.

The Panama Canal

A canal that cut across Central America and connected the Atlantic and Pacific oceans would allow ships to avoid the dangers of South America's Cape Horn and shorten sea voyages by several months and thousands of miles. But it wasn't until the Industrial Revolution that it became possible.

✳   ✳   ✳   ✳

FERDINAND DE LESSEPS, a French entrepreneur who had over-seen construction of the Suez Canal, first made an attempt to construct a Central American waterway in 1881. Heavy rain, intense heat, difficult terrain, and tropical diseases stymied de Lesseps's sea-level plan. Whenever the crews made progress, flooding would wash away all their work. When only 11 miles of the required 50 had been dug after seven years of work, almost $300 million, and the loss of 20,000 lives, de Lesseps gave up.

In 1901, President Theodore Roosevelt committed to building a canal, citing the need to join the U.S. Atlantic and Pacific fleets as a matter of national security. When negotiations with the government of Colombia (which owned the area considered most suit-able for a canal) broke down, the United States supported a group of local residents and businesspeople in seceding from Colombia and forming the nation of Panama. The new country promptly granted the U.S. the rights to build the canal.

Construction began in 1904, and the canal opened in 1914. Key to the success of the American effort was an organized and practi-cal "lake and lock" plan by chief engineer John Franks Stevens. A successful public-health campaign led by Dr. William Gorgas greatly reduced the scourges of malaria and yellow fever.

Alexander Bernard Heron, who died in 2000, was the last of the Panama Canal builders. He was born in 1894 and started work-ing at the Panama Canal's construction site at age 14. He ulti-mately spent a total of 48 years of his life working at the canal.

The Curse of King Tut's Tomb

The curse of King Tut's tomb is a classic tale, even if there's a lot of evidence that says there was never anything to worry about.

✳ ✳ ✳ ✳

Curse, Schmurse!

IN THE EARLY 1920S, English explorer and archaeologist Howard Carter led an expedition that was funded by the Fifth Earl of Carnarvon to unearth the tomb of Egyptian King Tutankhamun. Most of the tombs of the Egyptian kings had been ransacked long ago, but Carter had reason to believe that King Tut's 3,000-year-old tomb was still full of artifacts from the ancient world. He was right.

Within the king's burial chamber were vases, precious metals, statues, and even whole chariots—all buried with the king to aid him in the afterlife. Carter and his team excavated to their hearts' content, and due to their hard work, we now know a great deal more about the lives of ancient Egyptian people during King Tut's time than we did before.

Carter had been warned about the dangers of disrupting an ancient tomb, but he didn't buy into the rumors of curses and hexes. After opening the tomb, however, it was hard to deny that some strange, unpleasant events began to occur in the lives of those involved in the expedition.

Curse or Coincidence?

Several men involved in the excavation died shortly after entering King Tut's tomb. The first to go—the Fifth Earl of Carnarvon—died only a few months after the excavation. Legend has it that at the exact moment the earl died, all the lights in the city of Cairo mysteriously went out. That morning, his dog allegedly dropped dead, too.

Egyptologists claim that the spores and mold released by opening an ancient grave are often enough to make a person sick or worse. The earl had been suffering from a chronic illness before he left for Egypt, which could have made him more susceptible to the mold and, therefore, led to his death.

Other stories say that the earl was bitten by a mosquito. Considering the sanitary conditions in Egypt at the time, a mosquito bite in Cairo could have some serious consequences, including malaria and other deadly diseases. Some reports indicate that the bite became infected and he died as a result—not because an ancient pharaoh was annoyed with him.

There were other odd happenings surrounding the excavation, and the public, already interested in the discovery of the tomb itself, was hungry for details of "the curse of the pharaohs." Newspapers reported all kinds of "proof": The earl's younger brother died suddenly five months after the excavation, and on the morning of the opening of the tomb, Carter's pet bird was swallowed by a cobra—the same kind of vicious cobra depicted on the mask of King Tut. Two of the workers hired for the dig died after opening the tomb, though their deaths were likely due to malaria, not any curse.

Six of the 26 explorers involved in the excavations died within a decade, but many of those involved lived long, happy lives, including Carter. He never paid much attention to the "curse," and, apparently, it never paid much attention to him. In 1939, Carter died of natural causes at age 64, after working with King Tut and his treasures for more than 17 years.

Yeah, Right

King Tutankhamun's sarcophagus and treasures have toured the world on a nearly continual basis since their discovery and restoration. When the exhibit came to America in the 1970s, some people tried to revive awareness of the old curse. When a San Francisco police officer suffered a mild stroke while guarding a gold funeral mask, he unsuccessfully tried to collect compensation by claiming that his stroke was due to the pharaoh's curse.

What Really Caused the Chicago Fire?

*Mrs. O'Leary and her cow didn't really start the fire. So who—
or what—did?*

✳ ✳ ✳ ✳

THE STORY IS now legend: Old Mrs. O'Leary left a kerosene lantern burning in her barn. A cow kicked over the lantern, and the hay caught fire. The winds blew the flames, and they quickly spread through Chicago, destroying most of downtown.

The story that O'Leary and her cow were the culprits spread around Chicago before the flames had even stopped smoldering. The cow was butchered and served up as oxtail soup at the posh Royal Palm restaurant on Thanksgiving Day—a month after the fire. The O'Learys had to go into hiding because they feared being lynched.

But we now know that the story was total nonsense. While the fire did start in the O'Leary barn, the story that O'Leary or her cow was responsible was invented by a reporter who thought it made for a colorful story. The local press, much of which was openly anti-Irish, picked up on the story and ran with it; the *Chicago Tribune* spoke of O'Leary's "typical Irish know-nothingness." O'Leary was hounded on the anniversary of the fire for the rest of her life. Not surprisingly, she developed a lifelong hatred for reporters, and she never allowed herself to be photographed.

It Came from Outer Space?

So, if it wasn't O'Leary, what *did* cause the fire? Lots of stories have circulated through the years. A few people have sheepishly admitted to family members in their old age that they were the cause of the fire, whether by knocking burning ashes from a clay pipe into the hay or by sneaking into the barn to milk the cow on a dare, causing the irate cow to start kicking. They can't all have been telling the truth, though.

One strange theory alleges that the fire was actually caused by a meteor that crashed down on the day of the fire in Peshtigo, Wisconsin. The meteor certainly caused major fires throughout the state—some of which were actually even bigger than the one that blazed in Chicago. No one is entirely sure whether the fire could possibly have traveled all the way south to Chicago (and ended up localized in a barn southwest of the Loop, skipping everything west of Halsted and north of Taylor Street in the process). Still, it seems to be an awfully funny coincidence that major fires would have broken out in two nearby areas with totally different causes on the same day.

The Mysterious Theory of "Big Jim" O'Leary

O'Leary's son had an even wilder story. Young James grew up to be "Big Jim" O'Leary, a stockyards saloonkeeper and gambling king. When a statement was made in the *Tribune* that the fire had been started by two young men trying to milk the cow in order to make whiskey punch, he was outraged.

"The true cause of the fire has never been told," he said to a *Tribune* reporter in 1904. "But I'll speak out. That story about the cow kicking over the lamp was the monumental fake of the last century. I know what I'm talking about when I say that the fire was caused by spontaneous combustion!"

That's right. Spontaneous combustion.

According to Big Jim, his father had just purchased some mysterious green hay, which, he said, had spontaneously combusted in the hay loft.

The true origin of the fire will probably never be known, but most scientists agree that there is no type of hay that is known to spontaneously combust. In any case, in the late 20th century, the city finally issued an apology to the defamed woman.

Ghastly Medieval Torture Devices

The following devices—designed to maim, torture, and kill—prove that humankind has far to go in its quest for civility.

✳ ✳ ✳ ✳

The Rack

DURING MEDIEVAL TIMES, being interrogated meant experiencing excruciating pain as one's body was stretched on the infamous Rack. The operating premise was diabolically simple. Victims laid on their backs with arms extended while straps anchored the hands and feet to opposite ends of the table. The torture began when the operator rotated rollers at each end in opposing directions. At the very least, severe joint dislocations occurred; at worst, limbs were ripped clean off and death would result. Even if a tortured victim was subsequently released, he or she would often be incapable of standing erect since muscle fibers stretched beyond a certain point lose their ability to contract.

The Iron Maiden

The Iron Maiden torture device differs wildly from the popular heavy metal band of the same name, even if both could ultimately make one's ears bleed. Insidious in its intent, the sarcophagus-shaped container opened to allow the victim to step inside. Once there, protruding spikes on the front and back halves would spear the occupant as the door was closed. Agonies were prolonged because spikes were strategically positioned to find the eyes, chest, and back but not vital organs. As a result, death occurred only after the victim had bled out, an agonizing process that could last for days.

The Pear

Despite sharing its name with a sweet fruit, there was nothing at all tasty about the Pear. Designed to be inserted in the most sensitive of the body's orifices (i.e., mouth, rectum, vagina), the pear-shaped torture tool was used as a punishment for those who had committed sexual sins or blasphemy. Once put in place, a screw mechanism caused pointed outer leaves to expand ever wider, resulting in severe internal mutilation.

The Tongue-Tearer

Self-explanatory in name, the Tongue-Tearer worked precisely as advertised. Resembling a wire cutter with an eye bolt passing through its end grips, a victim's mouth was forced open as the Tongue-Tearer was employed. After finding purchase on its slippery quarry, the eye bolt at the opposite end of the device was tightened ever so slowly, until the tongue became completely detached from the horrified victim's mouth.

The Lead Sprinkler

With its innocuous sounding name, one might expect to find this item gracing a formal garden, not doing the devil's handiwork in a dank dungeon. Shaped like a maraca, the Lead Sprinkler held molten lead inside a perforated spherical head. The torturer would simply hold the device over the victim and give it a shake. The ensuing screams were the only music to come from this instrument.

✳ In 1890, Emperor Menelik II of Abyssinia became so enthralled with stories of the new "electric chair" from America that he ordered three of them, even though his country didn't have any electricity at the time. (He kept one unplugged death chair as his throne.)

The *Titanic's* Mystery Ship

Could there have been a ship close to the Titanic *when it sank?*

SEVERAL PASSENGERS AND crew members aboard the *Titanic* reported seeing the lights of another ship on the horizon as the *Titanic* was foundering. Modern *Titanic* buffs feud over this "mystery ship." What is known:

* The British freighter SS *Californian* was in the vicinity.

* *Titanic* survivors reported seeing some sort of ship nearby.

* *Californian* witnesses recall seeing white rockets at the right time to be *Titanic's* distress signals.

Solved? Not quite. Some doubt that the mystery ship was the *Californian*. Discrepancies in *Titanic* and *Californian* witnesses' stories raise reasonable doubt. It's hard to say whether the differences mean something or if they resulted from simple stress and distraction.

By the time *Californian* steamed over to help, no survivors remained. The *Californian's* captain later came under cold scrutiny for the tardy response. The question of his culpability boils enthusiasts' blood to this day.

---

* In the most famous nautical disaster in history, on April 15, 1912, the RMS *Titanic* struck an iceberg in the North Atlantic and sank in less than three hours. Of the 2,227 passengers and crew, only 705 survived. Of these, Britain's Millvina Dean— the youngest passenger on the ship at about ten weeks old— was the last survivor. Dean lost her father in the catastrophe. In 2009, actors Leonardo DiCaprio and Kate Winslet made generous donations to a fund to cover Dean's mounting nursing-home expenses. Dean passed away on May 31, 2009, at age 97.

Sailing from Slavery to Freedom

In the early hours of a spring morning, Robert Smalls became a fugitive slave—and a national hero.

❋ ❋ ❋ ❋

ROBERT SMALLS WAS born into slavery in 1839 in Beaufort, South Carolina, where he lived until age 12, when he moved with his owner to Charleston. There his owner allowed Smalls to hire himself out, with all but one dollar of his wages going to the owner. When he turned 18, Smalls renegotiated with his owner to pay him $15 per month and keep the rest of what he earned. He married Hannah Jones, who was owned by a different slaveholder. The next year, Hannah gave birth to a daughter, Elizabeth. Robert quickly made arrangements to buy Hannah and Elizabeth from their owner for $800; he had saved money toward that purpose when the Civil War broke out.

On the Job

Smalls held several jobs prior to the war, and through them he accumulated a variety of skills. Many of these skills were related to sailing, and in 1861, Smalls was hired as a sailor on the *Planter*, a cotton steamer, for $16 a month. After the war began, the *Planter* came under control of the Confederate government. It served General Roswell Ripley, commander of the Second Military District of South Carolina. Smalls's superiors recognized his skills: Although his status as a slave kept him from gaining an official title, for all intents and purposes, Smalls was made the *Planter*'s pilot.

An Early-Morning Excursion

With that position, Smalls was given details of the signals and pass codes necessary for a vessel to navigate the heavily fortified Charleston harbor. This information came in very handy in the early hours of May 13, 1862. The 23-year-old, illiterate Smalls did something that astonished everybody: He commandeered the *Planter*. The day before, the ship had been loaded with armaments to transport to forts where they were needed. While the white Confederate crew was ashore, leaving eight slaves aboard, Smalls took the ship and went to pick up their families. He then sailed toward the Union fleet blockading Charleston Harbor, giving the appropriate pass codes as he navigated past Confederate ships.

Smalls had carefully watched and studied the *Planter*'s captain, and this morning he wore the captain's hat and coat, mimicking his voice and movements as best he could to avoid suspicion from others in the Confederate fleet. As Smalls and his crew approached the nearest blockading ship, the USS *Onward,* they were aware that their Confederate steamer looked suspect. Smalls raised the white flag of surrender, and as the *Planter* drew near the Union vessel, the nattily dressed slave doffed his cap, shouting, "Good morning, sir! I have brought you some of the old United States' guns, sir!"

A New Life

For the capture of the *Planter,* President Lincoln awarded Smalls and his associates $1,500 in prize money. The Union press made Smalls a hero, and newspapers across the North told of his bravery, citing his actions as an argument against racial stereotypes. In one *New York Daily Tribune* editorial, the author wrote, "Is he not also a man—and is he not fit for freedom, since he made such a hazardous dash to gain it? . . . Perhaps [blacks are inferior to whites] but they seem to possess good material for improvement. Few white men have a better record than Robert Smalls."

In August 1862, Smalls and Mansfield French, a missionary, traveled to Washington, D.C., to meet with the president. Their objective: to ask the Union army to recruit 5,000 black soldiers to fight in the war. That request was granted, and black soldiers and all-black units were enrolled in the military for the first time in U.S. history.

Smalls remained active throughout the war. On April 7, 1863, he piloted the ironclad ship *Keokuk* during an attack on Fort Sumter; the ship was struck 19 times and sank the following morning. On December 1, 1863, Smalls became the first black captain of a U.S. vessel and was honored for his bravery under fire.

Going Home

After the war, Smalls entered the world of politics. Tutors had taught him to read and write, and he had always been regarded as well-spoken. In 1868, he contributed to drafting the South Carolina state constitution, and he went on to serve in the South Carolina House of 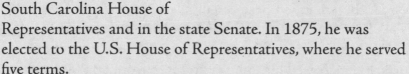 Representatives and in the state Senate. In 1875, he was elected to the U.S. House of Representatives, where he served five terms.

The man who was once a Beaufort slave boy had become the most powerful black man in South Carolina. Smalls moved back to Beaufort, where he bought the very house in which he had been a slave. Smalls died on February 23, 1915; his house is now a national historic landmark.

"I appear this evening as a thief and robber. I stole this head, these limbs, this body from my master and ran off with them."

—FREDERICK DOUGLASS

The Potsdam Giants

Ever heard of a Napoleon complex? Well, this short-statured Prussian king took a unique approach to building himself up.

✳ ✳ ✳ ✳

IT MAKES SENSE that the king of Prussia should want a decent military. And you can't really blame a guy for having specific tastes when it comes to soldiers. But you can definitely accuse Prussian King Frederick William I (also known as Fredrich Wilhelm) of being a little weird and a little too, well ... *particular.*

Assembling an Army

Prussia was a kingdom in north-central Europe that included present-day northern Germany and Poland. The state became a republic in 1918 and was formally abolished after World War II (though it had been essentially annexed by Germany in 1934). But prior to that, Frederick took the Prussian throne in 1713. His first order of business was to improve his military forces. He increased his military strength from 38,000 men to an army of 83,000. He outfitted them with better weapons and insisted that they train hard or face harsh consequences. Satisfied with his army, the king could move onto beefing up his own personal regiment—and that's where things got a little strange.

How's the Weather Up There?

Convinced that a tall soldier was a better soldier, the vertically challenged king (he stood at about 4′11″) began to recruit men with above-average height for his group of "Grand Grenadiers of Potsdam." He set a height requirement at 5′11″, which in the 18th century was seriously tall (the average height of a European man at the time was 5′6″). Most of the men were at least six feet tall, and many were nearly seven feet tall. Frederick became obsessed with "collecting" tall soldiers and was known to actually order the kidnapping of grown men who he felt fit the bill. Stories of Irish longshoremen being thrown into boxes and Austrian diplomats being snatched off the street were common.

Word got out about Frederick's need for hundreds of tall men. Neighboring countries would send their tallest male citizens (often against their wills) as gifts to keep relations with the king open and friendly. Frederick demanded that his army of giant men marry tall women in hopes of breeding tall children. One medical historian noted that Frederick's Potsdam Giants were "The tallest men ever assembled until the birth of professional basketball."

Unfortunately, the Potsdam Giants weren't actually fit for battle. Many of them were mentally disabled or somewhat mentally handicapped as a result of their gigantism. Most of them never used the weapons that the king provided. Even if they were able to fight, a lot of them wouldn't have wanted to. Living conditions in their camp were atrocious. Food and shelter shortages were common, and more than 200 of the Giants mutinied every year. The ones who were caught after they escaped had their noses and ears cut off before being thrown in jail. Frederick loved his soldiers—that is, as long as they were simply tall and not opinionated.

What Goes Up Must Come Down

In 1740, Frederick died, and with him perished his dream of the infantry of tall soldiers he had so brutally created. The crown prince, Frederick II, took the throne; interestingly, Frederick II was a small, sickly child, which had been quite a source of consternation to his father. Despite this tense relationship, Frederick II was a natural leader and became known as Frederick the Great. One of his first acts as king was to disband the Potsdam Giants. According to one biography of Frederick the Great, contemporary sources claim that after the Giants were released, "the roads to Paris were littered with halfwits trying to find their way back home."

The NAACP's Inspirational Origin

At the beginning of the 20th century, those who sought equal rights for African Americans lacked a powerful national organization to unite them. It took the initiative of one tenacious reformer to make the National Association for the Advancement of Colored People (NAACP) a reality.

✳ ✳ ✳ ✳

B Y THE TURN of the 20th century, African Americans had been free from the bonds of slavery for almost 40 years, but they still faced a culture, an economy, and a political process that made true equality seem like a distant dream. It was difficult for African Americans to vote, hold public office, or buy property; segregation laws made quality education nearly impossible; and lynchings and violence took the place of justice. Race riots spread like wildfire through the country.

The Niagara Movement

Organizations that sought to improve this situation existed but lacked money and political influence Nevertheless, in 1905, the illustrious scholar and political activist W.E.B. Du Bois called for a national meeting of black leaders. The result was the short-lived yet legendary Niagara Movement, during which black leaders and activists delineated their goals of equal economic opportunity, an end to disenfranchisement for blacks and women, an end to segregation and discrimination, and the abolition of injustice in legislation and judicial processes.

Meanwhile, Mary White Ovington, a white journalist and activist from New York City, was putting together a group of her own. In 1906, the *New York Evening Post* sent Ovington to cover the second annual meeting of the Niagara Movement, where she met and was inspired by many of the leaders. Two years later, Ovington read a harrowing newspaper article about violent race riots in Springfield, Illinois, during which several blacks were lynched at random for no reason. She responded by writing a letter to the author of the article, journalist William English Walling. Ovington suggested that she and Walling meet in New York City to discuss their common concern for the plight of African Americans in the United States.

In the first week of January 1909, the initial meeting of the fledgling NAACP took place in a tiny New York City apartment. Present were Ovington, Walling, and Henry Moskovitz, a Jewish social worker. The three agreed that on Abraham Lincoln's birthday, February 12, they would circulate an open call for a conference about the "Negro situation," as it was called at the time. They managed to get the resulting document signed by 60 prominent activists, black and white, including members of the Niagara Movement.

The National Negro Committee

The first conference convened on May 30, 1909. After three meetings, the participants began calling themselves the "National Negro Committee." After the Niagara Movement disbanded in 1910, many of their former members sought to solidify and strengthen the young National Negro Committee. The group's name was changed once again to its current moniker, the National Association for the Advancement of Colored People.

This new organization was potent for several reasons. It was biracial, so the power of black leaders was coupled with the legal and political clout that was largely reserved for whites. Further, the fundamental principle of the NAACP was to seek radical change through legal—never violent—means. By 1913, the NAACP had opened 24 branch offices in the United States. Over the years, it supported legal battles that advanced equality—most notably the *Brown v. Board of Education* decision, which favored desegregation in schools.

Today, the NAACP continues to employ its peaceful means to advance freedom and equality for all. And it all began with a small group of leaders who would not sit back and ignore violence and injustice.

The King Who Talked with Trees

Known as "Farmer George" for his love of rural life and "America's Last King" for losing the Revolutionary War, King George III could well have been "His Royal Madness-ty."

✳ ✳ ✳ ✳

A Fine Start

GEORGE III BECAME King of England in 1760 at age 22, after his grandfather, George II, died from an aneurism. (Young George's father, Frederick, who would have been heir to the throne, had died when the prince was 12.)

George III and his queen, Charlotte, had 15 children, and they became known for their model family life. To the chagrin of his family and other proper upper-class folks, George enjoyed many hobbies that were considered eccentric for a royal. Keenly interested in agriculture, he spent much time constructing and supervising model farms. He collected books (approximately 65,250) and tinkered endlessly with watches and other small gadgets. His subjects found these common traits endearing. But neither they nor the upper crust could have imagined the fate of their modest and generous leader.

Monarch Madness

Although he was a healthy devotee of exercise and simple food, George had his first mental breakdown in the fall of 1788, when a page observed him involved in a heated discussion with a tree. At other times, George would incoherently and incessantly babble. After he started foaming at the mouth, doctors bled him, put him in a straitjacket, and covered him with smelly hot compresses that were designed to draw out evil substances. Despite the archaic 18th-century medical practices, he miraculously recovered in about a year.

However, George's illness returned in 1811 after the death of his favorite daughter, Amelia, who had succumbed to tuberculosis. Then 72, George began acting with such "dreadful excitement" that the "mad doctor" had to be called back to the castle. The king continued to suffer frequent delusions, often declaring that Princess Amelia was alive and living in Germany. Eventually his aides had to bring out restraints again; George barely escaped a proposed shock treatment of being doused with buckets of cold water. His son, the future George IV, ruled in his father's stead as the official prince regent of Great Britain.

King George spent the last ten years of his life wandering in an enclosed area of Windsor Castle in a purple robe, blind and eventually deaf, his royal jewelry pinned to his chest. Some of his later delusions were that God had sent a second flood similar to Noah's, and that he himself was dead. Eventually, he was so far removed from reality that he didn't know that his wife had died in 1818. The king quietly passed away in his bed in 1820 at age 81.

Why the Weirdness?

Two doctors, Ida Macalpine and Richard Hunter, published an article in 1966 claiming that George III actually suffered from a complex of genetic disorders called *porphyria*, a medical condition that was not recognized in the monarch's time. Porphyria causes the blood to make too much red pigment, which affects the entire nervous system and brain and can cause a wide variety of symptoms, including bizarre behavior.

In 2003, an envelope containing a few strands of hair from George III was found buried in a London museum. Specialists tested the hair and found that the king's medicine contained high levels of arsenic—as much as 300 times the toxic level—which could have poisoned him and brought on the porphyric attacks. Ironically, the medicine was prescribed by the "mad doctor"— who was simply trying to cure the very illness he was creating.

The Diaries of Mary Chesnut

A Southern woman offered her views on war and slavery—and touched on women's rights while she was at it.

✳   ✳   ✳   ✳

THE DAUGHTER OF a U.S. senator and governor of South Carolina and the wife of another senator from South Carolina, Mary Boykin Miller Chesnut lived a life of privilege and wealth. Educated and well-read, she left history her Civil War diaries, some of the most fascinating primary sources from the time. When her husband, James Chesnut, resigned from the Senate to join the Confederate cause, in 1861, Mary put pen to paper and documented her experiences in a series of diaries that are still read more than a century after her death.

Scribbling Mania

"The scribbling mania is strong upon me," she wrote. That mania led to her writing several volumes of diaries written until 1865 that would shed light on the wartime Southern home front. Chesnut tried to aid the war effort by volunteering in hospitals like other women in the community, but she fainted and "deemed it wise to do my hospital work from the outside." But in writing about her brief time at the hospitals, she allows modern readers a glimpse into the lives of Southern women. "I cannot bear young girls to go to hospitals," she wrote, explaining how women were subjected to the leers and inappropriate comments of the soldiers—a situation that disgusted her.

In fact, Chesnut devoted much of her writing to the behavior and customs of Southern women. She was open about her feelings in her diaries, often penning scathing entries on women in her social circles and giving them nasty nicknames such as "Lucy Long-tongue." Chesnut was also appalled by the behavior of war widows who openly flirted. "As soon as she began whining about her dead beaux I knew she was after another one. She wouldn't lose any time."

No Subject Spared

No one was safe from Chesnut's poison pen. When fleeing from invading Union troops, she turned her weapon toward a Northern woman who offered hospitality: "She does not brush her teeth—the first evidence of civilization and lives amidst dirt in a way that would shame the poorest overseer's wife."

Though Chesnut had stated with Southern pride that she was never afraid of her slaves, her position changed after her cousin was smothered to death by her own servants. Although Chesnut wrote about slavery and her opposition to it, it is not in the way that modern readers might expect. Raised with slavery as a common part of her life, her issue with the institution was not its human indignity, but rather the practice of Southern men using their female slaves as sex objects, which she believed cheapened the men by giving them easy access to sex. Unfortunately, Chesnut neglected to record any thoughts that she may have had about how this affected the slave women.

Early Women's Lib

It can be said that Mary Chesnut struck some early blows for women's rights with her writing, especially when she addressed the social standing of women who took less-traditional routes with their lives. "South Carolina as a rule does not think it necessary for women to have any existence outside of their pantries or nurseries," Chesnut, who couldn't have children, wrote. "If they have not children, let them nurse the walls."

After the war, the Chesnuts returned to their ransacked plantation. In order to survive, Mary sold eggs and butter, but she never forgot her writing, selling stories to local newspapers and writing three novels. She also turned her eye to her diaries, editing them after the fact into a coherent story line. They were first published in 1905, almost 20 years after her death. The diaries would continue to be edited until 1981, when what is now considered the definitive version was released as *Mary Chesnut's Civil War* and awarded the Pulitzer Prize.

The Sheep Wars

Read on for a little-known tale of rivalry, property damage, and death in the Old West.

✳   ✳   ✳   ✳

THE REPUTATION OF the American cowboy dominates and defines the West: As anyone who's ever enjoyed a steak can attest, even the cattle that cowboys protected have become a part of America's identity. But before the cowboy—and the cattle he tended—ruled the landscape, they first had to do battle with another group that encroached upon their stomping grounds: shepherds and their flocks of sheep.

The shepherds seemed to have their own way of doing things. Whether riding a donkey or traveling on foot, shepherds moved slowly with their grazing flocks. Instead of turning away when they met with a fence, they would often simply cut it to cross ranch lands. It didn't help that the sheep competed with the cattle for food. In addition, many of the shepherds were Mexican.

The two groups continuously butted heads in what was to be known as the "Sheep Wars." The cowboys were swift to move when a flock was discovered crossing their land. In 1875, cowhands in western Texas drove 400 sheep into the Canadian River. In separate incidents in Wyoming, cattlemen drove 10,000 sheep into the mountains to die and more than 12,000 sheep off a cliff. In northern Arizona, issues between cattlemen and shepherds got hot enough to escalate into what is known as "The Graham-Tewksbury Feud"—a decadelong series of skirmishes between two families. Before it was over, 26 cattlemen and 6 shepherds were dead, along with scores of beasts.

The Sheep Wars ended around 1900 with the closing of the free prairie and the introduction of laws that made fence-cutting a felony. Though sheep and cattle coexist peacefully, today it is the cowboy who defines the American West.

The First Flag-Raising on Iwo Jima

Each year, more than 30,000 people compete in the Marine Corps Marathon, which finishes next to Arlington Cemetery, the resting place of American war dead, just outside of Washington, D.C. As runners climb the final hill near the finish line, they're inspired by the Marines Corps War Memorial—the 60-foot-tall statue of leathernecks raising a flag over Iwo Jima's Mount Suribachi. Few realize that the iconic image depicts not the first but the second flag that was hoisted at Iwo Jima.

✳ ✳ ✳ ✳

THE FAMOUS EMBLEM-RAISING took place shortly after noon on February 23, 1945; the first happened about 90 minutes earlier.

For four days, thousands of U.S. Marines were killed or wounded while battling toward the volcanic mount that dominates the island. "The Japs had the whole beach zeroed in. Most of the fire was coming from Suribachi," said Corporal Charles W. Lindberg, one of the first flag's raisers.

Capturing Mount Suribachi

Lieutenant Colonel Chandler W. Johnson commanded the 2nd Battalion, 28th Regiment, 5th Marine Division. His task was to capture the top of Suribachi. On the morning of February 23, after one of Johnson's companies scouted a path to the crest, he assigned the mission to 40 marines from 2nd Battalion's Company F. Johnson and gave the company commander, First Lieutenant Harold G. Schrier, an American flag that measured 54 inches by 28 inches. "If you get to the top," Johnson told Schrier, "raise it."

One of the patrol's members was Sergeant Ernest Ivy Thomas. On February 21, two days before the flag was raised, he had taken command of his rifle platoon after its leader was wounded. Armed with only a knife after his rifle was shot away, he repeatedly braved enemy mortars and machine guns while directing tanks against bunkers at the base of Suribachi. Led by Lieutenant Schrier, the patrol began its climb at 0800 hours on February 23, with Lindberg lugging a 72-pound flamethrower. They reached the top at 1015 and were attacked by a small Japanese force. While the skirmish continued, Thomas and another man scrounged up a 20-foot-long iron pipe. At 1020, recalled Lindberg, "We tied the flag to it, carried it to the highest spot we could find, and raised it." Hoisting the ensign was Lindberg, Thomas, and Schrier, along with Sergeant Henry "Hank" O. Hansen, Private First Class Louis C. Charlo, and Private First Class James Michels.

The First Photo

Sergeant Lou Lowery, a photographer for *Leatherneck* magazine, took a photo of the flag-raising. While he snapped the picture, a Japanese soldier tossed a grenade nearby. Just in time, Lowery threw himself over the crater's lip. He landed 50 feet below, his camera lens smashed, and yet his film remained intact.

After the raising, "All hell broke loose below," said Lindberg. "Troops cheered, ships blew horns and whistles, and some men openly wept."

"Make It a Bigger One!"

On the invasion beach was Secretary of the Navy James Forrestal, who'd arrived to watch the capture of Suribachi with Marine Commander General Holland "Howlin' Mad" Smith. Forrestal told Smith, "Holland, the raising of that flag on Suribachi means a Marine Corps for the next five hundred years." According to the book *Flags of Our Fathers* by James Bradley, Forrestal requested the emblem as a souvenir.

Battalion chief Johnson was not pleased. "The hell with that!" he said. He ordered an operations officer, Lieutenant Ted

Tuttle, to have another patrol secure the flag and replace it with another banner. "And make it a bigger one!" he told Tuttle.

Soon after, as Lowery climbed down from Suribachi, he ran into three fellow photographers going up: Marines Bob Campbell and Bill Genaust, and the Associated Press's Joe Rosenthal, whose photo of the second flag-raising garnered him the Pulitzer Prize. The three photographers were considering going back down the mount, but Lowery informed them the summit offered good views, so they kept trudging up. At the top, Schrier had the first flag lowered at the same time as the second flag—which measured 96 inches by 56 inches—was raised. Down on the sands, most didn't notice the switch.

The Battle Raged On

The terrible battle continued. Of the 40 marines in the first patrol, 36 were later killed or wounded on Iwo Jima. Private First Class Charlo, grandson of a noted Flathead Indian chief, was killed March 2, and Sergeant Thomas died on March 3. Photographer Genaust had taken a video of the second flag-raising, which featured imagery similar to Rosenthal's; his film was featured in a famous newsreel. He was killed March 4. Altogether, close to 7,000 marines died and 19,000 were wounded in the battle.

On March 1, Lindberg was wounded on Suribachi and was awarded the Silver Star. "I was after a mortar position up there, and I was shot, and it shattered my arm all to pieces," Lindberg said. Back in the States, he was angered by his patrol's lack of recognition. "I went on home and started talking about this," he stated. "I was called a liar and everything else. It was terrible." But "it was the truth," he said. "I mean, everyone says, 'Iwo Jima flag-raising,' they look at the other one."

The Japanese were also bitter, with good reason—they lost the battle. Iwo Jima was part of the prefecture of the city of Tokyo. Its loss meant that the U.S. Marines had secured their first piece of official Japanese soil. The rest of Japan beckoned.

One Lucky Lady

"Three strikes and you're out" apparently didn't apply to Violet Jessop. The lucky Englishwoman served aboard the infamous trio of White Star ships (Olympic, Titanic, Britannic) when accidents befell them—and lived to tell the tale each time.

IN 1911, VIOLET JESSOP was working as a first-class attendant on the *Olympic*. While cruising off the Isle of Wight, the British warship HMS *Hawke* collided with the vessel. Both ships sustained heavy damage, but they managed to stay afloat and no lives were lost. Strike one.

In 1912, Jessop was working aboard the luxury steamer *Titanic*. After the ship struck an iceberg, Jessop was ordered into a lifeboat. There, she watched the enormous ship plunge into infamy in the icy waters of the North Atlantic. Strike two.

Jessop's most harrowing moment came in 1916 while she was serving as a nurse aboard the doomed HMHS *Britannic*. Jessop was forced to abandon the swiftly sinking vessel after it had hit a mine. "I leapt into the water but was sucked under the ship's keel which struck my head," said Jessop. Somehow, she managed to survive. Strike three.

At this point an "out" would normally be called. But not for this plucky lass. She continued to sail and even served on the *Olympic* once again before retiring from sea duty. The unlikely tale of Violet Jessop, thrice saved by the fickle hand of fate, seems nothing short of miraculous.

✳ After the *Britannic* struck an underwater mine, its trip to the bottom of the Aegean Sea was extraordinarily swift. Contributing to the speed were scores of open portholes on the vessel's lower decks. Left open as a way to ventilate the ship's hospital wards, they allowed water to pour in when the vessel began to list. The ship slid beneath the waves less than an hour after the explosion.

Chapter 14

Around the World

Weird Hotels

What makes a hotel weird? It's not ugly curtains, a funny smell in the carpet, or wagon-wheel coffee tables in the lobby. To earn a spot on this list, a hotel has to be in a weird place, be made out of weird materials, or otherwise completely embody a weird ethos.

✳　✳　✳　✳

Icehotel (Jukkasjärvi, Sweden)

AT THIS HOTEL, guests sit on chairs made out of ice, drink from ice cups, and sleep on ice beds. With an average indoor temperature of 23 degrees Fahrenheit, it's just like any other hotel, only much colder. Guests can book a night in one of the ice rooms or ice-sculpture-filled suites, get married in the exquisite ice chapel, and meet friends at the ice bar. But make your reservation early: This hotel is only open from December to April, when it melts. It is rebuilt the next year.

Capsule Inn Akihabara (Tokyo, Japan)

Designed to be economical and space-saving, the concept of the capsule hotel is catching on in Japan and other parts of the world. For approximately $40 per night, guests get an 18-square-foot metal cubby (3 feet wide, 3 feet tall, 6 feet long)—just big enough for a body to get in and lie down. The "room" comes with a mattress, a pillow, a TV, a radio, an alarm clock, lights, curtains, and a control panel similar to an airplane's; everything is accessible from a prone position.

Jules's Undersea Lodge (Key Largo, Florida)

To enter this former research laboratory, guests must dive 21 feet underwater to reach the entrance. Built five feet above the ocean floor, this two-bedroom hotel has a view like no other, as guests are greeted by tropical fish peering in at them through the windows. With plenty to do and see, guests can remain underwater for days at a time.

Das Park Hotel (Linz, Austria)

Due to the strange nature and limited amenities of this hotel, guests are asked to "pay as you wish." Between May and October, guests can stay in three bedrooms made from concrete drainpipes outfitted with double beds and electricity. For luxuries such as restrooms and showers, guests must use the facilities at the public park surrounding the hotel.

The Old Jail (Mount Gambier, Australia)

This hotel/dorm/hostel operated as the South Australian State Prison from 1866 to 1995. Now it offers cheap beds in a gorgeous landscape, complete with gun turrets and barbed wire. Guests sleep tight behind bars in the original cells, shower in communal baths in the former women's cellblock, and can play soccer in the enclosed recreation yard.

Dog Bark Park Inn (Cottonwood, Idaho)

A bed-and-breakfast isn't complete without a friendly dog, but this B & B actually *looks* like a dog. It was created by chain saw artists whose carved wooden dogs are featured on the bedposts and in the gift shop. Up to four guests can sleep in Sweet Willy, which is currently the world's largest beagle at 30 feet tall, while the world's second-largest beagle, 12-foot-tall Toby, looks on.

Woodlyn Park (Otorohanga, New Zealand)

Would you rather sleep in a 1950s train car, a 1950s cargo plane, a World War II patrol boat, or the "world's first hobbit motel"? At Woodlyn Park, you can try them all out. Guests can also enjoy nearby glowworm caves and a Kiwi Culture show that includes sheep shearing, log sawing, and a dancing pig.

Harbour Crane Hotel (Harlingen, Netherlands)

This one-room hotel sits on a dock overlooking the Wadden Sea. Used as a crane until 1996, guests can still slip into the cabin to control the machine for panoramic views of the harbor from more than 50 feet above ground. Two elevators take guests and (very small) bags to the former machine room, where a minibar, a plasma TV, a rooftop patio, ultramodern decor, and a two-person shower with a light show await.

Controversy Tram Hotel (Hoogwoud, Netherlands)

The strange brainchild of Frank and Irma Appel, this hotel consists of two tram cars and a luxury train car. The two trams can sleep four couples and contain bathrooms and common rooms; the luxury train car also includes a whirlpool tub and can sleep four. A UFO-styled library floats nearby, overlooked by a MIG fighter jet. The hosts' house is built around a French van and a double-decker bus. Their pet donkeys sleep in the bus, but their cat prefers the roof of a Lamborghini.

Ariau Amazon Towers Hotel (Manaus, Brazil)

This eco-resort is built entirely at the level of the rainforest canopy. With seven towers linked by more than four miles of catwalks, this treehouse hotel offers eye-level views of monkeys, birds, and sloths, while piranhas circle below. Although it might seem more suited to Tarzan's tastes, it promises a glimpse of the Amazon that is usually off-limits to tourists.

Wigwam Village Motel, Various Locations

The first of several Wigwam Village motels was built in 1933 near Mammoth Cave in Kentucky. Seven more dotted the southern and southwestern United States by the early 1950s. Each wigwam features a guest room that is naturally suited to the southwestern stretch of Route 66. The Holbrook, Arizona; Rialto, California; and Cave City, Kentucky, locations are still in business; the marquee in front of the Holbrook location poses the question, "Have you slept in a wigwam lately?"

Shenzhen, China: The World's Fastest Growing City

Three decades ago, Shenzhen was sparsely populated farmland. Today, it's a city of more than ten million people.

✳   ✳   ✳   ✳

THE YEAR IS 1978, and the small area in southern China soon to be known as Shenzhen is thinly populated with farmers and fishermen. The inhabitants of this rural enclave don't know it yet, but something's astir in Chinese politics that will forever change their lives. Their rural land, a place that for centuries was just a minor producer of crops, also has some features that can't be found anywhere else. Shenzhen is located amid the Pearl River Delta, making it a prime spot for an export economy. It also happens to be just north of Hong Kong, the world's busiest port city.

The change began in the closed chamber rooms of the Chinese government. In the late 1970s, the economy of the People's Republic of China underwent a radical transformation. The new leader of the Republic, Deng Xiaoping (1904–1997), sought to modernize and globalize the country's economy. He appreciated the vast potential of China's massive size and population, as well as the necessity of international trade. Yet in order to be a global player, China needed something that it didn't yet have: the West's technology and industrial management skills.

Deng encouraged industries throughout China to open up to foreign partnerships, and he relaxed government restrictions on trade and financial institutions. As part of this plan, Deng established four Special Economic Zones in China. Different areas in Africa, Asia, and South America enjoy lower taxes and fewer commercial restrictions in order to encourage export-based economies. These specialized trade laws enable the developing world to supply the Western world's vast consumption. The Chinese government took many of the strategies of these duty-free international zones and applied them to their own economy as well.

China's State Council officially declared Shenzhen a Special Economic Zone in 1980. Shenzhen quickly proved to be the fastest growing and most influential of all the zones. Migrant workers from across China flocked to the area like moths to a lightbulb. In 1979, the population of the Shenzhen area was, at most, 300,000. Today, it is at least ten million—making it more populous than New York City. The Chinese government supplied some of the funds for infrastructure development, but this rapid transformation was mostly the result of joint ventures with foreign companies. Chinese companies supplied the cheap land, labor, and materials; the foreign investors supplied the technology and management.

Within ten years, Shenzhen was a full-fledged modern city, complete with skyscrapers, fancy hotels, and petty crime. Yet unlike other cities, Shenzhen was populated with such haste that it developed some unique characteristics. Shenzhen's population of migrant workers is unusually high. In 2006, poor migrant workers accounted for more than three-quarters of the population. Complex residency restrictions make it difficult for these laborers, who decades ago lived in small villages throughout China, to settle into Shenzhen for the long haul.

Watch What You Say!

These common American gestures can get you in hot water overseas!

＊　＊　＊　＊

When "OK" Is Not Okay

THE NEXT TIME you're visiting Brazil, refrain from giving the traditional "OK" sign by touching the tips of your thumb and forefinger. In South America's largest nation, that all-American gesture is considered obscene.

That's just one example of how body language can get you into trouble when traveling overseas. International travel experts note that a run-of-the-mill gesture in the United States may have a completely different meaning in other parts of the world.

Take, for example, the "V for victory" sign, which is made by raising the index and middle finger of one hand, palm facing out. (Since the '60s, this has also become known as the "peace" sign.) This gesture is perfectly acceptable in the United States, but in England, if the palm faces inward, it means "up yours."

Depending on what country you're visiting, the "thumbs up" gesture may have a different connotation. In the United States, it typically means that something is met with approval. In Australia, however, if you make a slight jerk upward with the extended digit, it means "up yours." The gesture is also an insult in Nigeria, so keep that in mind when hitchhiking there.

And then there's the common gesture of rotating a finger around one's ear, which in the United States usually signifies that someone is crazy. Not in Argentina, however. There, the gesture means that you have a telephone call.

Don't Get Me Wrong...

Two potentially problematic gestures are nodding and shaking your head. In the United States, of course, a nod means yes and a shake means no. One would think this is universal, but in parts of Greece, Bulgaria, Yugoslavia, Turkey, and Iran, the reverse is true—a shake of the head means yes and a nod means no.

Personal contact can be a particularly touchy issue when traveling abroad. In the United States, we're taught to give each other a little space when conversing, usually an arm's length. But in densely populated Japan, where personal space is appreciated and touching of any kind is considered taboo, the distance between two conversing individuals is typically greater.

Making Friends in the Middle East

However, the opposite is true in many Middle Eastern countries, where it's not uncommon to see two people standing almost toe to toe as they talk and touching each other's arms and bodies in the process. In the Middle East, it's also perfectly acceptable for two men to walk hand in hand in public. The gesture has no particular sexual connotation; it's just a sign of friendship.

When visiting the Middle East, you should also refrain from inadvertently showing others the soles of your shoes by crossing your legs. The gesture is considered extremely insulting because the soles of the shoes are viewed as dirty. Public displays of affection between men and women are also frowned upon, and in some very strict Muslim nations, it can even land you in jail. In addition, get used to using your right hand for everything, including eating, presenting gifts, and touching others. In Middle Eastern culture, the right hand is the "clean" hand, while the left hand is considered "unclean" and is used primarily for bodily hygiene. Sorry, southpaws.

World travel can be enlightening and a lot of fun, but make sure that you are well versed in the local customs of the regions you visit. Otherwise, an innocuous hand gesture could result in a punch to the nose!

Great Canadian Adventures

Everyone's heard that Montreal is charming, Victoria is quaint, and Blue Jays games are fun, but here are some Canadian roads (and places lacking roads) less traveled.

✳ ✳ ✳ ✳

Dempster Highway (Yukon Territory/Northwest Territory): This 417-mile gravel road will take you from Dawson, Yukon Territory (of gold rush fame), to Inuvik, Northwest Territory. You'll cross the Arctic Circle after the halfway point. Take two spare tires, fuel up at every opportunity, and go in late August or early September, when the mosquitoes and blackflies ease up and the fall colors are in their glory.

Churchill (Manitoba): For the moment, you can still see polar bears and beluga whales (July through August) near this Hudson Bay community and seaport. It's not a big place (pop. 923), and you'll have to fly or take a train from Winnipeg or Thompson because no road extends to Churchill. There's kayaking, an old Hudson's Bay Company stone fort, bird-watching, and more.

Jasper National Park (Alberta): Many people go to Banff yet never venture up the spine of the Canadian Rockies to take in this gem. Walk the trail of Maligne Canyon, golf at Lac Beauvert, take the tram up to Whistler's Mountain (not to be confused with Whistler ski resort in British Columbia), or just stare in awe at Stutfield Glacier.

Toronto (Ontario): Toronto is Canada's cleaner, smaller, and safer version of New York City, and a quarter of all Canadians live in the surrounding region. Highlights include the CN Tower, the Ontario Science Center, the Toronto Zoo, the Art Gallery of Ontario, the Gardiner Museum of Ceramics, Fort York historic site, and the Hockey Hall of Fame. Toronto offers plenty of quality hotels, good transportation, and fine dining—some with dress codes, so pack appropriately.

Charlottetown (Prince Edward Island): One used to have to boat or fly to Prince Edward Island, Canada's smallest province; now there's a toll bridge. If you love to golf, take your clubs: There are 20 courses within an hour of Charlottetown where you can slice tee shots into the rough. Otherwise, enjoy the historic architecture, friendly atmosphere, and unspoiled natural beauty.

Acadian Coast (New Brunswick): Acadian culture is not Quebecois culture, though it does involve the French language. New Brunswick's north coast is the Acadian part of the province. Walk the long, snaking boardwalk along the odd Dune de Bouctouche (a big sandbar with dunes), and go the short distance to Caraquet for immersion in Acadian history.

Quebec City Winter Carnival (Quebec): North America has one walled city: historic Quebec, a Francophone city that's far smaller than Montreal. Take the kids and go in February—that's Winter Carnival time. While French is the first language of locals, during le Carnaval de Québec, you can surely find English-speakers if needed. Take in the Ice Palace, competitions on the Plains of Abraham, and views of the lighted, snowy city at night.

Vancouver to Kamloops (British Columbia): The Canadian San Francisco, Vancouver is a large port city full of great dining and ethnic diversity. Take the freeway up the Fraser Valley past Bridal Veil Falls, past Hope (with its Hell's Gate tram across the river), and up the gorgeous Coquihalla Highway to the ruggedly Western 'Loops. Go in late September or October to catch a Kamloops Blazers game. Until you've seen a junior hockey game in a smaller town, you haven't quite touched the soul of Canada.

Cape Breton (Nova Scotia): For the full maritime experience, you must go to sea. Take a boat outing to the stony Bird Islands, where in season you'll see puffins, guillemots, cormorants, bald eagles, kittiwakes, and even gray seals. Stay in a seaside cottage where you can visit the Louisbourg fortress or explore the Cabot trail. Canada's great inventor, Alexander Graham Bell, did his later research in Cape Breton.

Labor of Love: The Taj Mahal

Known as one of the Wonders of the World, the Taj Mahal was a shrine to love and one man's obsession. Each year, around three million tourists travel to see what the United Nations has declared a World Heritage site.

✳   ✳   ✳   ✳

Taj Mahal: Foundations

THE MUGHAL (OR "Mogul") Empire occupied India from the mid-1500s to the early 1800s. In 1628, a young prince named Khurram took the throne. Six years prior, Khurram's father gave him the title Shah Jahan after a military victory. With much of the subcontinent at his feet, the title was apt: *Shah Jahan* is Persian for "King of the World."

When Khurram Met Arjumand

Being shah had a lot of fringe benefits—banquets, treasures, and multiple wives, among other things. Shah Jahan did have several wives, but one stood out from the rest. When he was 15, he was betrothed to 14-year-old Arjumand Banu Begam. Her beauty and compassion knocked the emperor-to-be off his feet; five years later, they were married. The bride took the title of *Mumtaz Mahal*, which means "Chosen One of the Palace," "Exalted One of the Palace," or "Beloved Ornament of the Palace."

Court historians have recorded the couple's close friendship, companionship, and intimate relationship. The couple traveled extensively together; Mumtaz even accompanied her husband on his military jaunts sometimes. But tragedy struck on one of these trips in 1631, when Mumtaz died giving birth to what would have been their 14th child.

Breaking Ground

Devastated, Shah Jahan began work on what would become the Taj Mahal, a palatial monument to his dead wife and their eternal love. While there were surely many hands on deck for the planning of the Taj, the architect who is most often credited with the building's design is Ustad Ahmad Lahori. The project took until 1648 to complete and enlisted the labor of 20,000 workers and 1,000 elephants. The structure and its surrounding grounds cover 42 acres. The following are the basic parts of Mumtaz's giant mausoleum.

The Gardens: To get to the main structure of the Taj Mahal, one must cross the enormous gardens surrounding it. In classic Persian garden design, the grounds to the south of the buildings are made up of four sections divided by marble canals (reflecting pools) with adjacent pathways. The gardens stretch from the main gateway to the foot of the Taj.

The Main Gateway: Made of red sandstone and standing approximately 100 feet high and 150 feet wide, the main gateway is composed of a central arch with towers attached to each of its corners. The walls are richly adorned with calligraphy and floral arabesques inlaid with gemstones.

The Tomb: Unlike most Mughal mausoleums, Mumtaz's tomb is placed at the north end of the Taj Mahal, above the river and between the mosque and the guesthouse. The tomb is entirely sheathed in white marble with an exterior dome that is almost 250 feet above ground level. Depending on the light at various times of the day, the tomb can appear pink, white, or gold.

The Mosque and the Jawab: On either side of the great tomb stand two smaller buildings. One is a mosque, and the other is called the *jawab*, or "answer." The mosque was used, of course, as a place of worship; the jawab was often used as a guesthouse. Both buildings are made of red sandstone, so as not to take away from the grandeur of the tomb. The shah's monument to the love of his life still stands, and still awes, more than 360 years later.

Hong Kong: The City That Isn't

Hong Kong was once a British crown colony. Now it's a Special Administrative Region of China. But the one thing that it has never been is a city.

✳ ✳ ✳ ✳

HONG KONG IS a dynamic Asian metropolis that seemingly rises directly from the waters of the South China Sea. Its main harbor is ringed by a captivating phalanx of skyscrapers and towers that illuminate the night with vivid neon hues.

Unquestionably, Hong Kong would rank with New York, London, and Rome as one of the world's great cities if not for a minor detail: Hong Kong is not a city. It is, officially, the Hong Kong Special Administrative Region (SAR), a territory governed by China that enjoys autonomy in its political and economic affairs (aside from foreign relations and defense). The Hong Kong SAR came into being in 1997, when Great Britain transferred control of Hong Kong to China after ruling the territory as a crown colony since 1842. Initially, the colony consisted of only Hong Kong Island, but in the span of 55 years, its territory expanded to include the Kowloon Peninsula, Lantau Island, and 260 smaller islands.

Within Hong Kong are several cities and towns, which are concentrated mainly on the Kowloon Peninsula and the northern shore of Hong Kong Island. These settlements have no formal boundaries and mesh into one urbanized area that's crammed with most of Hong Kong's seven million residents—making Hong Kong seem like one big city.

✳ Another popular misconception about Hong Kong is that it is a monolithic urban jungle. In reality, it is a very green place— less than a quarter of its 426-square-mile area is developed. The rest is protected parklands, nature reserves, beaches, and lushly forested hillsides and mountains.

In English, *Por Favor*

It may be just a common misconception, but is English really the most widely spoken language on Earth?

✳   ✳   ✳   ✳

THERE ARE BETWEEN 5,000 and 6,000 languages spoken around the world today, depending on the criteria used to differentiate *language* from *dialect*. Approximately a third of these languages are spoken by no more than 1,000 people, but 200 are used by more than a million native speakers each. Of course, we don't need to concern ourselves too much with most of these languages because, as English speakers, we're already fluent in the most commonly spoken language in the world—or at least that's what many of us mistakenly believe.

Estimates of how many people speak a language tend to be imprecise, but there is no doubt that Mandarin Chinese has the most native speakers—about 1.1 billion compared with approximately 330 million native English speakers. Mandarin developed from the principal dialect that was spoken in the Beijing area of China and is now the country's official language. It is also spoken outside of China in countries such as Brunei, Cambodia, Indonesia, Malaysia, Mongolia, the Philippines, Singapore, and Thailand.

English, on the other hand, is spoken in 115 countries around the world, more than any other language. Many people speak English as a secondary language, and it is considered the primary language of international diplomacy. However, if we include all the people who use English as a secondary language, the number of speakers swells to approximately 480 million—still a distant second to Mandarin Chinese.

So if you plan on traveling abroad, we recommend that you at least master the following Mandarin phrase: *Ni hui jiang yingyu ma?* ("Do you speak English?")

Odd and Unusual Structures

Whether they're made out of bones or toilet paper tubes, the following structures rank among some of the most creative and outlandish in the world.

✳ ✳ ✳ ✳

Sedlec Ossuary: The Skeleton Sanctuary

A CHANDELIER MADE OF every bone in the human skeleton, a heap of 14th-century skulls with arrow wounds, and a skull and crossbones atop a tower all make the little chapel known as the Sedlec Ossuary a most unusual church.

Located just outside the medieval silver-mining center of Kutna Hora, Czech Republic, in a suburb now called Sedlec, the chapel was predated by a cemetery that was made famous in 1278 after a church official sprinkled it with soil from the Holy Land. The chapel was constructed in 1400, but by 1511, the cemetery was so overcrowded that bones were dug up and stored inside the building. In 1870, a woodcarver named Frantisek Rindt was hired to organize the bones of the 40,000 people stashed in the ossuary; he decided to assemble them into a fantastic assortment of altars, sculptures, and other furnishings and decorations.

Shigeru Ban's Houses of Cardboard

In the 1980s, Japanese architect Shigeru Ban began making buildings from cardboard tubes because he felt that good structures should be affordable to anyone and that a new kind of architecture could arise from so-called "weak materials." Ban used his idea to develop inexpensive yet durable shelters for refugees from natural disasters. He has also designed spectacular buildings from what he calls "improved wood." One of the most famous is his Nomadic Museum, which is made out of approximately 150 shipping containers that can be sent to and reassembled anywhere in the world. Inside, it features giant curtains made from thousands of recycled tea bags.

Dancing House of Prague: A Building that Boogies

Originally named the "Astaire and Rogers" building after the famous dance duo, this modern structure's swaying towers suggest a dancing couple. The building, which houses a popular restaurant, was designed by architects Vlado Milunic and Frank Gehry and built in the mid-1990s. Set in a traditional neighborhood of Baroque, Gothic, and Art Nouveau buildings, the crunched appearance of the Dancing House inspired great local controversy when it was first proposed.

Bangkok Robot Building: Banking on Robots

The world's first robot-shaped building houses the United Overseas Bank headquarters in Bangkok, Thailand, and was designed to reflect the hope that robots will someday release humanity from the burden of drudge work. Designed by Dr. Sumet Jumsai and built mostly of native materials, the energy-efficient, 20-story "robot" includes a day-care center and an 18th-floor dining room with a sweeping view of the city. The building, which was completed in 1986, has reflective glass eyes, lightning rod "antennae," and bright blue walls.

Topsy-Turvy in Tennessee

Imagine that a colonnaded building from some genteel laboratory was transported in a windstorm and landed upside down, where it remained . . . chock-full of strange anomalies for visitors to explore. That's the scenario that operators of Wonder Works in Pigeon Forge, Tennessee, and Orlando, Florida, hope paying guests will believe. Both places appear to be upside down on the inside too, and patrons can experience a re-creation of the 1989 San Francisco earthquake, among other interactive displays. The Orlando attraction was built in 1998, and its sister structure plopped in Pigeon Forge in 2006.

Poland's Crooked House: Wavy-Walled Wonder

Tipsy bar patrons who wander outside the Crooked House in Sopot, Poland, may rightly wonder if their eyes deceive them. But whether viewed with or without an alcoholic haze, the curving facade is purposely out-of-kilter, and its green shingles are intended to look like the scales of a dragon. The building is home to several pubs, coffee shops, and other businesses and has become a popular attraction since it opened in 2004.

The Haines Shoe House: A "Boot"-iful Home

Anyone wanting a place to just kick back should focus solely on a boot-shape home near Hellam, Pennsylvania. The 25-foot-tall building was created in 1948 by shoe magnate Mahlon Haines to help publicize his wares. The house was designed to function as an actual house with a kitchen, two bathrooms, three bedrooms, and a living room. It's currently an ice cream shop but has also served as an inn.

Futuro Flying Saucer Homes

In the 1960s, people were looking forward to a futuristic tomorrow, while at the same time, waves of UFO sightings swept the world. Finnish architect Matti Suuronen combined the two social phenomena to create houses that resembled flying saucers. Suuronen's Futuro units were made from fiberglass and measured 26 feet in diameter and 11 feet high. Like any good spaceship, the Futuro homes had windows that looked like portholes, and visitors entered units through a lower hatch. The insides featured a set of built-in plastic chairs, a kitchenette, and a tiny bedroom and bathroom arranged around a fireplace. Suuronen envisioned the Futuro units as the ultimate mobile home, so they were designed to be moved easily. Although the oil embargo of the early 1970s drove up petroleum prices so high that plastic houses were no longer affordable, about 100 models were produced, and several still exist.

Houston Beer Can House: The Six-Pack Shack

The ultimate in recycling projects, John Milkovisch's Houston home, which is covered with siding made from smashed beer cans, has proven that even a six-pack-a-day habit (which is about what it took to provide materials over the course of 20 years) can be eco-friendly. After his retirement from the railroad in 1968, Milkovisch started the project just to have something to do and drank some 40,000 cans of Coors, Bud Light, and Texas Pride to create his unique habitat. For a bit of flair, he added wind-chime curtains made from pull tabs and arranged horizontal rows of cans to provide decorative fences. Both Milkovisch and his wife, Mary, are now deceased, so an arts preservation foundation now owns the property.

South Korea Toilet House: The Potty House

South Korean lawmaker Sim Jae-Duck is flush with pride over the house he built in Suwon in 2007—the home is shaped like a giant toilet. Sim Jae-Duck, who built the home for a meeting of the World Toilet Association, said that he hoped to persuade people to think of toilets not just as a "place of defecation" but also a "place of culture." The two-story home is outfitted with four actual toilets, including one encased in motion-sensitive glass that fogs up for privacy. The house's staircase spirals where the drain would be if the structure were actually a toilet. The 4,520-square-foot house stands on the site of Sim Jae-Duck's former home, which was tanked for the new project.

A Tisket, a Tasket, That Building's a Basket!

One of the most creative corporate headquarters anywhere is the Longaberger building in Newark, Ohio. Built to mimic the company's product—woven baskets—the structure includes a giant version of the Longaberger brand tag and trademark "bent wood" handles balancing over the top. The building stands seven stories high with windows set in indentations of the synthetic basket-weave exterior.

The City That Rose from the Sand

Over the past 50 years, Dubai has transformed from a small trading outpost to one of the most powerful cities in the world. The area's exponential growth can be traced to the investment sense and tenacious imagination of the city's ruling royal family.

✳ ✳ ✳ ✳

THE STUNNING CITYSCAPE of Dubai, a Middle Eastern city perched on the Persian Gulf, seems almost surreal in its splendor. Dubai is home to the world's largest indoor ski resort, the only seven-star hotel on the planet, the world's largest mall, and a set of artificial islands in the shape of a palm tree that can be seen from space. In the making is a cluster of islands shaped like a map of the world. And let's not forget the Burj Khalifa, the tallest building in the world—which surpasses the second-tallest by more than 1,000 feet. Investment bankers and real estate moguls from both the West and the Middle East are drawn to this bastion of tourism and wealth.

Descended from the Desert

The area around Dubai is a sparse desert landscape. Over the millennia, the various nomadic tribes that inhabited the area were involved in trade, especially with nearby India. The modern historical trajectory of Dubai didn't begin until 1833, when 800 members of the Bani Yas tribe made the city an outpost for sea and pearl trading. Descendants of this same tribe still rule the city, but now the family is a constitutional monarchy.

Dubai is one emirate in a nation known as the United Arab Emirates (UAE); Dubai City lies within the Dubai emirate. The UAE was formed in 1971, when the British officially abandoned their colonial influence over the region. The wealth of Dubai is inextricably linked with the oil wealth of Abu Dhabi, another of the UAE's emirates. Dubai traded in oil extracted from Abu Dhabi until 1966, when oil was discovered in Dubai. In the following years, Dubai's population grew immensely.

Transformation Takes Flight

But Dubai's recent exponential growth comes from investment income, not oil wealth. In 1985, Dubai became a pet project of the UAE's then-prince (now prime minister) Sheik Mohammed bin Rashid Al  Maktoum. It all started with the well-known plight of travel inconvenience: One day, Maktoum's flight out of Dubai was cancelled. Because of this, Maktoum decided to invest his money in a new international airline. Today, this airline is known simply as Emirates, and it is the world's second-most-profitable airline.

The royal family realized the potential of investing their own money to make Dubai a center of finance and tourism. They hosted international sporting events, established zones with special tax-free trading and finance laws, and sought foreign capital in real estate and other ventures. To make all this happen, an influx of immigrant construction and service workers inundated the city. Liberal laws were established so that foreign residents of Dubai would not be subject to the region's religious traditionalism. (Islam is the official religion of the UAE, but immigrant workers are allowed religious freedom.)

The royal family's plans were immediately successful. In 1985, the population of Dubai was 370,800; today it is 1.4 million. As with other cities that were created out of nothingness through sudden wealth investment, the population of Dubai is oddly lopsided: Almost 90 percent of Dubai's population comes from outside the UAE. Still, UAE nationals are granted special rights and privileges, especially when it comes to land ownership. Living side by side with the affluent nationals and foreign investors that come to Dubai for its promise of wealth is a much larger population of poor migrant families who built the city up from the desert in a blink of the metaphorical eye.

Unusual World Customs

Table Manners

✳ In Italy, eat spaghetti as the Romans do: with a fork only. Using a spoon to help collect the pasta is considered uncouth.

✳ It is considered improper and impolite to use silverware to eat chicken in Turkey.

✳ Keep your right elbow off the table when eating in Chile.

✳ In the United States, one should never butter an entire piece of bread before eating it. The proper, albeit impractical, way to eat bread is to pull off a small bit from your larger piece and butter it before popping it into your mouth.

✳ In China, slurping one's food and belching at the end of a meal are considered acceptable and even polite.

✳ Tipping is uncommon, and even considered rude, in many Asian countries.

Personal Care

✳ Up until the 19th century, long fingernails were considered a symbol of gentility and affluence among the Chinese aristocracy. Wealthy Chinese often sported fingernails that were several inches in length and protected them by wearing special coverings made of gold.

✳ In ancient Rome, urine was commonly used to whiten teeth.

✳ In many countries, toilet paper is unheard of. Instead, people use their left hand to wash themselves after going to the bathroom. For this reason it is considered rude to use the left hand in many social situations.

Body Language

✳ When conversing in Quebec, keep your hands where they can be seen. Talking with your hands in your pockets is considered rude.

* In Chile, pounding your left palm with your right fist is considered vulgar.

* In Thailand, feet are considered unclean. Using one's foot to move an object or gesture toward somebody is considered the height of rudeness. Similarly, one should never cross their legs when in the presence of elders.

* Make sure to get enough sleep when traveling through Ecuador. Yawning in public is tacky.

Rituals and Traditions

* In China, small, dainty feet were considered a symbol of status. Until being banned in 1912, foot binding was common in China. The practice, which involved breaking girls' toes and wrapping them tightly in cloth, prevented women's feet from growing normally.

* In parts of India, some women still perform *sati*, an ancient custom in which a widow commits suicide by throwing herself on the funeral pyre of her deceased husband.

* For almost 500 years, a form of conflict resolution known as "dueling" took place in western Europe and the United States. The highly ritualistic tradition began with the offended party throwing down his glove at the foot of another and ended with a sword or pistol fight—often to the death.

Down Under

When the going got rough, these cities rose above—literally.

✳ ✳ ✳ ✳

Seattle, Washington

YOU MAY THINK that Seattle was built on Microsoft, but it was actually built on mud. In 1852, the first neighborhood of Seattle was settled in the area known today as Pioneer Square. Even though the builders brought in fill from surrounding areas, the city's sea-level status still posed some problems. When the tide came in, the streets would fill with mud and toilets would turn into gushing fountains. Sounds pleasant!

In 1889, a fire destroyed the young city, and a new Seattle was built on top of the ashes of the old. The city planners knew that this was their chance to fix the mistakes of the past, so with the relief money they received, they built retaining walls on the sides of the existing streets, filled the space, and raised the streets eight feet higher than they previously were. In effect, the first (that is, original) floors of stores and homes became the underground. Today, humorous tours are available to those who are looking to explore Seattle's underside.

Edinburgh, Scotland

Take a tour of Edinburgh's underground and you'll find plenty of fables and good old-fashioned scares. But the history of the city's underground is not quite so lighthearted. In Edinburgh's early days, the rich and the poor shared space. But as Old Edinburgh became overcrowded in the 18th century, the more fortunate folks moved to the north and south ends of the city. The North and South bridges were built to link the two wealthy areas. The South Bridge was built over a large ravine, and the area below the bridge was excavated. Rooms and chambers were created; these became known as the Edinburgh Vaults. The area was used primarily for storage, but over the years, it became home to a number of businesses.

When the bridge began to leak, the businesses moved out and the city's poorest and most destitute residents—including victims of the bubonic plague—moved into the vaults. The vaults were eventually filled and forgotten until they were rediscovered in 1985. Today, visitors can take tours of the vaults.

Rome, Italy

Visitors to this ancient city are familiar with the typical haunts: the museums, the Coliseum, and the Pantheon. But below the well-known attractions lies a labyrinth of catacombs, roads, temples, and houses that date back to the fall of the Roman Empire. Because of frequent floods, the citizens of Rome began to build upward and continued to do so for some 3,000 years, covering previous cities as well as their own history. Though a few archeologists have explored the underground in hopes of finding clues about the ancient city, much of it has remained untouched. According to *National Geographic* writer Paul Bennett, you can dig a hole just about anywhere within the walls that once surrounded the city and find something of historical significance. In fact, many of the discoveries to date have been made by accident. In 2008, while a modern rugby stadium was being renovated, an ancient cemetery was found underneath it.

Cappadocia, Turkey

This ancient region located in present-day Turkey features a number of cities that sit atop buried historic towns. In all, there are thought to be 150 underground dwellings lurking beneath the area's rocky landscape. Perhaps the best known of these, Derinkuyu, is eight stories deep—about 279 feet beneath the surface. These underground cities are thought to have been originally dug out upwards of 2,000 years ago and served as hiding places for early Christians who were escaping persecution from Romans and Muslims. Many of these refuges feature churches, trapdoors, and remnants of kitchens and shops, and were large enough to hold thousands of people. Today, a handful of these subterranean cities have been turned into swanky hotels and homes for affluent Turks.

Thomas Tresham Tied Triangles Together

Nothing prompts a person to build a bizarre, triangular stone lodge in the middle of nowhere quite like religious persecution. At least that's how Sir Thomas Tresham felt back in 1593 when he began work on what would become the mysterious Rushton Triangular Lodge, a structure that he hoped would be much more than a place to hang his hat.

✳   ✳   ✳   ✳

But Why?

A T THE END of the 16th century, life wasn't much fun for Catholics living in England, as Tresham could attest. As a devout Catholic, he'd spent 15 years in prison because his faith had made him a criminal in the eyes of the law. After he was released, Tresham figured he'd better keep his mouth shut about his religion, but that didn't stop him from professing his faith in other ways.

Tresham decided to build a secretly Catholic monument near Rushton that would encompass encoded messages to keep it safe from Protestant adversaries. The bricks and mortar of his lodge would showcase aspects of his faith without threatening his freedom. The man stuck to his plan so diligently that the details of Rushton Triangular Lodge are downright weird.

The Rule of Three

To represent the Holy Trinity, Tresham designed the building with only three walls. The structure is itself a perfect equilateral triangle, with its walls meeting at 60-degree angles. Glorifying the Holy Trinity via the rule of three is repeated (and repeated and repeated) throughout the building. Check this out:

✳ Each of the three walls is 33.3 feet.

✳ Each floor of the three-story building has three windows.

✳ Each wall has three gables. Each gable is 3 feet by 3 feet with three-sided pinnacles.

✳ There are nine gargoyles (three sets of three).

✳ Friezes run along the walls on each side of the building—each contains a phrase written in Latin, and each phrase has exactly 33 letters.

✳ Though the main room on each floor is hexagonal in shape, if you draw three bisecting lines through a hexagon, you get six more equilateral triangles.

✳ And if all those threes are making you a little dizzy, just wait—Tresham was far from content with a few triangular tricks. The building's ornaments were where Tresham spared no expense to work in secret codes for the glory of the Lord.

Gables, and Windows, and Math, Oh My!

Two of the gables of Tresham's lodge are inscribed with the dates 1641 and 1626. Indeed, Tresham carved future dates into the side of the building for a reason: If you subtract 1593 (the year he started building) from 1626, you get 33. Subtract 1593 from 1641, and you get 48. Both numbers are divisible by three—no big surprise there—but there's something more. If you add the Anno Domani (commonly known as "A.D."), you get the years of Jesus' death and the Virgin Mary's death, respectively. The second gable shows the dates 3898 B.C. and 3509 B.C., dates that are said to be the years of the Great Flood and the call of Abraham.

The windows provided another place for Tresham to work his code magic. The three windows on the first floor are in the shape of a Gothic trefoil, a vaguely triangular Christian symbol that also happened to be the Tresham family crest. The trefoil is carried through to the basement windows as well, all of which are, of course, repeated in threes.

Double Entendres

* As mentioned before, there were three 33-character-long inscriptions on the Rushton Lodge, one on each side. The inscriptions and their respective translations read as follows:

    * *Aperiatur terra & germinet salvatorem* means "Let the earth open and let them bring forth a Savior."

    * *Quis seperabit nos a charitate Christi* means "Who shall separate us from the love of Christ?"

    * *Consideravi opera tua domine at expavi* means "I have considered your works and am sorely afraid."

* In addition, if you inspect the waterspouts at Tresham's place, you'll find a letter above each one. Together, they create an acronym for the first three letters of a Latin mass. An inscription above the main door to the lodge reads *Tres Testiminium Dant*, which means "these three bear witness." But Tresham's wife is said to have called him "Tres" for short; knowing that, one might interpret this as "Tresham bears witness," which was certainly the point of all his obsessive building.

Even More Hidden Meaning?

Tresham got away with his secretly Catholic building, although it certainly raised a few eyebrows. In fact, the building (which is now maintained as a historical site by the English Heritage organization) is still a source of much discussion. Some people think that Tresham wasn't über-Catholic, but rather all those numbers and all that funky math were rooted in black magic.

Either way, the building is a great example of the era's love of allegory—using something to represent something else entirely. After the Triangular Lodge was completed, Tresham started *another* building full of secret codes and mysterious math called Lyveden New Bield, but he died before it was finished. It still stands in England exactly as it was left: half-built and full of its own mystery.

Chapter 15

Technology and Inventions

Learning to Fly

Back in the 18th century, all you needed to reach the sky was a silk bag, hot air, and a lot of guts.

※　※　※　※

The Heat Is On

AS THE NAME implies, hot-air balloons rely on heat: As heat is produced and ultimately rises, it gets caught within the lightweight balloon. Because the heated air is less dense, it causes the balloon to rise. Winds push the balloon along, but its operator can also control the device manually by increasing or decreasing the heat to raise and lower the balloon. It's a seemingly simple concept—but one that wasn't discovered until the late 1700s.

When Sheep Fly...

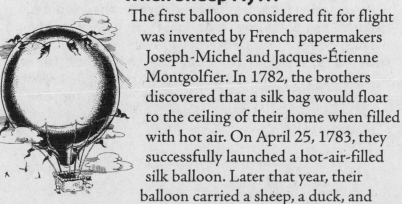

The first balloon considered fit for flight was invented by French papermakers Joseph-Michel and Jacques-Étienne Montgolfier. In 1782, the brothers discovered that a silk bag would float to the ceiling of their home when filled with hot air. On April 25, 1783, they successfully launched a hot-air-filled silk balloon. Later that year, their balloon carried a sheep, a duck, and a rooster into the air. The balloon landed safely, and none of the animals were the worse for the experience.

October 15 marked another landmark for hot-air-balloon flight in France. The *Aerostat Reveillion* balloon carried scientist Jean François Pilâtre de Rozier 250 feet into the air, although it remained tethered to the ground by a rope. It floated for about 15 minutes, and then landed safely in a nearby clearing.

But the real moment of glory came on November 21, 1783, at the Bois de Boulogne in Paris: A 70-foot silk-and-paper balloon made by the Montgolfier brothers was launched without a tether, carrying its first human passengers: de Rozier and François Laurent, the Marquis d'Arlandes. The balloon rose to around 500 feet and flew a distance of 5 1/2 miles, remaining aloft for 25 minutes before the straw that was used to stoke the hot-air pit set fire to the balloon. Although legend tells of the lofty gentlemen handing bottles of champagne to startled farmers upon landing, the real story is that they landed in a deserted farming area just outside of Paris—with no spectators nearby.

✳ On September 16, 1979, Peter Strelzyk, Günter Wetzel, their wives, and their four children dropped from the night sky onto a field in West Germany in a homemade hot-air balloon. Strelzyk, an electrician, and Wetzel, a bricklayer, built the balloon's platform and burners in one of their basements. Their wives sewed together curtains, bedsheets, shower liners, and whatever other fabric was on hand to make the 75-foot-high balloon. Their bid to escape communist East Germany during the days of the Iron Curtain was two years in the making, spanned 15 miles, and took 28 minutes to complete. Unsure whether they had reached freedom, the two families spent the next morning hiding in a barn, until they saw an Audi driving down a nearby road and realized that they had indeed made it to West Germany.

A Perfect Container...
But How Do You Open It?

The can opener was a sharp invention that was long overdue.

✳ ✳ ✳ ✳

BEFORE THE CAN opener, there was a revolutionary (albeit somewhat half-baked) invention called the can. The process for canning food was patented by Peter Durand of Britain in 1810; the first commercial canning factory opened three years later. The British Army quickly became a leading customer for the innovative product—after all, the can greatly simplified the logistics of keeping the nation's soldiers fed. In 1846, a new machine that could produce 60 cans per hour increased the rate of production tenfold.

Unfortunately, in all that time, no one came up with a way to address the most serious drawback of this perfectly sealed container: how to open it. In fact, some cans actually came with instructions that read, "Cut 'round the top near the outer edge with a chisel and hammer."

It wasn't until the mid-1800s—when manufacturers devised methods for producing thinner cans made of steel—that there was any hope of creating a simple and safe way to open them. In 1858, Ezra Warner of Connecticut patented the first functional can opener—a bulky thing resembling a bent bayonet that you shoved through the top of the can and then carefully forced around the lid. It was an improvement, no doubt, but still a potentially dangerous way to get at your potted meat.

Finally, in 1870, William Lyman designed a wheeled blade that could cut a can open as it rolled around the edge of the lid (similar to what we use today). The next big inventions in can-opening technology were a long time coming: The electric can opener was invented in the 1930s, and pull-open cans arrived in 1966.

ARPANET: The Grandfather of the Internet

Rumors abound that ARPANET was originally designed as a communications network that would withstand nuclear attacks. That simply isn't true. The creators of ARPANET weren't seeking invulnerability, but rather reliability—in order to fulfill one man's vision of an "inter-galactic" computer network.

✳ ✳ ✳ ✳

O N OCTOBER 4, 1957, the Soviet Union launched the world's first artificial satellite, *Sputnik I.* It was a clear message that Russian technology was more advanced than American technology. To amend this oversight, the Advanced Research Projects Agency (ARPA) was formed to fund technical research. The United States already had a substantial financial investment in computer tech—the initial purpose of ARPA was to figure out the best way to put that to use. Though it fell under the auspices of the U.S. Department of Defense (and was renamed DARPA), the research was never intended to be used solely for military purposes. Instead, the agency's purpose was to develop technology that would benefit the world in general.

Not Connected to Other Galaxies—Yet

The expert chosen to head ARPA's initial effort was Joseph Licklider, a leading computer scientist. Licklider envisioned a worldwide communications network connected by computers, which he referred to as the "inter-galactic computer network." He departed ARPA in 1965, before his plan was implemented, but he left a lasting impression on his successor, Bob Taylor.

Taylor selected a new leader for the system design team that would make Licklider's vision a reality: Dr. Lawrence G. Roberts, an MIT researcher. He became one of the four people most closely associated with the birth of the Internet. (The other three are Vinton Cerf, Leonard Kleinrock, and Robert

Kahn.) Roberts had gained experience in computer-linking while at MIT, having linked computers using the old-fashioned telephone method of circuit switching. The concept of packet switching was at first controversial, but it proved to be one of the key factors in linking multiple computers to form a network. The other important technical achievement was the use of small computers, known as interface message processors (IMPs), to store and handle the data packets.

By 1968, the concept for ARPANET was in place, and invitations to bid on the project were sent to 140 institutions; only 12 replied. The others apparently believed the concept to be impractical, even bizarre, and never bothered to bid. In the end, BBN Technologies—Licklider's former employer—got the nod.

A Hesitant Start

The first piece went to UCLA, thanks to the reputation of Professor Kleinrock, an expert in computer statistical analysis and measurement. The first IMP link was with Stanford Research Institute (SRI), and the first message was sent on October 29, 1969. By December 5, four IMPs were linked: UCLA, SRI, the University of California at Santa Barbara, and the University of Utah.

ARPANET was finally a reality. Its growth during the 1970s was phenomenal, as newer and better protocols were designed. In 1971, e-mail was born; in 1972, telnet was developed; and in 1973, file transfer protocol was created.

By 1986, ARPANET had serious competition from the National Science Foundation Network (NSFNET), which became the true backbone of the Internet. ARPANET closed up shop in 1990. In 1991, NSFNET opened to the public, introducing the Internet we know today. Within four years, more than 50 million people had traveled the information superhighway. As of 2011, worldwide Internet usage stood at more than 2 billion—and that's only the beginning.

Faulty Forecasts

Predicting the future is seldom easy, even for people who are widely respected for their expertise and authority. Here are some serious statements that were way off the mark.

✳ ✳ ✳ ✳

"I think there is a world market for maybe five computers."

—THOMAS WATSON, CHAIRMAN OF IBM, 1943

"The Americans have need of the telephone, but we do not. We have plenty of messenger boys."

—SIR WILLIAM PREECE, CHIEF ENGINEER OF THE BRITISH POST OFFICE, 1876

"Who the hell wants to hear actors talk?"

—H. M. WARNER, WARNER BROTHERS, 1927

"We don't like their sound, and guitar music is on the way out."

—DECCA RECORDING COMPANY, REJECTING THE BEATLES, 1962

"It will be years—not in my time—before a woman will become prime minister."

—MARGARET THATCHER, 1974

"I see no good reasons why the views given in this volume should shock the religious feelings of anyone."

—CHARLES DARWIN, *THE ORIGIN OF SPECIES*, 1869

"With over 50 foreign cars already on sale here, the Japanese auto industry isn't likely to carve out a big slice of the U.S. market."

—*BUSINESS WEEK*, 1968

"The bomb will never go off. I speak as an expert in explosives."

—ADMIRAL WILLIAM LEAHY, UNITED STATES ATOMIC BOMB PROJECT

"Airplanes are interesting toys but of no military value."

—MARECHAL FERDINAND FOCH, PROFESSOR OF STRATEGY,
L'ÉCOLE SUPÉRIEURE DE GUERRE, 1911

"Louis Pasteur's theory of germs is ridiculous fiction."

—PIERRE PACHET, PROFESSOR OF PHYSIOLOGY AT TOULOUSE, 1872

*"The abdomen, the chest, and the brain will forever be shut from the
intrusion of the wise and humane surgeon."*

—SIR JOHN ERIC ERICKSEN, SURGEON-EXTRAORDINARY TO QUEEN VICTORIA, 1873

*"What? Men dodging this way for single bullets? What will you do
when they open fire along the whole line? I am ashamed of you.
They couldn't hit an elephant at this distance!"*

—LAST WORDS OF GENERAL JOHN SEDGWICK, SPOKEN DURING THE BATTLE OF
SPOTSYLVANIA, MOMENTS BEFORE HE WAS KILLED BY A BULLET TO THE EYE

"Reagan doesn't have that presidential look."

—UNITED ARTISTS EXECUTIVE, REJECTING REAGAN FOR THE LEAD ROLE IN THE
1964 FILM *THE BEST MAN*

*"Capitalist production begets, with the inexorability of a law of
nature, its own negation."*

—KARL MARX, *DAS KAPITAL*

*"Stick to driving a truck, because you're never going to make it as a
singer."*

—MEMPHIS MUSICIAN EDDIE BOND, AFTER HEARING ELVIS PRESLEY, 1954

Dem Bones: X-ray Shoe-Fitting Machines

During the 1940s, people were particularly concerned about their feet—mothers, fathers, even the U.S. Army. As a result, the guardian of modern foot care was created: the Adrian X-ray Shoe Fitting Machine.

✳ ✳ ✳ ✳

A Star Is Born

ALTHOUGH THERE ARE conflicting stories about its origin, the first X-ray shoe-fitting machine has generally been attributed to Dr. Jacob Lowe, a Boston physician who was looking for a fast and efficient way to analyze soldiers' feet during World War I. Dr. Lowe was concerned with the poorly fitting boots worn by many military recruits and was interested in a way to reduce their foot-related injuries. In addition to providing the doctor with a superior view of the foot, the X-ray shoe-fitting machine allowed him to speed up production, since soldiers didn't have to remove their boots.

The machine was a simple design: A fluoroscope that sent X-rays upward toward a fluorescent screen was mounted on the base of a wooden platform. The client placed his or her foot inside and the image would be directed up to a reflector, where three viewing scopes displayed the foot's image. The entire area was sealed within a lead-shielded area for protection of the client. Unlike X-rays that are captured on film, the machine displayed a real-time image of the client's foot—shoes and all.

After the war, Dr. Lowe starting making the rounds to retail shoe stores, and in 1927, he sold the patent for the device to the Adrian Company of Milwaukee. At about the same time, a similar patent was granted in Great Britain for the Pedoscope, although the Pedoscope Company claimed that its instrument had already been in use for more than five years.

Better Shoe Fitting Through Science

The public went wild over this new way to be "scientifically" fitted for shoes. Concerned mothers were grateful that someone had finally come up with a method for accurately fitting their children's shoes; the manufacturer claimed that by being able to view the foot inside the shoe, a child's shoe would last longer and promote the child's foot health and comfort. By the 1950s, more than 10,000 Adrian X-ray Shoe Fitting Machines had been installed all over the country.

As the machines gained popularity, so did the government's concern over their safety. When the machines were first introduced, little was known about continued exposure to radiation. As a result, children and adults were repeatedly exposed to X-ray radiation with little concern over its ill effects. The average shoe salesman might expose himself to 20 to 30 doses in a single shift. In 1946, the American Standards Association defined a "safe and tolerable dosage" of radiation and began regulating how the X-ray machines were used. Ultimately, the American Conference of Governmental Industrial Hygienists issued uniform guidance standards.

The Adrian Company assured parents that the machines were safe and could "just as easily be operated by 'old timers' with more than 20 years of shoe fitting experience as 'Saturday extras' who only had their jobs for a few weeks."

The End of an (X-ray) Era

By the early '50s, a number of medical societies became concerned that nonmedical personnel were operating the fluoroscopes. Eventually they issued warnings suggesting that only licensed physiotherapists should operate the machines. A number of states began requiring that only licensed physicians use the X-ray fitting machines. In 1957, Pennsylvania became the first state to ban the use of the shoe-fitting fluoroscopes; it didn't take long for the rest of the country to follow suit.

8 Things Invented or Discovered by Accident

We tend to hold inventors in high esteem, but often their discoveries are accidental or twists of fate. This is true of many everyday items, including the following surprise inventions.

✳   ✳   ✳   ✳

1. **Play-Doh:** One smell most people remember from childhood is the odor of Play-Doh, the brightly colored, nontoxic modeling clay. It was accidentally invented in 1955 by Joseph and Noah McVicker while they were trying to make a wallpaper cleaner. It was marketed a year later by Rainbow Crafts. More than 700 million pounds of Play-Doh have sold since then, but the recipe remains a secret.

2. **Fireworks:** Fireworks originated in China some 2,000 years ago, and legend has it that they were accidentally invented by a cook who mixed together charcoal, sulfur, and saltpeter—all items commonly found in kitchens in those days. The mixture burned, and when it was compressed in a bamboo tube, it exploded. There's no record of whether it was the cook's last day on the job.

3. **Potato Chips:** If you can't eat just one potato chip, blame it on chef George Crum. He reportedly created the salty snack in 1853 at Moon's Lake House near Saratoga Springs, New York. Fed up with a customer who continuously sent his fried potatoes back, complaining that they were soggy and not crunchy enough, Crum sliced the potatoes as thin as possible, fried them in hot grease, then doused them with salt. The customer loved them and "Saratoga Chips" quickly became a popular item at the lodge and throughout New England. Eventually, the chips were mass-produced for home consumption, but since they were stored in barrels or tins, they quickly went stale. Then, in the 1920s,

Laura Scudder invented the airtight bag by ironing together two pieces of waxed paper, which kept the chips fresh longer. Today, chips are packaged in plastic or foil bags and come in a variety of flavors, including sour cream and onion, barbecue, and salt and vinegar.

4. **Slinky:** In 1943, naval engineer Richard James was trying to develop a spring that would support and stabilize sensitive equipment on ships. When one of his prototypes accidentally fell off a shelf, it continued moving, and James got the idea for a toy. His wife Betty came up with the name, and when the Slinky made its debut in late 1945, James sold 400 of the bouncy toys in 90 minutes. Since then, more than 250 million Slinkys have been sold worldwide.

5. **Post-it Notes:** A Post-it Note is a small piece of paper with a strip of low-tack adhesive on the back that allows it to be temporarily attached to documents, computer monitors, and just about anything else. The idea for the Post-it Note was conceived in 1974 by 3M employee Arthur Fry as a way of holding bookmarks in his hymnal while he was singing in the church choir. He was aware of an adhesive that had been accidentally developed in 1968 by coworker Spencer Silver, but no application for the lightly sticky stuff was apparent until Fry's idea. The 3M company was initially skeptical about the product's potential profitability, but in 1980, the product was introduced around the world. Today, Post-it Notes are sold in more than 100 countries.

6. **Silly Putty:** It bounces, it stretches, it breaks—it's Silly Putty, the silicone-based plastic clay marketed as a children's toy. During World War II, while attempting to create a synthetic rubber substitute, James Wright dropped boric acid into silicone oil. The result was a polymerized substance that bounced, but it took several years to find a use for the product. Finally, in 1950, marketing expert Peter Hodgson saw its potential as a toy, renamed it Silly Putty, and a

classic toy was born. Not only is it fun, but Silly Putty also has practical uses—it picks up dirt, lint, and pet hair; can stabilize wobbly furniture; and is useful in stress reduction and in medical and scientific simulations. It was even used by the crew of *Apollo 8* to secure tools in zero gravity.

7. Microwave Ovens: The microwave oven is a standard appliance in most American households, but it has only been around since the late 1940s. In 1945, Percy Spencer was experimenting with a new vacuum tube called a magnetron while doing research for the Raytheon Corporation. He was intrigued when the candy bar in his pocket began to melt, and he immediately saw the potential in this revolutionary process. In 1947, Raytheon built the first microwave oven, the Radarange, which weighed 750 pounds, was 5½ feet tall, and cost about $5,000. When the Radarange became available for home use in the early 1950s, its bulky size and expensive price tag made it unpopular with consumers. But in 1967, a much more popular countertop version was introduced at a price of $495.

8. Corn Flakes: In 1894, Dr. John Harvey Kellogg was the superintendent of the Battle Creek Sanitarium in Michigan. He and his brother, Will Keith Kellogg, were Seventh Day Adventists, and they were searching for wholesome foods to feed patients that also complied with their religion's strict vegetarian diet. Will accidentally left some boiled wheat sitting out, and it was stale by the time he returned. Rather than throw it away, the brothers sent it through rollers, hoping to make long sheets of dough, but they got flakes instead. They toasted the flakes, which were a big hit with patients, and patented them under the name Granose. The brothers experimented with other grains, including corn, and in 1906, Will created the Kellogg's company to sell the corn flakes. On principle, John refused to join the company because Will lowered the health benefits of the cereal by adding sugar.

America's First Skyscraper

Chicago's Home Insurance Building certainly earned the grandiose nickname "the father of the skyscraper"—its creation set off the trend toward today's array of sky-climbing constructs.

❋   ❋   ❋   ❋

COMPLETED IN 1885, William LeBaron Jenney's architectural innovation was the first building ever to feature a structural steel frame. It was erected at ten stories tall, or about 138 feet. Even with two additional stories tacked on five years later, the structure known as "America's First Skyscraper" would, at 180 feet, still be tiny when compared to Chicago's current skyscraper giant, the Willis Tower, which boasts a rooftop height of approximately 1,450 feet. Stack up eight Home Insurance Buildings, and the Willis Tower would still soar overhead.

Vertical Construction

The Home Insurance Building led the way toward the upward urban sprawl that now defines major U.S. cities. Prior to its creation, architects had only plotted building growth in a horizontal fashion. But with this new breed of building, the future of architecture was all upward. After all, why rely on mere land when the sky's the limit?

Torn down in 1931—less than 50 years after its construction— the Home Insurance Building was replaced just three years later by the still-standing, 45-story LaSalle National Bank Building (originally called the Field Building). The Home Insurance Building had served its purpose, and the city was ready to ring in the next round of architectural advancements. Just as the Home Insurance Building initiated a new era in architecture, its replacement structure was significant in its own way: It capitalized on the architectural trend of the time by capturing the eloquence of the Art Deco movement. But as significant as the Field Building may have been, it didn't break new ground in the same way that the Home Insurance Building had.

The Real Origin of the Barbie Doll

It didn't take long for the Barbie doll to become not only an iconic toy, but also a symbol of adult style and ambition. Where exactly did Barbie come from?

✳ ✳ ✳ ✳

WHEN BARBIE APPEARED in 1959, she was the first adult-styled toy for young girls. Previously, girls were offered stuffed animals and vinyl baby dolls, which were presumably meant to evoke their maternal instinct. But Barbie was a creature to which a young girl could *aspire*—albeit one that was, er, rather statuesque.

German Influences

Before Barbie, there Bild Lilli, an alluring model of a cartoon character from German newspaper *Bild*. Clad in a short skirt and tight sweater, Bild Lilli was sold in tobacco shops and bars.

Ruth Handler, the genius behind Barbie, understood that young girls were in a hurry to grow up. Handler had noticed her own daughter, Barbara, assigning her paper dolls grown-up roles while playing with them. When Handler spotted Bild Lilli during a family vacation in Europe, she recognized the doll's commercial potential and purchased several in different outfits.

Barbie's earliest appearance, although toned down from that of Bild Lilli, still owed much to her forerunner: She was pale, with pouty red lips and heavily made-up eyes that cast a knowing sidelong glance, and her measurements (which, on a human counterpart, would equal 38"-18"-32") were absurd.

Barbie was also a protofeminist icon, holding at least 95 different careers. Her original job description as "a teenage model dressed in the latest style" (even though she looked at least a decade older) later evolved into the nonspecific "career girl," with astronaut, ballerina, doctor, pilot, attorney, paleontologist, and presidential candidate versions, to name a few.

The Barbie Empire Grows

Millions of baby-boom children grew up imagining themselves as Barbie, who went on to acquire boyfriend Ken, close friend Midge, little sister Skipper, and a wardrobe that any movie star would envy. Actually, "acquire" could have been Barbie's middle name (although it was really Millicent): At first, it was just a trunk to hold her elegant clothes and teeny accessories. Then came furniture and a beach house, accessed via luxury cars (pink Cadillac, red Porsche, etc.). Her features and hair softened; her other friends (Stacey, a British pal, and Christie, her African American friend) came and went; and, after the longest engagement in history, she finally dumped Ken (who was named after Handler's son).

During the past half-century, Barbie has reflected both fashion and social history. Charlotte Johnson, who originally dressed the doll, took her job seriously, interpreting the most important clothing trends of each successive era. Barbie also inspired top designers—including Armani, Christian Dior, John Galliano, Dolce & Gabbana, Gucci, Alexander McQueen, and Vivienne Westwood—to create exclusive originals for her. Philip Treacy designed her hats and Manolo Blahnik her shoes.

The Decline of the Barbie Era

But what was once cutting-edge has now become rather staid. What excites little girls nowadays is the hypersexualized, even more cartoonish-looking Bratz dolls, which come off as even poorer role models. In comparison, Barbie looks square and has accordingly lost market share. Her future seems to be more as a nostalgic collectible for adults. Indeed, a complete collection of early Barbies, plus friends and outfits, was auctioned for record prices at the venerable Christie's in late 2006; a single doll, Barbie in Midnight Red, sold for $17,000.

In an odd way, Barbie has come full circle: From her sordid, bar-hopping start as Bild Lilli, Barbie has once again become a toy for grown-ups.

Smile—You're on Camera Obscura

Photography didn't begin with the daguerreotype (1839), but Louis Daguerre's gadget was the breakthrough step.

❋ ❋ ❋ ❋

Pinhole Cameras

IRONICALLY, NO PHOTOS exist of the earliest known camera pioneer, a Chinese man named Mo Ti. It would be nice if his likeness had been captured for the world to remember, but alas, technology in China circa 400 B.C. was not quite there yet. History credits Mo with uncovering a simple key principle of photography: Light traveling through a small hole will cast an image onto a surface. Unfortunately, there was no efficient way to capture the image—which explains the lack of pics of Mo.

Camera Obscuras

In about A.D. 1015, it came to light (so to speak) that a closed box placed behind a pinhole camera makes a *camera obscura* (Latin for "dark chamber"). At that time, an Arab scholar published a landmark optics text describing how the device could project an image onto paper, enabling one to sit and trace/draw the image. Limited, but useful.

The situation improved in the 1500s, when clever pioneers added a magnifying lens to the pinhole. It inverted the image, but one could later flip the sketch upright and paint in the colors.

Fixing the Image

During the 1700s, researchers discovered that light could change certain substances, particularly silver compounds. In 1826, Frenchman Joseph Niépce used a camera obscura and about eight hours of exposure to fix an image onto bitumen (natural asphalt). Where the light hit the bitumen, it hardened, and Niépce dissolved away the unfixed remainder. The image wasn't very clear, but it was a start. The oldest surviving example of what's called a *heliograph* evidently dates to about 1826, when the image was much improved. But a clever chap named Louis Daguerre grabbed onto Niépce's coattails a few years before Niépce's death in 1833. Daguerre also grabbed the credit, so few people outside of France have ever heard of Niépce.

Daguerreotyping

Daguerre kept experimenting with chemicals until 1837, when he found a suitable light-sensitive combination involving iodine and mercury vapor. Thanks to this breakthrough, the subjects had to stand still for "only" half an hour, which seems like forever—until you consider that the alternative was a much longer portrait sitting. Nearly all the earliest surviving photos of American icons such as Dolley Madison and young Horatio Alger at Harvard are daguerreotypes.

Daguerre's camera was to photography what the first self-assembled mail-order Altair 8800 PC would become to young computer nerds in the late 1970s: It set numerous clever souls to fiddling with camera technology. With the invention of negatives in 1835, multiple copies of photos became feasible. By 1871, image plates no longer required immediate development.

From Mainstream to Mass Market

George Eastman, whom you may connect with Eastman Kodak, patented photo emulsion-coated paper and rollers in 1880; after that, shutterbugs everywhere said farewell to cumbersome plates. Photography was about to go mainstream.

In 1900, Kodak's Brownie, a $1 mass-market camera, hit the shelves. The 1900s would become photography's democratic age, in which anyone could preserve on film whatever he or she wished. Before long, camera nuts would keep whole families waiting in the Model T so this or that fascinating item could be recorded for all posterity in the family album.

For whatever reason, the paparazzi wouldn't get saddled up until the 1930s. Sure took them long enough.

Inventors Killed by Their Inventions

The success of history's most famous inventors rested not only upon brilliant ideas, but also upon having the dedication and confidence to pursue those ideas in the face of public doubt. Unfortunately, inventors have sometimes been too confident in their work—with disastrous consequences. Here are six inventors whose inventions got the better of them.

✳ ✳ ✳ ✳

Henry Winstanley, Lighthouse Architect

WHILE 17TH-CENTURY lighthouse-smith Henry Winstanley didn't invent lighthouses, he did design a new kind of lighthouse—the Eddystone Lighthouse, an octagonal-shaped structure built to withstand treacherous conditions on tenuous ground. Despite observers' doubts that the lighthouse would stand up to serious meteorological assault, Hank believed in his design—so much so that he insisted on taking shelter in it during a terrible storm in November 1703. It was a poor decision—the lighthouse collapsed, ending Winstanley's life.

Marie Curie, Radiation Pioneer

Marie Curie is known to schoolchildren as the discoverer of the elements radium and polonium, the first woman to win a Nobel Prize, a pioneer in the field of radioactivity, and the inventor of a method for isolating radioactive isotopes. Unfortunately, she is also known as a pioneer in the field of radiation-induced cancer. Curie, who was working with radioactive isotopes well before the dangers of radiation were fully known, contracted leukemia from radiation exposure and died at age 66.

Karel Soucek, Inventor of the "Stunt Capsule"

Soucek shot to fame in 1984 by designing a special capsule that he used to plunge over Niagara Falls. Seeking to capitalize on his newfound popularity, Soucek decided to repeat the stunt in 1985—only this time from an artificial waterfall running from the top of the Houston Astrodome down to a tank of water. It seemed like a bad idea, and it was: The capsule exploded upon impact, and Soucek suffered fatal injuries.

William Bullock, Inventor of the Web Rotary Printing Press

Before the 19th century, the printing press hadn't advanced much beyond Gutenberg's first effort from the 15th century. But in 1863, William Bullock changed everything by coming up with the idea of a web rotary press—a self-feeding, high-speed press that could print as many as 10,000 pages per hour. Unfortunately, Bullock forgot a basic rule of printing presses: Don't stick your foot into the rotating gears. In 1867, Bullock got tangled up in his invention, severely injuring his foot. Gangrene set in, and he died shortly afterward.

Otto Lilienthal, Inventor of the Hang Glider

Until the late 19th century, human flight was little more than a pipe dream. Otto Lilienthal changed all that with his hang glider, and his successful glides made him famous the world over. Unfortunately, in 1896, Lilienthal plunged more than 50 feet during one of his test runs. The fall broke his spine, and he died shortly thereafter.

Cowper Phipps Coles, Inventor of the Rotating Ship Turret

The splendidly named Cowper Phipps Coles was a captain in the British navy who invented a "rotating gun turret" for British naval vessels during the Crimean War. After the war, Coles patented his invention and set about building ships equipped with his new turret. Unfortunately, the first ship that he built, the HMS *Captain*, turned into the HMS Capsized. In order to accommodate his turret design, shipbuilders were forced to make odd adjustments to the rest of the ship, which seriously raised its center of gravity. The end result? The ship sank on one of its first voyages, killing Coles and much of his crew.

The Frisbee: From Pie Plate to Toy

For something that now seems so familiar, the flying disc has an interesting story.

✳   ✳   ✳   ✳

William Russell Frisbie

IT ALL STARTED with W. R. Frisbie, a bakery manager. In 1871, he bought a Connecticut baking business, renamed the firm the Frisbie Pie Co., and got to work baking pies. After he passed away in 1903, his son Joseph P. Frisbie donned the baker's hat. By the time of Joe's death in 1940, the Frisbie Pie Co. had become a great regional success. One example of Joe's savvy marketing: metal pie plates stamped with "Frisbie's Pies," so that anyone keeping the plates remembered the name.

Recreational Yalies

Yale University is a quick jog away from the Frisbie Pie Co.'s old New Haven bakery. Like any self-respecting college students, Yalies both loved a good feed and liked to find creative ways to amuse themselves. After porking out on Frisbie's pies, they flung the empty tins around for fun, quickly noting that they flew better if thrown with a quick flick of the wrist to impart spin. If you threw them with a spin, you could play catch with them. You could also accidentally bonk Professor Stuffshirt upside the head, so they learned to yell "Frisbie!"—as golfers holler "Fore!"—to warn of an incoming flying object.

Morrison & Franscioni

Walter "Fred" Morrison was a World War II air combat veteran working for a bottled gas company in California in the late 1940s. He brought the idea for the flying plate to his employer, fellow vet Warren Franscioni. They started experimenting on the side.

The partners soon learned that a streamlined plastic disk was the ideal configuration. They began to make and market the

Flyin' Saucer—attempting to capitalize on budding American interest in UFOs—but the product didn't take off. In the meantime, the gas company tanked. Franscioni rejoined the Air Force and was relocated to South Dakota.

Wham-O

Morrison renamed the Flyin' Saucer the Pluto Platter. In 1957, he was demonstrating the projectile and caught the attention of a small slingshot company named Wham-O. Impressed, the Wham-O people offered to market the Pluto Platter, and they knew what they were doing: The toy sold well.

Not long thereafter, Wham-O cofounder Rich Knerr was giving out Pluto Platters at East Coast universities to build brand awareness and demand. At Yale, he saw students chucking metal pie tins around. When a tin was headed for a noggin, the Yalies yelled "Frisbie!," as tradition and safety demanded. (Ironically, Frisbie's Pies shut down that same year.)

Knerr soon renamed the Pluto Platter the Frisbee. Along with the Hula Hoop, the Frisbee became a Wham-O cash cow.

Morrison got royalties; Franscioni did not. Franscioni died in 1974 while still considering legal action. Wham-O's official history of the Frisbee does not mention him.

✳ Wham-O was founded by Richard Knerr and Arthur "Spud" Melin in 1948. Their first product offering was a slingshot specifically designed to hurl bits of meat into the air to feed hawks and falcons.

✳ Wham-O began marketing the Hula Hoop in 1957 and the Super Ball in 1964.

✳ You can also thank Wham-O for the Slip-N-Slide, Silly String, and the Air Blaster—which is guaranteed to blow out a candle from 20 feet away.

Listerine: Inventing the Need

Some 40 years after Listerine was invented, marketing gurus finally figured out how to sell it to Jane and Joe Average: Tell Joe that if he didn't cure his toxic breath, Jane wouldn't kiss him.

❉　❉　❉　❉

Not by Lister?

SURGERY WAS A filthy business in the 19th century. President James Garfield died from an infection due to the lack of sterilization practices—and Garfield had the finest medical care of the day. For the less fortunate, healthcare was a Dante's *Inferno* of squalor, amputation, infection, and gangrene. In wartime, a bullet to the gut generally meant an agonizing death that surgery would merely hasten.

Into that breach stepped Dr. Joseph Lister, an English doctor who advised his fellow physicians to wear gloves, wash their hands, and sanitize their instruments. It was that simple. The mortality rate dropped so far that a grateful Queen Victoria made Lister a baron. But later generations would come to associate him with a mouthwash that he didn't invent and probably never used.

Then Who Did?

Dr. Joseph Lawrence and Jordan Lambert first whipped up a batch of surgical disinfectant in 1879. It was a blend of thymol, menthol, eucalyptol, and methyl salicylate mixed in grain alcohol. Did the inventors name it after Lister to honor him or to ride on his famous coattails? That depends on your opinion of their motives, but profit surely drove Listerine's invention and marketing. There's no evidence that Dr. Lister ever received a penny for the appropriation of his name, nor that he complained or cared. A devout man of science and medicine, Lister didn't seek riches.

Here Comes the Marketing

In 1884, Lambert formed the Lambert Company to sell Listerine. By 1895, it was clear that Listerine did a quick, safe job of wiping out oral bacteria, so Lambert started hawking it to dentists as well. One might say that Western medicine washed away centuries of sin with a baptism in Listerine.

If Duels Were Still Legal, Your Breath Could Win

By 1914, Lambert's son Gerard was calling the shots for Listerine. In an era of snake-oil concoctions with names like Spurrier's Powders, Kvak's Pills, and Boga's Tonic, which promised to cure everything from baldness and "female complaints" to rheumatism and colds yet conferred no actual health benefits, Listerine was the exception—*it actually did something useful.*

Gerard started selling Listerine over the counter as a mouthwash, thus inventing the category... and he soon realized that he also had to invent the problem. His advertising focused on halitosis (bad breath), and played on the idea (often wellfounded, granted) that oblivious persons with poor dental habits might be offending everyone with their dragon breath.

The cure? Americans started rinsing their mouths with medical disinfectant. The Lambert Company's annual revenues went from $115,000 to $8 million in seven years.

Today, Listerine comes in a variety of flavors, but you can still buy the old-school version. It hasn't changed much, if at all, since Jordan Lambert started selling it as a surgical sanitizer.

＊ Listerine tastes nasty, so even during Prohibition (1920–1933), you had to be a pretty desperate alcoholic to swallow the stuff.

＊ One claim that Listerine eventually had to back off of was that it helped fight colds. Its makers got away with that one for 50 years before the Federal Trade Commission made them drop it.

Amazing Amazon

A good idea and a better name resulted in one of the Web's greatest success stories.

✻ ✻ ✻ ✻

JEFF BEZOS WAS never a slouch. After graduating summa cum laude from Princeton in 1986, Bezos distinguished himself, first at a high-tech firm, then with financial-services companies. But it wasn't until Bezos applied his skills as a computer nerd and financial whiz that he really found his niche: He decided that e-commerce would be the next big thing.

But what to sell? With a list of 20 possible products, Bezos eventually settled on books, a market that comprised more than one million products but no single company dominated.

In 1994, Bezos launched what would eventually become the world's largest online store—Cadabra.com, which was supposed to sound like "abracadabra." Unfortunately, it also sounded like "cadaver"—not a great association for a new startup.

Bezos decided he needed a new name, and he wanted one that would come up first in an alphabetical search. In 1995, he changed his site's name to Amazon.com, naming it after the largest river in the world.

The very first book sold on Amazon was *Fluid Concepts and Creative Analogies: Computer Models of the Fundamental Mechanisms of Thought*. Nevertheless, during its first month of business, Amazon.com filled orders in all 50 states and 45 countries from the garage of Bezos's Seattle home.

Amazon soon became one of the most-visited sites on the Web, though the company didn't actually make a profit until the end of 2003—Bezos was too busy plowing money back into the company and buying new businesses. Today, the site sells everything from lawn mowers to watches.

The Segway

✳ ✳ ✳ ✳

✳ In 2001, mastermind Dean Kamen, a self-taught physicist and established inventor, developed the Segway, a product that he claimed would make walking "obsolete." Kamen had previously patented a phonebook-size portable dialysis machine, a nonpolluting engine, and more than 150 other contraptions.

✳ The Segway is officially referred to as a "human transport device."

✳ Segway PTs ("personal transporters") have electric motors that drive the apparatus at speeds up to 12.5 miles per hour.

✳ The Segway is designed with "redundant technology." This means that the device features duplicates of its important pieces of hardware. If one function fails, an internal computer uses the duplicate function to keep the machine stable long enough for the rider to hop off safely.

✳ Segways respond as if they're controlled by the riders' thoughts alone. The secret of newer Segways is in the control shaft, which sways in sync with the rider if he or she wants to turn.

✳ You operate a Segway like this: Turn it on, step onto the two-wheeled platform, grip the waist-high handle, lean forward, and off you go.

✳ It is illegal to use a Segway on streets, roads, or highways. They are allowed on sidewalks and in bike lanes.

✳ Most Segways can travel about 12 miles before they need to be recharged.

✳ Despite a lot of media attention, Segways haven't sold very well. As of 2010, only about 50,000 units had been sold.

Sorry, Mates, but Aussies Didn't Invent the Boomerang

Contrary to popular myth, lore, and Australian drinking songs, boomerangs—or "The Throwing Wood," as proponents prefer to call them—did not originate down under.

✳ ✳ ✳ ✳

THE COLONISTS, ADVENTURERS, prisoners, and explorers who ventured into the heart of the Australian wilderness may be excused for believing that the local aborigines created these little aerodynamic marvels, considering the proficiency with which they used them to bring down wild game and wilder colonials. The gyroscopic precision with which boomerangs were (and still are) crafted by primitive peoples continues to intrigue and astonish those who come in contact with the lightweight, spinning missiles, which—if thrown correctly—actually will return to their throwers.

Many Happy Returns

As a weapon of war and especially as a tool for hunting small game, the boomerang has been around for nearly 10,000 years. In fact, evidence of boomerangs has been discovered in almost every nook and cranny in the world. Pictures of boomerangs can be found in Neolithic-era cave drawings in France, Spain, and Poland. The *lagobolon*—or "hare club," as it was called—was commonly used by nobles in Crete around 2000 B.C. And King Tutankhamun, ruler of Egypt around 1350 B.C., had a large collection of boomerangs—several of which were found when his tomb was discovered in the 1920s.

On pottery made during the Homeric era, Greek mythological hero Hercules is depicted tossing around a curved *clava*, or "throwing stick." Carthaginian invaders in the second century B.C. were bombarded by Gallic warriors who rained *catela*, or "throwing clubs," on them. The Roman historian Horace describes a flexible wooden *caia* used by German tribes, saying "if thrown by a master, it returns to the one who threw it." Roman Emperor Caesar Augustus' favorite contemporary author, Virgil, also describes a similar curved missile weapon used by natives of the province of Hispania.

However, Europeans can no more claim the invention of the boomerang than their Australian cousins can. Archaeologists have unearthed evidence of boomerang use throughout Neolithic-era Africa, from Sudan to Niger and Cameroon to Morocco. Tribes in southern India, the American southwest, Mexico, and Java all used the boomerang, or something very similar, and for the same purposes.

Australians, however, can be credited with bringing the boomerang to the attention of the modern world. They helped popularize it both as a child's toy and as an item for sport. The Boomerang World Cup is held every other year, and enthusiasts and scientists still compete to design, construct, and throw the perfect boomerang.

Though the boomerang did not originate in Australia—or at least did not originate exclusively in Australia—the word itself is Australian. *Boomerang* is a blending of the words *woomerang* and *bumarin*, which are used by different groups of Australian aborigines to describe their little wooden wonders.

✳ **From boomerang to helicopter? David Unaipon, an Australian Aboriginal inventor, drew preliminary mechanical illustrations of a helicopter based on the principles of boomerang flight. Today, Unaipon appears on the Australian $50 note.**

✳ **While an officer in the South Pacific during World War II, future president Richard Nixon set up a hamburger stand to provide sandwiches and Australian beer to returning flight crews.**

Slippery Slopes

From our nation's first oil well came a clear slime with healing properties.

✳   ✳   ✳   ✳

SOME OF THE best inventions are the ones that come about accidentally. In 1859, while visiting the oil-boom town of Titusville, Pennsylvania, chemist Robert Chesebrough literally stumbled upon petroleum jelly.

Chesebrough was in the Titusville region, home of the famous Drake Well, to strike crude-oil deals with drilling bigwigs. As the 22-year-old owner of an oil business, Chesebrough had his entrepreneurial eyes set firmly on future prizes. The young man wasn't aware of it yet, but he was about to realize success beyond his wildest dreams.

While touring Titusville's oil fields, Chesebrough noticed a worker scraping waxlike goo from a pump rod. The worker told him he was scraping off "rod wax." It turns out that the substance was a by-product of the drilling process that would cause the pumps to foul if allowed to accumulate.

Then the man told Chesebrough something that piqued the businessman's interest. When a worker would burn or cut himself, a dollop of this stuff applied to the wound would "fix it right up." Chesebrough was intrigued. A new business venture was about to emerge.

Over the next decade, Chesebrough modified naturally occurring rod wax and used himself as a guinea pig to test its effectiveness. He became convinced of the product's curative properties and brought it to market. By 1875, the invention was selling at the rate of one jar per minute. By the 1880s, the substance was a staple in American homes. These days, people refer to it not as rod wax but as petroleum jelly. Chesebrough had found success in the slimiest of places. We all share in his rewards.

The Cat Who Came in from the Cold

Ed Lowe's innovation led to a new kind of pet—the indoor cat.

✳   ✳   ✳   ✳

CATS HAVE BEEN beloved pets for thousands of years. For much of that time, though, felines were considered outdoor animals. They would often spend time in the house, but their owners usually put them out at night. Cat lovers who kept their pets indoors paid a rather smelly price because commonly used cat box fillers such as sand, sawdust, and ashes did little to combat the notoriously rank odors that little Fluffy left behind.

In 1947, Kay Draper of Cassopolis, Michigan, found herself short of cat box filler and went to a neighbor's to see if he might have something that she could use. Lucky for Kay and for cat lovers everywhere, her neighbor Ed Lowe sold industrial absorbents, and he suggested that she try some fuller's earth—small granules of dried clay used for soaking up oil spills and such. After trying the clay, Draper raved that it was not only cleaner than other fillers she had used, but it also helped control odors.

Smelling an opportunity, Lowe filled several paper bags with the stuff, scrawled the name "Kitty Litter" on the side, and headed to a local pet store. The owner was skeptical of the idea; who would pay 65 cents for a bag of dirt when sand and sawdust were virtually free? Undaunted, Lowe told him to give the bags away to anyone willing to try it. Within a short time, those customers came back saying they would gladly pay for more.

Lowe spent the next few years driving around to pet stores and cat shows to promote his product; his diligence paid off. Americans now spend nearly $800 million a year on clay cat box filler, and millions of pampered felines enjoy the luxury of indoor living.

The "Fax" Are In

The fax machine of the 1980s actually got its start in the 1840s.

✳   ✳   ✳   ✳

VISIT ALMOST ANY office supply store and you'll find that fax machines are in plentiful supply. Considered *de rigueur* in the workplace during the business boom of the 1980s, the fax—or, more correctly, "facsimile"—machine continues to hold its own in this day of lightning-quick e-mailing and dexterity-challenging text messaging. And why shouldn't it? The ability to "fax" a reproduced sheet of paper from here to virtually anywhere within a telephone's reach has changed the business landscape.

But this breakthrough invention doesn't actually hail from modern times. In fact, this gem came along when Abraham Lincoln was still a lawyer and the telephone—the yin to the fax machine's yang—had not yet been invented. But how is that possible?

Scottish mechanic Alexander Bain used existing telegraph technology paired with a stylus functioning as a pendulum to work such magic. The stylus picked up images from a rotating metal surface (a precursor of modern-day scanning), then converted them to electrical impulses for transmission. These, in turn, were sent to chemically treated paper. Bain was granted a British patent for his groundbreaking invention on May 27, 1843—some 33 years before the telephone was patented.

The process underwent various improvements over the years to reach its present form. By the 1920s, an updated scanning system allowed the paper original to remain in a fixed position, like many modern-day machines. During that time, the American Telephone & Telegraph Company (AT&T) became involved in fax development, thus ushering the fax machine into modern times. On March 4, 1955, the very first fax transmission was sent via radio waves across the continent. Not bad for a pre–Civil War machine produced by an amateur.

The Stats on Web Surfing

The Internet began in the United States in 1969 as a network of computers designed to connect government and defense installations in the event of a telecommunications failure. Even with the 1991 introduction of the World Wide Web, the Internet initially had limited uses and few users. These days, you can practically live your life online—and some people do. Here's a snapshot of the ever-evolving Internet world.

✻ ✻ ✻ ✻

✻ By 2008, Google had identified one trillion individual Web pages. The Microsoft Bing team estimates it would take six million years to read everything available online.

✻ According to Internet World Stats, there were more than two billion people online as of 2011. That seems like a big number, but that's actually only about 30 percent of the world's population.

✻ In 2007, a 75-year-old Swedish woman got the world's fastest home Internet connection. Her son, an optical Internet expert, set up a 40-gigabits-per-second connection to demonstrate fiber-optic technology. Her setup could download an entire movie in two seconds, but she mainly used it to dry laundry. (The routing system would get very hot.)

✻ Between 2004 and 2009, the online population of American kids grew by 18 percent. Their time online grew 63 percent, to an average of 11 hours per month.

✻ As of October 2011, Wikipedia boasted more than 19 million articles in 282 languages, including over 3.7 million articles in English.

✻ In July 2009, Americans racked up 21.4 billion online video views.

The Little Disc that Could

Digital downloads may have hurt the traditional record store, but the CD remains the world's most popular medium for audio recordings. Not only that, but the technology also changed the way digital data is stored.

✳   ✳   ✳   ✳

COMPACT DISCS FIRST hit the market in 1982. After little more than a decade, the format dominated the market, quickly replacing the vinyl records and cassette tapes that had been the standard for years. Although music purists and audiophiles maintain that vinyl has a richer, warmer sound, the compact disc has superior advantages—notably that repeated playback causes virtually no wear to the disc's surface; the recordings sound the same after 10 plays or 10,000 plays.

The compact disc was invented by James T. Russell in the late 1960s. Russell patented the technology in 1970.

The first CD player, Sony's CDP-101, was released in Japan in October 1982 and arrived in the United States the following spring. The compact disc quickly became a breakthrough format for popular music. The 1985 album *Brothers in Arms* by Dire Straits was the first compact disc to sell a million copies.

The CD ultimately lent itself to more than just music. The same digitally encoded data that made music storage possible also worked as a storage device for computers. In 1990, the CD-ROM (Read-Only Memory) and recordable CDs (for music and data) were introduced. And while the MP3 may be stealing the spotlight from the music CD, the discs will live on through volume, if nothing else: To date, more than 200 billion CDs have been sold worldwide.

The Way the Future Wasn't: Cities Under Glass

The vision of cities flourishing under glass domes dates back to the science fiction of the late 19th century. During the early 20th century, futuristic pictures of urban life frequently depicted gigantic overturned bowls protecting people and buildings from the elements. But it was little Winooski, Vermont, that almost became the world's first real domed city.

✳ ✳ ✳ ✳

I N 1979, RISING oil prices gave Winooski residents a chill. One fateful night, Mark Tisan, a 32-year-old community development planner, came up with a novel idea: Why not enclose the entire city under a glass dome?

Though it seems unbelievable, city officials encouraged Tisan to pursue federal funding for the so-called Winooski Dome Project. Tisan held a press conference, and the story went viral. Soon, he received a flood of mail from people who were eager to help.

Tisan didn't even know what the dome would look like, so he hired conceptual architect John Anderson, who quickly whipped up a design for a clear vinyl structure supported by a web of metal struts. Shaped rather like a hamburger bun with a wide, flat top, the dome would be 250 feet tall and encompass an area of one square mile—most of the city. Of course, there were still a few kinks in the plan—like how to get rid of auto exhaust and other fumes. Tisan assured skeptics that electric cars and similar innovations would solve any problems.

All systems were go for Winooski to become the city of the future. Engineer Buckminster Fuller, designer of the geodesic dome, endorsed the project, as did President Jimmy Carter. But Carter lost his bid for reelection to conservative Ronald Reagan. In 1980, the government turned down Tisan's request for funds, and the dome disappeared into the annals of history.

The Early Days of Video Games

Today's die-hard video gamers might chuckle at the thought of playing a simple game of table tennis on a TV screen. But without Pong, there might not be Grand Theft Auto. Read on to learn about the early history of the video game.

✳   ✳   ✳   ✳

Spacewar

At MIT in 1962, Steve Russell programmed the world's first video game on a bulky computer known as the DEC PDP-1. *Spacewar* featured spaceships fighting amid an astronomically correct screen full of stars. The technological fever spread quickly, and by the end of the decade, nearly every research computer in the United States had a copy of *Spacewar* on it.

Pong

In 1972, Nolan Bushnell founded Atari, taking the company's name from the Japanese word for the chess term "check." Atari released the coin-operated *Pong* later that year, and its simple, addictive action of bouncing a pixel ball between two paddles became an instant arcade hit. In 1975, the TV-console version of *Pong* was released. It was received with great enthusiasm by people who could play hours of the tennislike game in the comfort of their homes.

Tetris

After runaway success in the Soviet Union, *Tetris* jumped the Bering Strait and took over the U.S. market in 1986. Invented by Soviet mathematician Alexi Pajitnov, the game features simple play—turning and dropping geometric shapes into tightly packed rows—that drew avid fans in both countries. Many gamers call *Tetris* the most addictive game of all time.

Space Invaders

Released in 1978, Midway's *Space Invaders* was the arcade equivalent of *Star Wars*: a ubiquitous hit that generated a lot of money. It also introduced the "high score" concept. A year later, Atari released *Asteroids* and outdid *Space Invaders* by enabling the high scorer to enter his or her initials for posterity.

Pac-Man

This 1980 Midway classic is the world's most successful arcade game, selling some 99,000 units. Featuring the yellow maw of the title character, a maze of dots, and four colorful ghosts, the game inspired rap songs, Saturday morning cartoons, and a slew of sequels, including *Ms. Pac-Man*.

Donkey Kong

In 1980, Nintendo's first game marked the debut of Mario, who was soon to become one of the most recognizable fictional characters in the world. Originally dubbed Jumpman, Mario was named for Mario Segali, the onetime owner of Nintendo's warehouse in Seattle.

Q*bert

Released by Gottlieb in 1982, this game featured the title character jumping around on a pyramid of cubes, squashing and dodging enemies. Designers originally wanted Q*bert to shoot slime from his nose, but that was deemed too gross.

✳ From 1988 to 1990, Nintendo sold roughly 50 million home-entertainment systems. In 1996, the company sold its billionth video game cartridge for home systems.

✳ In 1981, 15-year-old Steve Juraszek set a world record on Williams Electronics' *Defender*. His score of 15,963,100 landed his picture in *Time* magazine; it also got him suspended from school. He played part of his 16-hour game when he should have been attending class.

✳ In 1977, Atari opened the first Chuck E. Cheese's pizzeria/arcade establishment in San Jose, California. Nolan Bushnell bought the rights to the pizza business when he parted ways with Atari in 1978, then he turned it into a nationwide phenomenon. It was later acquired by its primary competitor, ShowBiz Pizza.

The Firefighter's Best Friend

A device that was first conceived in the 1600s saves countless lives and millions of dollars in property every year.

✳ ✳ ✳ ✳

FIRE HYDRANTS ARE among the most ubiquitous fixtures in U.S. cities. Squat, brightly painted, and immediately recognizable, two or three of them adorn virtually every city block in the country. New York City alone has more than 100,000 hydrants within its city limits.

Fire hydrants as we know them today have been around for more than 200 years, but their predecessors first appeared in London in the 1600s. At that time, Britain's capital had an impressive municipal water system that consisted of networks of wooden pipes—essentially hollowed-out tree trunks—that snaked beneath the cobblestone streets. During large blazes, firefighters would dig through the street and cut into the pipe, allowing the hole they had dug to fill with water. That created an instant cistern that provided a supply of water for the bucket brigade. After extinguishing the blaze, they would drain the hole and plug the wooden pipe—which is the origin of the term *fireplug*—and then mark the spot for the next time a fire broke out in the vicinity.

Fanning the Flames of Invention

In 1666, a terrible fire raged through nearly three-quarters of London. As the city was rebuilt, the wooden pipes beneath the streets were redesigned to include predrilled plugs that rose to ground level. The following century, these crude fireplugs were improved with the addition of valves that allowed firefighters to insert portable standpipes that reached down to the mains. Many European countries use systems of similar design today.

With the advent of metal piping, it became possible to install valve-controlled pipes that rose above street level. Frederick Graff Sr., the chief engineer of the Philadelphia Water Works department, is generally credited with designing the first hydrant of this type in 1801, as part of the city's effort to revamp its water system. A scant two years later, pumping systems became available, and Graff retrofitted Philly's hydrants with nozzles to accommodate the new fire hoses, giving hydrants essentially the same appearance that they have today. By 1811, the city boasted 185 cast-iron hydrants, along with 230 wooden ones.

Hydrants Spread Like Wildfire

Over the next 50 years, hydrants became commonplace in all major American cities. But many communities in the north faced a serious fire-safety problem during the bitter winters: The mains were usually placed well below the frost line, allowing the free-flow of water year-round, but the aboveground hydrants were prone to freezing, rendering them useless. Some cities tried putting wooden casings filled with sawdust or other insulating material—such as manure—around the hydrants, but that wasn't enough to stave off the cold. Others sent out armies of workers on the worst winter nights to turn the hydrants on for a few minutes each hour and let the water flow.

The freezing problem wasn't fully solved until the 1850s, with the development of dry-barrel hydrants. These use a dual-valve system that keeps water out of the hydrant until it is needed. Firefighters turn a nut on the top of the hydrant to open a valve where the hydrant meets the main, letting the water rise to street level. When this main valve is closed, a drainage valve automatically opens so that any water remaining in the hydrant can flow out. Very little about the design of fire hydrants has changed since. In fact, some cities are still using ones that were installed in the early 1900s.

Invented by Canadians

Even Canadians might be surprised by how much has been discovered by sons and daughters of the North. It's a lot more than Alexander Graham Bell's telephone.

✻ ✻ ✻ ✻

Automatic postal sorter (Dr. Maurice Levy, 1956): Revolutionizing mail delivery, Levy's first model processed 30,000 letters an hour with an average of three errors. By 1957, a model was processing approximately 200,000 letters per hour.

Basketball (James Naismith, 1891): Most people don't know that the Ontarian invented the game to keep kids out of trouble. Teaching at a Massachusetts YMCA school, the Canadian pedagogue believed that a noncontact indoor sport would do wonders for kids' behavior.

Electric car heater (Thomas Ahearn, 1890): Like many Canadian inventions, this was a natural, considering the chilly Canadian winters. It was one of this Ottawan's many patents relating to electric heat, including a hot-water heater and an iron.

Electric light bulb (Henry Woodward and Mathew Evans, 1874): These Canadians paved the way for Thomas Edison's later improvements to the bulb.

Rotary snowplow (J. W. Elliott, 1869): Consider the difficulties in building a railroad across a country with snowy winters. How many people had to shovel off several thousand miles of track several months per year? Elliott, a dentist by trade, devised a fan that enabled a locomotive to plow the snow itself.

Hydrofoil boat (Alexander Graham Bell and Casey Baldwin, 1908): Fact: Water's drag slows down boats. In the early days of aviation, Bell and Baldwin wondered, "What if you could mount a wing under a boat?" The wing lifted the hull out of the water, improving both the speed and the ride.

Insulin process (Sir Frederick Banting, et. al., 1922): Banting's team pioneered the understanding of insulin's central role in diabetes, then learned to produce it from the pancreas of a cow. Through injections, diabetics could regulate their conditions. Banting's group deserves special credit for placing the invention into the public domain rather than making a lot of money with it.

Flight suit (Dr. Wilbur Franks, 1941): This antigravity suit enabled combat pilots to withstand G-force pressure and extreme acceleration.

Java (James Gosling, 1994): This Alberta native created the Java programming language. Most people think of Java as it relates to the Internet, but it's also found in other diverse devices, including the Mars Rover, toasters, cars, and industrial-inventory tagging machines.

Paint roller (Norman Breakey, 1940): Breakey was a victim of the legal system: He invented this useful painting method but didn't have the capital to defend his patent from those eager to make money from his idea. This may explain why he also invented a beer keg tap.

Snowblower (Arthur Sicard, 1925): As a boy, Sicard saw that snowstorm-blocked roads caused a lot of dairy spoilage. He unveiled the first "snowblowing" device in 1925—a truck with his fanlike snowblower in front. The walk-behind snowblower didn't come along until the 1950s, by which time Sicard's machines were already busily working on the streets of Canada.

Wonderbra (Louise Poirier, 1964): Poirier developed this item while working for Canadelle, a company well known to Canadian women for more than half a century of firm support. The Wonderbra name dates back to the 1930s, but Poirier pushed the bustline-maximizer version to market in 1964.

When Broken Dishes Mean Business

Not every tale of invention is a rags-to-riches story. This one is more of a "dishes-to-riches" story.

✳ ✳ ✳ ✳

LIKE MANY SOCIETY women of her time, Josephine Cochran liked to entertain. What she didn't like was the way her servants handled her good china while washing it after parties, often chipping or breaking it.

Mrs. Cochran decided that the world needed a mechanical dishwasher. Patents had already been issued for such devices, first to Joel Houghton in 1850 for a hand-cranked model and then to L. A. Alexander in 1865. But these were clumsy models that didn't do a very good job of washing dishes. "If nobody else is going to invent a dishwashing machine," Mrs. Cochran is reported to have said, "I'll do it myself."

And so she did. In a shed behind her home, she measured her plates, cups, and saucers, then fashioned wire baskets to fit them. The baskets were loaded into a wheel inside a copper boiler, where hot water rained down upon them as the wheel turned.

Mrs. Cochran decided to start her own company—Cochran's Crescent Washing Machine Company—and in 1886, she received a patent for her "dish-washing machine."

Her first customer was the Palmer House Hotel in Chicago, the city where her invention also took a first prize at the World's Columbian Exposition of 1893.

Mrs. Cochran's dishwashers quickly became popular with restaurants and hotels, but they were too expensive for most homeowners. It wasn't until the middle of the 20th century that electric dishwashers caught on. Mrs. Cochran's company changed names over the years, but it lives on as KitchenAid.

Really Wrong Predictions

Getting it right isn't always easy, even if you're Albert Einstein. Of course, when Einstein gets it wrong, it goes down in history. But he's not the only one to make a major slip-up when it comes to science and technology predictions. Here are some doozies:

✳ ✳ ✳ ✳

"I have traveled the length and breadth of this country and talked with the best people, and I can assure you that data processing is a fad that won't last out the year."

—PRENTICE HALL BUSINESS BOOKS EDITOR, 1957

"There is no reason anyone would want a computer in their home."

—DIGITAL EQUIPMENT CORP. PRESIDENT KEN OLSON, 1977

"640K ought to be enough for anybody."

—ATTRIBUTED TO FORMER MICROSOFT CHAIRMAN BILL GATES, 1981

"Where a calculator on the ENIAC is equipped with 18,000 vacuum tubes and weighs 30 tons, computers in the future may have only 1,000 vacuum tubes and weigh only 1.5 tons."

—POPULAR MECHANICS, 1949

"While theoretically and technically television may be feasible, commercially and financially, it is an impossibility."

—RADIO PIONEER LEE DEFOREST, 1926

"This 'telephone' has too many shortcomings to be seriously considered as a means of communication. The device is inherently of no value to us."

—WESTERN UNION INTERNAL MEMO, 1876

Chapter 16

Death and the Macabre

True Tales of Being Buried Alive

Generally, burial is something that happens after death. But whether by accident, intent, or diabolical design, there are some who settle down for a "dirt nap" a little prematurely.

※ ※ ※ ※

David Blaine

O N APRIL 5, 1999, some 75,000 people gathered on Manhattan's Upper West Side to watch magician David Blaine voluntarily be buried alive—albeit with a small air tube. After being placed inside a transparent plastic coffin with a mere six inches of headroom and two inches on either side, the illusionist was lowered into his burial pit. Next, a three-ton water tank was lowered on top of his tomb. The magician ate nothing during his stunt and reportedly only sipped two to three tablespoons of water per day—a fasting schedule likely designed to keep bodily wastes to a minimum. After seven days of self-entombment, Blaine popped out, none the worse for wear. Blaine called it a "test of endurance of . . . the human body and mind . . ." Even famous debunker and fellow magician James Randi praised him.

Barbara Jane Mackle

Could there be anything more horrifying than a forced burial? Emory University student Barbara Jane Mackle suffered through such an ordeal and miraculously lived to tell the tale.

Abducted from a motel room on December 17, 1968, the 20-year-old daughter of business mogul Robert Mackle was whisked off to a remote location and buried under 18 inches of earth. Her coffinlike box was sparsely equipped with food, water, a pair of vent tubes, and a light. After receiving their $500,000 ransom, the kidnappers informed the FBI of the girl's location. Some 80 hours after Mackle entered her underground prison, she was discovered in a wooded area roughly 20 miles northeast of Atlanta. Her light source had failed just a few hours after her burial, and she was severely dehydrated, but otherwise the young woman was in good condition. Her kidnappers were eventually apprehended.

School Bus Kidnapping and Live Burial

In 1976, a school bus filled with children was hijacked in Chowchilla, California. The driver and 26 children were removed from the bus, placed into vans, and driven 100 miles to a quarry in the town of Livermore. There they were forced into a moving van that was buried several feet below ground, where they had limited survival supplies and minimal air vents. After 12 hours, the students feverishly started looking for ways out. Using old mattresses, the bus driver and a group of boys climbed to the top of the moving van, where they had originally entered. They used a wooden beam to slowly pry off the heavy metal lid that separated them from sweet freedom. After 16 hours below ground, the prisoners emerged. Despite their harrowing experience, all survived. A ransom note was eventually traced to the quarry owner's son. He and two accomplices were sentenced to life in prison.

Toltecs

The Toltecs, a pre-Columbian Native American people, practiced a ceremony called "Burial of the Warrior." During this rite of passage, a young man would enter a forest and bury himself in a shallow grave for a day. In a modern-day incarnation of the practice, people voluntarily go underground, claiming that burial beneath Mother Earth is a way of returning to the womb.

Lying in Wait

There are those who are famous in life, and there are those who are more famous in death. "Eugene" is certainly the latter. An unclaimed body in the town of Sabina, Ohio, turned a funeral home into a tourist attraction.

✳ ✳ ✳ ✳

WHEN MOST PEOPLE die, their remains are quickly taken care of. Not so for the mystery man known as Eugene. After being embalmed, he hung around for nearly 35 years before finally being laid to rest.

Eugene's decades-long saga began and ended in the tiny town of Sabina, which is where he died along the highway on June 6, 1929. Several people reported seeing him walk listlessly through town, as if he were ill, but he didn't stop or ask for help.

A Man Without a Name

No identification was found on the man's body except for a slip of paper with a Cincinnati address. But when the Cincinnati police went to investigate, they found only a vacant lot at the address. Police talked to a man living next door named Eugene Johnson, who became the unknown dead man's namesake.

All that is known about Eugene is that he was African American, was thought to be between 50 and 80 years old, and died of natural causes. Authorities took him to the Littleton Funeral Home, where he was embalmed. But rather than bury Eugene in a pauper's grave, the funeral home's owners decided to wait on the off chance that his survivors might be located.

On Display

Eugene spent the next 35 years at the funeral home, lying in state in a small house in the side yard. As his legend grew, curiosity seekers started dropping by to see the enbalmed man—so many, in fact, that the funeral home's owners erected a screen across the room to protect him from grabby souvenir seekers. Out of respect, the funeral home provided Eugene with a brand-new suit every year.

Before long, Eugene had become a bona fide tourist attraction. Buses passing through town would stop by the funeral home so travelers could stretch their legs and take a peek at the "Sabina mummy." On holidays and summer weekends, lines of people would form waiting to pass by Eugene's resting place, which shows how hungry the citizens of Sabina were for entertainment on Saturday nights.

While Eugene lay in state, an estimated one and a half million people paid their respects—a million of them, including several celebrities, signed his voluminous guest books. Sadly, none of the visitors ever claimed to recognize him.

Eugene was sometimes the victim of pranksters during his years at the funeral home. On a number of occasions he was kidnapped but was quickly recovered. Once, members of a fraternity drove him all the way to the Ohio State campus in Columbus.

Resting in Peace

In 1964, Eugene was finally put to rest. It was evident that no one was going to claim him, and many found the pranks played on him demeaning and harmful to business. Rather than bury Eugene in a potter's field, the owners of the funeral home purchased a plot in Sabina Cemetery and paid for his burial expenses, including another brand-new suit.

Only a handful of people attended the service for Eugene on October 21, 1964, and employees of the Sabina Cemetery, Spurgeon Vault Company, and Littleton Funeral Home acted as pallbearers. A local Methodist minister offered a few words and a prayer, and Eugene was finally able to rest in peace.

Memorable Epitaphs

They might be six feet under, but a good epitaph means they'll never been forgotten. Here are some favorite gravestone inscriptions.

✳   ✳   ✳   ✳

Mel Blanc: "That's all folks!"—Arguably the world's most famous voice actor, Mel Blanc's characters included Bugs Bunny, Porky Pig, Yosemite Sam, and Sylvester the Cat. When he died of heart disease and emphysema in 1989 at age 81, his epitaph was his best-known line.

Spike Milligan: *"Dúirt mé leat go raibh mé breoite."*— The Gaelic epitaph for this Irish comedian translates as, "I told you I was ill." Milligan, who died of liver failure in 2002 at age 83, was famous for his irreverent humor, which was showcased on TV and in films such as *Monty Python's Life of Brian*.

Joan Hackett: "Go away—I'm asleep."—The actor—who was a regular on TV throughout the 1960s and '70s, appearing on shows such as *The Twilight Zone* and *Bonanza*—died in 1983 of ovarian cancer at age 49. Her epitaph was copied from the note she hung on her dressing-room door when she didn't want to be disturbed.

Rodney Dangerfield: "There goes the neighborhood."— This comedian and actor died in 2004 from complications following heart surgery at age 82. His epitaph is fitting for this master of self-deprecating one-liners, who was best known for his catchphrase, "I don't get no respect."

Dee Dee Ramone: "OK...I gotta go now."—The bassist of the punk rock band The Ramones died of a drug overdose in 2002 at age 49. His epitaph is a reference to one of the group's hits, "Blitzkreig Bop."

Ludolph van Ceulen: "3.14159265358979323846264339283 27950288 . . ."—The life's work of van Ceulen, who died from unknown causes in 1610 at age 70, was to calculate the value of the mathematical constant pi to 35 digits. He was so proud of this achievement that he asked that the number be engraved on his tombstone.

George Johnson: "Here lies George Johnson, hanged by mistake 1882. He was right, we was wrong, but we strung him up and now he's gone."—Johnson bought a stolen horse in good faith, but the court didn't buy his story and sentenced him to hang. His final resting place is Boot Hill Cemetery in Tombstone, Arizona, which is also "home" to many notorious characters of the Wild West, including Billy Clanton and the McLaury brothers, who died in the infamous gunfight at the O.K. Corral.

John Yeast: "Here lies Johnny Yeast. Pardon me for not rising."—History hasn't recorded the date or cause of John Yeast's death, or even his profession. We can only hope that he was a baker.

Lester Moore: "Here lies Lester Moore. Four slugs from a 44, no Les, no more."—The date of birth of this Wells Fargo agent is not recorded, but the cause of his death in 1880 couldn't be clearer.

Jack Lemmon: "Jack Lemmon in . . ."—The star of *Some Like It Hot* (1959), *The Odd Couple* (1968), and *Grumpy Old Men* (1993) died of bladder cancer in 2001 at age 76.

Hank Williams: "I'll never get out of this world alive."—The gravestone of the legendary country singer, who died of a heart attack in 1953 at age 29, is inscribed with several of his song titles, of which this is the most apt.

Millions of Mummies

It sounds like the premise of a horror movie—millions of excess mummies just piling up. But for the Egyptians, this was simply an excuse to get a little creative.

✳ ✳ ✳ ✳

THE ANCIENT EGYPTIANS took death seriously. Their culture believed that the afterlife was a dark and tumultuous place where departed souls (*ka*) needed protection throughout eternity. By preserving their bodies as mummies, Egyptians provided their souls with a resting place—without which they would wander the afterlife forever.

Starting roughly around 3000 B.C., Egyptian morticians began the mummy trade. On receiving a corpse, they would first remove the brain and internal organs and store them in canopic jars. Next, they would stuff the body with straw to preserve its shape, cover it in salt and oils to prevent it from rotting, and then wrap it in linens—a process that could take up to 70 days. Finally, the finished mummies would be placed in a decorated sarcophagus, ready to face eternity.

Mummies have always been sources of great mystery and fascination. Tales of mummy curses were wildly popular in their time, and people still flock to horror movies involving vengeful mummies. Museum displays, especially those of King Tutankhamun or Ramses II, remain sure-fire draws, allowing patrons the chance for a remarkably preserved glimpse of ancient Egypt.

At first, mummification was so costly that it remained the exclusive domain of the wealthy, usually royalty. However, when the middle class began adopting the procedure, the mummy population exploded. Soon people were mummifying everything. The practice of mummifying the family cat was common as an offering to the cat goddess Bast.

Even those who could not afford to properly mummify their loved ones unknowingly contributed to the growing number of mummies. These folks buried their deceased in the Egyptian desert, where the hot, arid conditions dried out the bodies, inadvertently creating natural mummies. When you consider that this burial art was used for more than 3,000 years, it's not surprising that over time the bodies began piling up—literally.

So, with millions of mummies lying around, local entrepreneurs began looking for ways to cash in on these buried treasures. To them, mummies were a natural resource, not unlike oil, which could be extracted from the ground and sold at a heavy profit to eager buyers around the world.

Mummy Medicine

In medieval times, Egyptians began touting mummies for their secret medicinal qualities. European doctors began importing mummies, boiling off their oils and prescribing it to patients as treatment for sore throats, coughs, epilepsy, and skin disorders. Contemporary apothecaries also got into the act, marketing pulverized mummies to noblemen as a cure for nausea.

However, the medical community wasn't completely sold on the benefits of mummy medicine. Several doctors voiced their opinions against the practice; one wrote, "It ought to be rejected as loathsome and offensive," and another claimed, "This wicked kind of drugge doth nothing to help the diseased." A cholera epidemic that broke out in Europe was blamed on mummy bandages, and the use of mummy medicine was soon abandoned.

Mummy Merchants

Grave robbers, a common feature of 19th-century Egypt, made a huge profit from mummies. Arab traders would raid ancient tombs, sometimes making off with hundreds of bodies. These would be sold to visiting English merchants who, on returning to England, could resell them to wealthy buyers. Victorian socialites would buy mummies and hold parties so that their friends could come over and view their Egyptian prizes.

Mummies in Museums

By the mid-19th century, museums were becoming common in Europe, and mummies were prized exhibits. Hoping to make names for their museums, curators would travel to Egypt and purchase mummies to display back home. This provided a steady stream of revenue for the unscrupulous mummy merchants. In the 1850s, the Egyptian government finally stopped the looting of its priceless heritage. Laws were passed allowing only certified archaeologists access to mummy tombs, effectively putting the grave robbers out of business.

Mummy Myths

There are so many stories regarding the uses of mummies that it's often hard to separate fact from fiction. Some historians suggest the linens that comprised mummy wrappings were used by 19th-century American and Canadian industrialists to manufacture paper. At the time, there was a huge demand for paper, and suppliers often ran short of cotton rags—a key ingredient in the paper-making process. Although there's no concrete proof of this, some historians claim that when paper manufacturers ran out of rags, they imported mummies to use in their place.

Another curious claim comes courtesy of Mark Twain. In his popular 1869 travelogue *The Innocents Abroad*, Twain wrote: "The fuel [Egyptian train operators] use for the locomotive is composed of mummies three thousand years old, purchased by the ton or by the graveyard for that purpose." This item, almost assuredly meant as satire, was taken as fact by readers and survives as a legend to this day. However, there is no historical record of Egyptian trains running on burnt mummies. Besides, the mischievous Twain was never one to let a few facts get in the way of a good story. Perhaps those who believe the humorist's outlandish claim might offset it with another of his famous quotes: "A lie can travel halfway around the world while the truth is putting on its shoes."

Fast Facts on Mummification

❋ In your kitchen, you probably have the materials needed to make the starting element for mummification, but please don't try it at home. Natron is a naturally occurring combination of sodium bicarbonate and sodium chloride— baking soda and salt—dissolved in water.

❋ Egyptian embalmers were like modern funeral homes— they'd show you samples until you found a process that fit your price range.

❋ An "el cheapo" mummification consisted of cleaning out the intestines, then soaking the body in natron for ten weeks. After that, the body was returned to the family.

❋ Embalmers set canopic jars (often made of ornate alabaster) that were supposed to contain the mummy's entrails near the mummy. However, some lazy or unscrupulous embalmers just filled the jars with mud and cedar pitch. It was probably easy to get away with: The only people inspecting the jars were looters, who didn't care.

❋ In A.D. 392, Theodosius II, emperor of Rome, banned the practice of Egyptian mummification.

❋ By studying mummies, historians learned that it wasn't odd for Egyptians to have parasites that were almost large enough to merit separate mummification. Many had Guinea worms, horrible pests that can grow to three feet long.

❋ Bone tumors have been found in Egyptian mummies, but none with cancer of the internal organs have been unearthed.

❋ By the 1800s, mummy unwrappings rivaled garden parties and fox hunts as highlights of English social seasons. Lords would send out printed invitations to their friends: "A Mummy from Thebes to be unrolled at half past two, 10th June 1850."

Strange Ways to Die

Most of us strive to lead interesting lives, but some stake a place in history by dying in an unusual way.

✳ ✳ ✳ ✳

So funny it hurts: A fatal guffaw struck Alex Mitchell, a 50-year-old English bricklayer on March 24, 1975, while he and his wife watched his favorite TV sitcom, *The Goodies*. Mitchell found a sketch so hilarious that he laughed for 25 minutes straight until his heart gave out and he died. Mitchell's wife sent the show a letter thanking the producers and performers for making her husband's last moments so enjoyable.

The tortoise in the air: Those flying monkeys in *The Wizard of Oz* were scary enough, but did you ever think that you'd have to worry about flying tortoises? According to legend, Greek playwright Aeschylus was killed when an eagle or a bearded vulture mistook his noggin for a stone and dropped a tortoise on his bald head in an attempt to crack open the tortoise's shell.

Deadly twist: Isadora Duncan was one of the most famous dancers of her time. Her fans marveled at her artistic spirit and expressive moves, and she is credited with creating modern dance. But it was another modern creation that prematurely ended her life. She was leaving an appearance on September 14, 1927, when her trademark long scarf got caught in the wheel axle of her new convertible. She died of strangulation and a broken neck at age 50.

A terrible taste: All war is hell, but ancient wars were especially brutal. After the Persians captured the Roman emperor Valerian during battle around A.D. 260, Persia's King Shapur I is said to have humiliated Valerian by using him as a footstool. But it only got worse for the Roman. After Valerian offered a king's ransom for his release, Shapur responded by forcing molten gold down his prisoner's throat, stuffing him with straw, and then putting him on display, where he stayed for a few hundred years.

Too long in the tooth: Sigurd I of Orkney was a successful soldier who conquered most of northern Scotland in the 9th century. Following a fever-pitched victory in A.D. 892 over Maelbrigte of Moray and his army, Sigurd decapitated Maelbrigte and stuck his opponent's head on his saddle as a trophy. As Sigurd rode with the head, his leg kept rubbing against his foe's teeth. The choppers opened a cut on Sigurd's leg that became infected and led to blood poisoning. He died shortly thereafter.

An unfair way to go: Mark Twain once said, "Golf is a good walk spoiled," and although many a duffer has spent a frustrating couple of hours on the links, few actually die as a result. In 1997, Irishman David Bailey was not so lucky. Bailey was retrieving an errant shot from a ditch when a frightened rat ran up his pant leg and urinated on him. The rat didn't bite or scratch the golfer, so even though his friends kept telling him to shower, Bailey didn't think much of the encounter and kept playing. His kidneys failed two weeks later, and he died. The cause was leptospirosis, a bacterial infection spread by rodents, dogs, and livestock that is usually mild but can cause meningitis, pneumonia, liver disease, and kidney disease.

Fantasy meets harsh reality: Many people who like playing video games do so to escape the pressures of the real world for a bit. But when that escapism is taken too far, gamers can leave the real world altogether. That's what happened to South Korean Lee Seung Seop in August 2005. Lee was an industrial repair technician, but he had quit his job to spend more time playing games. Lee set himself up at a local Internet café and played *StarCraft* for nearly 50 hours straight, taking only brief breaks to go to the bathroom or nap. Exhaustion, dehydration, and heart failure caused Lee to collapse, and he died shortly thereafter at age 28.

The Mysterious 27 Club

If you're a rock star approaching your 27th birthday, perhaps you should take a year-long hiatus. The curse known as "the 27 Club" is a relatively new one, but that doesn't make it any less freaky. For those about to blow out 27 candles... good luck.

✳ ✳ ✳ ✳

Founding Members

KEITH RICHARDS AND Eric Clapton both cite guitarist Robert Johnson as a major musical influence. Born on May 8, 1911, Johnson played guitar so well at such a young age that some said he must have made a deal with the devil. Those spooky speculations have survived in part due to Johnson's untimely death. The blues legend died on August 16, 1938, at age 27, when the husband of a woman with whom Johnson was involved allegedly poisoned him.

After Johnson, the next musician to join the 27 Club was Brian Jones, one of the founding members of the Rolling Stones. Jones was a lifelong asthma sufferer, so his descent into drug and alcohol addiction was probably not the wisest choice. Still, the sex and drugs inherent in the music biz proved to be too much for Jones to pass up. Some believe he committed suicide because his time with the Stones had recently come to an end. Due to his enlarged liver, autopsy reports led others to believe that he overdosed. Either way, the 27-year-old rocker was found lifeless in a swimming pool in 1969. Jones, another person who cited Johnson as a musical influence, was unfortunately following in his idol's footsteps—and he would soon have company.

A Trio of Inductees

About a year later, the 27 Club would claim its biggest star yet. The counterculture of the late 1960s had embraced the incredibly talented Jimi Hendrix. Legions of fans worshipped the man and his music and sang along to "Purple Haze" at Woodstock.

On September 18, 1970, the rock star—who, like so many before him and since, had an affinity for drugs and alcohol—died in London at age 27. Hendrix choked on his own vomit after taking too many sleeping pills.

Texas-born singer-songwriter Janis Joplin was another megastar at the time and a friend of Hendrix. Widely regarded as one of the most influential artists in American history, Joplin's gravelly voice and vocal stylings were unique and incredibly popular. She screeched, growled, and strutted through numbers such as "Me and Bobby McGee" and "Piece of My Heart." She also tended to play as hard as she worked, typically with the aid of drugs (including psychedelics and methamphetamines) and her signature drink, Southern Comfort whiskey.

On October 4, 1970, when Joplin failed to show up for a recording session for her upcoming album, *Pearl*, one of her managers became worried and went to her motel room to check on her. He found the singer dead—at age 27—from a heroin overdose. After Joplin's death, rumors about this strange "club" began to take hold in the superstitious minds of the general public. Another tragic death less than a year later didn't help quell the story.

Florida-born Jim Morrison was yet another hard-living, super-famous, devil-may-care rock star. He skyrocketed to fame as the front man for the band The Doors. The young musician was known for his roguish good looks, his curly dark hair, and his charismatic and mysterious attitude. But his fans didn't have much time to love him. The Doors hit their peak in the late 1960s, and Morrison died (at age 27) from an overdose on July 3, 1971.

Another Inductee

If you were a fan of rock 'n' roll music in 1994 (especially if you were younger than 30), you probably remember where you were when you heard that Kurt Cobain had died. The tormented lead singer of the incredibly popular alternative rock band Nirvana committed suicide after a lifelong battle with drug addiction, chronic pain, and debilitating depression.

At the tender age of—you guessed it—27, Cobain ended his life and became another member of the Club. Cobain seemed to have known about the "elite" group of young, dead rock musicians: His mother told reporters, "Now he's gone and joined that stupid club. I told him not to join that stupid club."

The Latest Victim

Amy Winehouse was the most recent name added to the 27 Club's roster. The talented yet troubled songstress quickly rose to fame in the mid-2000s, winning five Grammy awards in 2006. But all the while, the sultry singer battled serious addiction to drugs and alcohol, which ultimately led to her premature death. On July 23, 2011, Winehouse died of alcohol poisoning; her blood-alcohol level was five times the legal limit.

Rock Steady? Probably Not

It is odd that these incredibly influential, iconic figures in the music biz would all die at age 27, well before their times. But when you think about all of the other rock stars who *didn't* die—Keith Richards, Paul McCartney, and Ozzy Osbourne, to name a few—the odds don't seem so bad. Plus, when you consider how hard these individuals lived while they were alive, it seems extraordinary that they survived as long as they did.

Rock stars are shrouded in speculation as well as the powerful effects of idol worship, so it's no wonder that fans have elevated what's probably just a strange coincidence into the stuff of legends or curses. Whether you believe in the 27 Club or not, you can still rock out to the music that these tragic stars left behind.

Chaplin's Coffin Held for Ransom

In the silent-film era, he entertained millions without uttering a single word. Perhaps that's why two men thought Charlie Chaplin's casket was worth more than half a million dollars in ransom.

✳ ✳ ✳ ✳

The Life and Death of a Silent Star

BORN IN LONDON on April 16, 1889, Charles Spencer Chaplin had his first taste of show business at age 5 when his mother, a failing music-hall entertainer, could not continue her act and little Charlie stepped up and finished her show. In February 1914, Chaplin's first movie, *Making a Living*, premiered; it would be the first of more than 30 shorts that Chaplin made in 1914 alone. In fact, from 1915 until the end of his career, Chaplin was featured in nearly 100 movies, mostly shorts. That was no small feat considering that he wrote, starred in, directed, produced, and even scored all of his own movies. All of this made Charlie Chaplin a household name as well as a very rich man.

But not everything was wine and roses for Chaplin. After three failed marriages, the 54-year-old actor caused quite a stir in 1943 when he married his fourth wife, 17-year-old Oona O'Neill. More scandal arose in the early 1950s when the U.S. government began to suspect that he and his family might be Communist sympathizers. There seemed to be very little to support these suspicions other than the fact that Chaplin had chosen to live in the United States without declaring U.S. citizenship. Regardless, Chaplin soon tired of what he deemed harassment and moved to Switzerland, where he lived with his wife and their eight children until his death on Christmas Day 1977. Shortly thereafter, perhaps the strangest chapter in Charlie Chaplin's saga began.

Grave Robbers

On March 2, 1978, visitors to Charlie Chaplin's grave were shocked to discover a massive hole where the actor's coffin had been. It soon became clear that sometime overnight, someone had dug up and stolen Chaplin's entire casket. But who would do such a thing? And why? It didn't take long to find the answer. Several days later, Oona began receiving phone calls from people who claimed to have stolen the body and demanded a portion of Chaplin's millions in exchange for the casket. Oona dismissed most of the callers as crackpots, with the exception of one: This mysterious male caller seemed to know an awful lot about what Chaplin's coffin looked like. But because he was demanding the equivalent of $600,000 U.S. dollars in exchange for the coffin's safe return, Oona told the caller she needed more proof. Several days later, a photo of Chaplin's unearthed casket arrived in her mailbox, so Oona alerted Swiss police.

The Arrests

When Oona first met with Swiss authorities, she could only show them the photo and tell them that the caller was male and that he spoke with a Slavic accent. She also told them that she had no intention of paying the ransom. But the police convinced Oona that the longer she pretended to be willing to pay the ransom, the better chance they had of catching the thief. Oona was emotionally unable to deal with the fiasco, so Chaplin's daughter Geraldine complied with the investigators' request.

During the next few weeks, Geraldine did such a convincing job that she talked the caller's ransom price down from $600,000 to $250,000, though the Chaplin family still did not plan to pay it. In the meantime, Swiss police were desperately trying to trace the calls. Their first big break came when they established that the calls were coming from a pay phone in Lausanne, Switzerland. However, there were more than 200 pay phones in the town. Undaunted, police began staking out all of them. Their hard work paid off when they arrested 24-year-old Roman Wardas, who admitted to stealing Chaplin's coffin, stating that he had gotten the idea after reading about a similar body "kidnapping" in an Italian newspaper. Based on information Wardas provided police, a second man, 38-year-old Gantcho Ganev, was also arrested. Like Wardas, Ganev admitted to helping take the casket but claimed that it was all Wardas's idea and that he just helped out.

So Where's the Body?

Of course, once the two suspects were in custody, the question on everyone's minds was, "Where was Chaplin's casket?" Ganev and Wardas claimed that after stealing the casket from the cemetery, they drove it to a field and buried it in a shallow hole.

Following directions provided by both suspects, Swiss police descended upon a farm about 12 miles from the Chaplin estate. Spotting a mound of what appeared to be freshly moved dirt, they began digging, and on May 17, 1978, in the middle of a cornfield, Chaplin's unopened coffin was recovered.

After word of the casket's location got out, people began flocking to the farm, so the farmer placed a small wooden cross, ornamented with a cane, over the hole where the casket had been buried. For several weeks, people brought flowers and paid their respects to the empty hole.

The Aftermath

After a very short trial, both men were convicted of extortion and disturbing the peace of the dead. As the admitted mastermind of the crime, Wardas was sentenced to nearly five years of hard labor. Ganev received only an 18-month suspended sentence.

As for Chaplin's unopened casket, it was returned to Corsier-Sur-Vevey Cemetery and was reburied in the exact spot where it had originally been interred. Only this time, to deter any future grave robbers, Oona ordered it buried under six feet of solid concrete.

Curious Classifieds

✳ "Mixing bowl set designed to please a cook with round bottom for efficient beating."

✳ "The fact that those we have served return once again, and recommend us to their friends, is a high endorsement of the service we render. Village Funeral Home."

✳ "Joining nudist colony, must sell washer and dryer—$300"

✳ "A superb and inexpensive restaurant. Fine food expertly served by waitresses in appetizing forms."

✳ "Christmas tag sale. Handmade gifts for the hard-to-find person."

✳ "Our bikinis are exciting. They are simply the tops."

✳ "Full-size mattress. Royal Tonic, 20-year warranty. Like new. Slight urine smell. $40."

✳ "For a successful affair, it's the Empire Hotel."

✳ "Tickle Me Elmo. New in box. Hardly tickled. $700."

✳ "Tired of cleaning yourself? Let me do it."

✳ "For sale—Diamonds, $20; microscopes, $15."

✳ "Will swap white satin wedding gown (worn once) for 50 pounds fresh Gravy Train."

✳ "Modular sofas. Only $299. For rest or fore play."

✳ "Used tombstone, perfect for someone named Homer Hendel Bergen Heinzel. One only."

✳ "Save regularly in our bank. You'll never reget it."

✳ "Springmaid sheets are known as America's favorite playground."

✳ "Illiterate? Write today for free help."

✳ "Used cars: Why go elsewhere to be cheated? Come here first!"

✳ "Cuisine of India—Nashville's BEST Italian Restaurant."

✳ "Oscar Mayer Bilingual Turkey Franks 3 for $4."

✳ "Free Puppies—Half cocker spaniel, half sneaky neighbor's dog."

Chapter 17

Politics

America's Only King

Few people would believe that a separate empire with its own king once existed within the United States of America. But James Jesse Strang was indeed crowned ruler of a Lake Michigan island kingdom.

✳ ✳ ✳ ✳

Growing the Garden

JAMES JESSE STRANG moved to Wisconsin in 1843 with his wife, Mary. There they bought a large parcel of land just west of what would become the city of Burlington.

Strang met the Mormon prophet Joseph Smith on a trip to Nauvoo, Illinois. Smith immediately appointed him an elder in the faith and authorized him to start a Mormon "stake" in Wisconsin named Voree, which meant "Garden of Peace."

Mormons from around the country flocked to Voree to build homes on the rolling, forested tract along the White River. A few months after Strang became a Mormon, Smith was killed. To everyone's amazement, Strang produced a letter naming him as the church's next prophet, which appeared to have been written and signed by Smith.

However, another leader named Brigham Young also claimed that title. Young eventually won, and Strang broke away to form his own branch of Mormonism.

Secrets from the Soil

In September 1845, Strang made a stunning announcement:
A divine revelation had told him to dig under an oak tree in
Voree located on a low rise called the "Hill of Promise." Armed
with shovels, four followers unearthed a box that contained some
small brass plates, each only a few inches tall.

The plates were covered with hieroglyphics, crude drawings of
the White River settlement area, and a vaguely Native American
human figure holding a scepter. Strang claimed to be able to
translate them using special stones, like the ones Smith had
used to translate similar buried plates in New York. The writing,
Strang said, was from a lost tribe of Israel that had made it to
North America. He showed the plates to hundreds of people.

He created subgroups among his followers. There was the
commune-style Order of Enoch, as well as the secretive
Illuminati, who pledged their allegiance to Strang as "sovereign
Lord and King on earth." Infighting developed within the ranks,
and area non-Mormons raised objections to the community.

In 1849, Strang received a second set of divine messages called
the Plates of Laban, which he said were originally in the Ark of
the Covenant. They contained instructions called *The Book of the
Law of the Lord*, which Strang again translated with his helpful
stones. The plates were not shown to the group at that time, but
they did eventually yield support, some would say rather conveniently, for the controversial practice of polygamy.

Polygamy Problems

Strang had personal reasons for getting divine approval to have
multiple wives. In July 1849, a 19-year-old woman named
Elvira Field secretly became Strang's second wife. There was
just one problem—he hadn't divorced his first one. Soon, Field
began traveling with him posing as a young man named Charlie
Douglas, with her hair cut short and wearing a man's black suit.
Yet the "clever" disguise did little to hide Field's ample figure.

At about that time, Strang claimed that another angel told him it was time to get out of Voree. Strang was to lead his people to a land surrounded by water and covered in timber. This land, according to Strang, was Beaver Island, the largest of a group of islands in the Beaver Archipelago north of Charlevoix, Michigan. It had recently been opened to settlement, and the Strangites moved there in the late 1840s.

The Promised Island

On July 8, 1950, Strang donned a crown and a red cape as his followers officially dubbed him King of Beaver Island. Falling short of becoming King of the United States, he was later elected to Michigan's state legislature, thanks to strong voter turnout among his followers. Perhaps reveling in his new power, he eventually took three more young wives, for a total of five.

On the island, Strang's divine revelations dictated every aspect of daily life. He mandated that women wear bloomers and that their skirts measure a certain length. He required severe lashings for adultery and forbade cigarettes and alcohol. Under this strict rule, some followers began to rebel. In addition, relations with local fishermen soured as the colony's businesses prospered.

On June 16, 1856, a colony member named Thomas Bedford, who had previously been publicly whipped, recruited an accomplice and then shot Strang. The king survived for several weeks and was eventually taken back to Voree by his young wives. He died in his parents' stone house—which still stands near Mormon Road on State Highway 11. At the time, four of his wives were pregnant. And back in Michigan, it wasn't long before local enemies and mobs of vigilantes from the mainland forcibly removed his followers from Beaver Island.

Strang was buried in Voree, but his remains were later moved to a cemetery in Burlington. Strang's memory also lives on in a religious group formed by several of his followers, the Reorganized Church of Jesus Christ of Latter-Day Saints.

Fast Facts

* The record for the longest filibuster is held by the late South Carolina senator Strom Thurmond, who spoke against the 1957 Civil Rights Act for 24 hours and 18 minutes.

* The first female senator, Rebecca Felton, was 87 years old when she was appointed to the Senate by the governor of Georgia. She served for one day: November 21, 1922.

* The halls of Congress are reputed to be haunted by a multitude of deceased politicians, among them John Quincy Adams and Daniel Webster. More surprising, perhaps, are the periodic sightings of a demonic cat.

* In the spring of 1930, the Senate almost voted to ban all dial telephones from the Senate wing of the Capitol, as the technophobic older senators found them too complicated to use.

* Morocco was the first country to recognize the United States as a sovereign nation, in 1777.

* The proud American motto *E pluribus unum*—out of many, one—was originally used by Virgil to describe salad dressing.

* Benjamin Franklin considered the bald eagle to be a "bird of bad moral character" and resented its being chosen to represent the United States of America.

* The founder of the Smithsonian, James Smithson, who in 1826 willed a then-staggering $508,318 to the United States, never set foot in America.

* On September 13, 1859, California senator David Broderick became the first sitting senator to be killed in a duel.

* Officially, the United States of America comprises only 46 states: Kentucky, Massachusetts, Pennsylvania, and Virginia designate themselves commonwealths.

Sitting on the Laps of Power

Former Secretary of State Henry Kissinger called power "the ultimate aphrodisiac." For centuries, influential—and married— men have been attracting women who are drawn by power.

✳ ✳ ✳ ✳

Presidential Follies

AMERICAN PRESIDENTIAL DALLIANCES seemed almost commonplace in the 20th century. Bill Clinton had Monica Lewinsky. Franklin D. Roosevelt had a decades-long affair with Lucy Mercer (later Rutherfurd)—in fact, it was she, and not his wife Eleanor, who was with him when he died at Warm Springs, Georgia, in 1945. John Kennedy allegedly had Angie Dickinson, Marilyn Monroe, Jayne Mansfield, and Judith Campbell Exner, among others.

Lucy Mercer

Warren G. Harding worked as a newspaper editor and Ohio state senator before he decided to capitalize on America's war-weariness by running for president. He became the nation's 29th president in 1920, and he promised America a "return to normalcy." However, despite being married to a woman he called "Duchess," Harding had previously carried on a 15-year-long affair with Carrie Phillips, the wife of a good friend.

When that affair ended, he took up with the much-younger Nan Britton. The innocence didn't last long—on one particular night of passion in the Senate Office Building in January 1919, she conceived a child. After Harding's death in 1927, Britton published a tell-all book about their trysts, *The President's Daughter*. So much for "normalcy."

Amorous Also-Rans

Sometimes a man lusting after both the presidency and women finds the two desires don't mix well. Such was the case in July

1791, when U.S. Treasury Secretary Alexander Hamilton began an affair with Maria Reynolds, a pretty 23-year-old woman who tearfully implored him for help after her husband left her. A few months later, Reynolds's husband, a professional con man, mysteriously returned and blackmailed Hamilton. Although he paid $1,750 to keep the affair quiet, Hamilton ultimately learned that blackmailers are never satisfied. In 1797, the affair came to light, creating one of the first sex scandals in American politics. Although Hamilton apologized, many historians believe the damage to his reputation cost him the presidency he so coveted.

Another man who saw his presidential chances wrecked on the rocks of infidelity was Gary Hart. The odds-on favorite to win the 1988 Democratic presidential nomination, the married senator from Colorado was caught by the press in the company of Donna Rice, a blonde 29-year-old actress and model, in April 1987. One of the places they had allegedly been together was on a yacht called, appropriately enough, *Monkey Business*. After several days of feverish headlines, Hart withdrew from the presidential race. Although he reentered the race later that year, Hart's monkey business had finished him as a force in national politics.

Political cost was likely not on the mind of Thomas Jefferson if, and when, he began an affair with Sally Hemings, a slave at Monticello. Although he was a powerful political figure, Jefferson was also a lonely widower who had promised his dying wife in 1782 that he would never remarry. At the time that he allegedly began the affair with Hemings, the presidency must have seemed a distant dream. In 1800, however, Jefferson became president. Two years later, a newspaper editor named James Callendar first published the charge that Jefferson and Hemings were an item, thus igniting a historical controversy that still rages today. Even modern methods such as DNA testing have failed to positively identify Thomas Jefferson as the father of Hemings's children. The only thing certain is that this story is far from over.

Foreign Affairs

Of course, American political figures aren't the only ones with roving eyes. Charles Stewart Parnell was a leader of the Ireland's Independence Movement in the 1880s. It seemed as if British Prime Minister William Gladstone was about to support Parnell and finally give Ireland its freedom, but in November 1890, it was revealed that Parnell had long been involved with Kitty O'Shea, the wife of Irish MP William Henry O'Shea. The disclosure rocked prim-and-proper England, causing Gladstone to distance himself from Parnell and pull back from endorsing Irish independence. Thus not just a political career but also the fate of an entire nation was affected by one man's indiscretion.

Of course, there's often much more at stake than a political career. Claretta Pettachi was a beautiful young Italian girl who became Italian dictator Benito Mussolini's lover. To her credit (or discredit), she stayed loyal to him to the end. In April 1945, she and Mussolini were captured as they tried to flee Italy. According to legend, Pettachi was offered her freedom but refused, and she threw her body in front of Mussolini's in a vain attempt to shield him from a firing squad's bullets. Photos show their bodies, which were subsequently hung upside down in a public square.

Pettachi's devotion to her fascist lover is perhaps only topped by that of Prince Pedro of Portugal. Pedro had an affair with one of his wife's maids, Inês de Castro, who bore him two children. His wife died in 1349, and de Castro was put to death in 1355. When Pedro became king in 1357, he had his mistress' body exhumed, married the corpse, and forced his entire court to honor her remains.

History is filled with many other examples of famous men and the women they attracted. However, this doesn't seem to work in the opposite direction—stories of powerful women and the men they attracted are much less common. Perhaps it is as Eleanor Roosevelt once observed: "[How] men despise women who have real power."

Cleveland's Secret Cancer

President Grover Cleveland was known as "Grover the Good" for his honesty. However, there was one time when Cleveland was not very forthcoming.

✳ ✳ ✳ ✳

The Painful Truth

IN 1893, CLEVELAND had recently returned to the White House, this time as the country's 24th president. America had just entered a national economic depression, the Panic of 1893. To help the economy recover, Cleveland wanted to repeal the Sherman Silver Purchase Act and maintain the gold standard.

On June 13, 1893, Cleveland felt pain on the left—or "cigar-chewing"—side of the roof of his mouth. A few days later, he noticed a rough spot in the area. He asked White House physician R. M. O'Reilly to take a look at it, and the doctor discovered a cauliflower-like area the size of a quarter. Samples were sent to several doctors for biopsy, and the results showed a malignancy.

We've Got a Secret

But having surgery wasn't that easy. With the economy on the skids, the country jittery, and a pro-silver vice president (Adlai Stevenson), Cleveland did not want to further upset the nation by announcing that his health was at risk. So he ordered his condition be kept secret from the public.

However, his medical team was racing against a deadline. It was now late June. Congress would reconvene on August 7, and Cleveland wanted to have recovered enough to personally persuade wavering congressmen to vote for free silver repeal. Meanwhile, the doctors were desperate to perform the surgery and stop the cancer from spreading.

It was agreed the surgery was to be conducted on July 1. For maximum secrecy, the operation was to be held on the yacht *Oneida*, which was owned by the president's friend Commodore

Elias Benedict. This was logical: Cleveland was often seen on Benedict's yacht, and with the July 4th holiday approaching, it was an entirely plausible place for him to be.

Sink or Swim

The doctors had already slipped on board when Cleveland joined them on June 30. The yacht was piloted to Bellevue Bay in New York City's East River—near Bellevue Hospital. On July 1, Cleveland went below deck and was propped up in a chair that was lashed to the mast. Fifty-six years old and corpulent, Cleveland was considered to be at high risk for a stroke if ether was used as an anesthetic, so Dr. Joseph D. Bryant decided to use nitrous oxide and hope for the best. The doctor was so anxious about the operation and Cleveland's condition that he told the captain that if he hit a rock, he should *really* hit it, so that they would all go to the bottom.

Over the next few hours, doctors removed Cleveland's entire left upper jaw, from the first bicuspid to past the last molar. Cleveland recovered in splendid isolation aboard the yacht for a few days, and then for a few weeks at his family's Cape Cod vacation home. A plug of vulcanized rubber was made to fill the huge hole in Cleveland's jaw. The press was told that the president had been treated aboard the ship for two ulcerated teeth and rheumatism and was recovering nicely.

(Oh, and by the way, the free silver repeal did indeed pass Congress later that year, just as Cleveland had planned.)

Nixon Would Have Been Proud

That's where matters stayed for two decades. Even though a gabby doctor spilled the beans to the *Philadelphia Press* in late August 1898, Cleveland's doctors and close friends managed to deny the story out of existence. Finally, in 1917—nine years after Cleveland's death—the full story was finally revealed in *The Saturday Evening Post*. Today, part of Cleveland's jaw can be seen at the Mütter Museum in Philadelphia.

Drama in Government: 10 Actors Turned Politicians

Government's nothing if not drama-filled. Who better, then, to lead us than someone straight out of Hollywood?

✳   ✳   ✳   ✳

Arnold Schwarzenegger

PERHAPS THE BIGGEST actor (literally) to transition into politics, Arnold Schwarzenegger raised more than a few eyebrows when he announced his run for governor of California in 2003. Going up against such reputable contenders as porn star Mary Carey and actor Gary Coleman, Schwarzenegger quickly moved to the head of the pack. Then-governor Gray Davis was out and the "Governator" was in.

Al Franken

Saturday Night Live alum Al Franken started making real headlines when he ran for a Minnesota Senate seat against incumbent Norm Coleman in 2008. A too-close-to-call race led to a seemingly endless string of recounts and court challenges. Franken finally won the seat on June 30, 2009.

Fred Thompson

Law & Order vet Fred Thompson tossed his hat into the ring for the 2008 presidential election. (If we have to tell you that he didn't win, you're probably reading the wrong book.) Thompson had previously served in the U.S. Senate from 1993 to 2003.

Ronald Reagan

Ronald Reagan progressed from president of the Screen Actors Guild to president of the United States. Reagan's early career saw him appear in dozens of films and television shows. In 1966, he made it to the California governor's mansion, paving the way for his two terms in the White House during the 1980s.

Jesse Ventura

Known to wrestling fans as "The Body," Jesse Ventura brought his bulk to the Minnesota governor's office for one term from 1999 to 2003, sailing past more traditionally qualified candidates to win the election and get the gig.

Sonny Bono

Cher's former partner-in-crime started his political career in 1988 when he was elected mayor of Palm Springs, California. From there, Bono served in the U.S. House of Representatives, where he served for two terms before his death in 1998.

Alan Autry

Alan Autry did it all: He played in the NFL, played Captain Bubba Skinner in TV's *In the Heat of the Night*, and then played politics for the city of Fresno, California. He was elected mayor in 2000 and served two terms.

Shirley Temple Black

Former child star Shirley Temple Black lost her 1967 run for a Congressional seat in California, but she didn't let that end her political aspirations. Black went on to hold numerous diplomatic posts, including delegate to the United Nations and U.S. Ambassador to Ghana and Czechoslovakia.

Ben Jones

Who would have thought that a guy known as "Cooter" could become a member of Congress? Ben Jones, the actor who was famous for his role as the mechanic in *The Dukes of Hazzard*, served two terms after being elected in Georgia.

Clint Eastwood

This cowboy wrangled his way into the political arena for a brief stint as mayor of Carmel, California. Eastwood ran for office in 1986 after the city's planning board prevented him from remodeling a piece of his property. Once mayor, Eastwood fired the board members who voted against his proposal.

The Lincoln–Kennedy Debates

After President John F. Kennedy's assassination in 1963, history buffs began finding peculiar parallels between Kennedy and President Abraham Lincoln.

✳ ✳ ✳ ✳

Honest-Abe Truths

Fact: Lincoln and Kennedy were elected to Congress 100 years apart (1846 and 1946, respectively). They were also elected to the presidency 100 years apart (1860, 1960) and had successors who were born 100 years apart (1808, 1908).

Fact: Both men were shot in the head.

Fact: Both had successors with the last name Johnson (Andrew and Lyndon).

Fact: The names Lincoln and Kennedy both contain seven letters.

None of these commonalities are particularly remarkable. The head is a logical target for a would-be assassin; Johnson is a common last name. Some people get excited about numeric coincidences while conveniently ignoring countless differences, such as the men's ages at the times of their deaths, years of their births, and the fact that their first names have different numbers of characters (and Lincoln didn't even have a middle name). If the 100-year connection seems uncanny, perhaps it's because round numbers are given undue importance (think of the excitement and foreboding that accompany the end of a decade, century, or millennium).

Semi-Truths

Coincidence: Both first ladies lost children while in the White House.

Fact: Although this is technically true, the situations were quite different. The Kennedys' child died a few days after a premature birth, and two older children (a boy and a girl) survived to adulthood. The Lincolns lost an 11-year-old son to typhoid while they were in the White House; they had three other sons, two of whom died before they reached adulthood.

Coincidence: Lincoln and Kennedy were both assassinated by and succeeded by Southerners.

Fact: Both men chose Southern running mates to help balance their tickets (Lincoln was from Illinois and Kennedy was from Massachusetts). That their assassins were both Southerners is debatable. Although Lincoln's assassin, John Wilkes Booth, sympathized with the South, he spent most of his life in the North and thought of himself as a Northerner. Kennedy's assassin, Lee Harvey Oswald, was born in New Orleans but moved around so much that he didn't identify as Northern or Southern.

Coincidence: Both assassins were known by three names.

Fact: Booth used the names "J. Wilkes Booth" and "John Wilkes" equally, whereas Oswald went primarily by "Lee Oswald" when he was not using one of his many aliases. Only after the assassination did police and the media use his full name to ensure proper identification.

Coincidence: Booth and Oswald were both assassinated before their trials.

Fact: Although both men were killed before going to trial, the details of their deaths were different. Two days after Oswald was captured by police, nightclub owner Jack Ruby shot him in the abdomen while he was in transit under police custody. A lawfully armed federal officer named Boston Corbett shot Booth in the neck as he was attempting to evade arrest (which hardly constitutes an assassination).

Coincidence: Lincoln and Kennedy were both assassinated on a Friday before a holiday.

Fact: Lincoln was shot on April 14, the Friday before Easter (also known as Good Friday). Kennedy was shot on November 22, the Friday before Thanksgiving. Lincoln lived until the next day, whereas Kennedy was declared dead shortly after he was shot.

Coincidence: Both presidents had special concerns for civil rights.

Fact: It is accurate to say that monumental events in civil rights occurred during each presidency—for example, Lincoln signed the Emancipation Proclamation in 1863, and the March on Washington took place in 1963, during Kennedy's term. But Lincoln was actually not a proponent of civil rights in the modern sense. In fact, during one of his famous debates with Stephen Douglas, he said, "I will say that I am not, nor ever have been, in favor of bringing about in any way the social and political equality of the white and black races." He maintained that blacks should not hold office, intermarry, vote, or sit on juries—rights we now consider inalienable.

Flat-Out Fibs

Myth: Lincoln's secretary, Kennedy, warned him not to go to Ford's Theatre; Kennedy's secretary, Lincoln, warned him not to go to Dallas.

Fact: Kennedy did have a secretary named Evelyn Lincoln, but there is no evidence that she warned him not to go to Dallas (he was, however, frequently informed of assassination plots). Lincoln, who was the subject of as many as 80 known plots, was also used to these warnings. But he never had a secretary by the name of Kennedy; his secretaries' surnames were Nicolay and Hay.

Myth: John Wilkes Booth and Lee Harvey Oswald were born 100 years apart, in 1839 and 1939, respectively.

Fact: Even if it were true, it would simply be another instance of giving false meaning to a round number. But it's not true: Booth was born in 1838.

Myth: Booth ran from a theater and was caught in a warehouse; Oswald ran from a warehouse and was caught in a theater.

Fact: Booth shot Lincoln in a theater during a performance of a play. He evaded police for 12 days, during which time he left the state and was captured in a tobacco shed on a farm—not what most would consider a warehouse. Oswald (allegedly) shot Kennedy from the window of a book depository and was captured in a movie theater a few hours later.

Myth: A month before Lincoln was assassinated, he visited Monroe, Maryland; a month before Kennedy was assassinated, he visited Marilyn Monroe.

Fact: Tee-hee. Marilyn Monroe, one of Kennedy's supposed lovers, died more than a year before his death. In addition, there is no Monroe in Maryland, unless one counts the 217 feet of Monroe Street that connects Commerce Street to Clay Street in the town of Point of Rocks.

Common Connections

The violent and untimely deaths of two popular presidents is certainly tragic. It's human nature to try to make meaning out of tragedy, and drawing these tenuous connections is an attempt at assigning order to a chaotic world. In 1992, the magazine *Skeptical Inquirer* asked for submissions of similar lists of coincidences between presidents. One of the winners found connections between 21 different pairs of U.S. presidents, while another found 16 shocking connections between Kennedy and former Mexican President Alvaro Obregón. However, connections can be found between almost any two people, given enough data and research. Comb through and compare siblings, parents, teachers, likes, dislikes, ages, employers, cars, pets, etc., and you'll eventually find a surprising number of matches. Don't have time for that? Good for you.

Bounced by the System

New York City' schoolteachers who have committed offenses have to spend time in reassignment centers, aka "rubber rooms" where they're paid to do nothing.

✳ ✳ ✳ ✳

Clock In, Hang Out

EVERY MORNING, HUNDREDS of New York City schoolteachers clock in and take their places—but not in classrooms. These teachers are expected to simply sit around and wait out their time in one of the many "reassignment centers" run by New York City's school systems. Otherwise known as "rubber rooms," these large, sometimes windowless rooms are where teachers (who have been taken off duty for offenses and are waiting to find out their fate) spend the school day playing games, sleeping, knitting, reading, or watching DVDs. Not one of them is ever found doing what they were hired to do: *teaching*.

A Divisive Situation

As of 2007, more than 750 teachers did time in 12 rubber rooms. The thing is, nobody's really quite sure what the end game is supposed to be or exactly what offenses qualify as actionable.

The Board of Education will tell you that the rubber rooms are a necessary evil—due to tenure and a strong teachers' union, bad or subpar teachers are almost impossible to fire. In addition, a teacher's contract requires a hearing before any action is taken.

Jeanne Allen—president of Washington, D.C.'s, Center for Education Reform—sums up this divisive issue. According to her, rubber rooms exist "because of worn-out and, quite frankly, irrelevant union contracts that do more to protect people's jobs than they do to protect kids."

Teachers, however, would probably respond that the rubber rooms are a heavy-handed tactic devised to break their wills, push them out the door for political reasons, or punish them for infractions that are slight or sometimes even imagined.

In 2008, David Pakter was subjected to a full year of rubber-room "duty" for purchasing an unauthorized plant for his school and giving students gifts for getting good grades. Now he's taken a stand as one of the more outspoken opponents of the rubber room system. Previously named a NYC "teacher of the year," Pakter used his salary from his stay in the rubber room—doing nothing—to purchase a Jaguar automobile. In a *New York Post* interview, he took the opportunity to rail against a system that is sorely in need of overhaul. "It's a present from [Schools Chancellor] Joel Klein," said Pakter of his expensive new ride. "I want to teach, they won't let me teach, but they'll pay me enough to buy a car. Can someone explain this to me?"

Nobody Wins

With New York City teaching salaries running between $42,500 and $93,400 in 2007, it's the taxpayers who lose. Throw in the exorbitant cost of hiring substitute teachers to replace those stuck in the rubber rooms, and the leasing of 12 rooms, and the problem only gets worse. According to the *Post*, the current rubber room policy costs about $40 million a year, based on the median teacher salary. For their parts, the offending teachers speak of depression brought on by the dehumanizing effects of this "guilty until proven innocent" scenario and the boring downtime that awaits them each weekday.

In the end, it's the schoolchildren who are getting the rawest deal of all. With regular teachers being held indefinitely and substitute teachers coming and going, the children's progress is prohibited by bureaucratic red tape. New Yorkers hope their kids' education can bounce back from the effects of the rubber rooms.

The Capital(s) of the United States

Washington, D.C., wasn't always our nation's capital: Since America's inception, no less than nine cities have served in that illustrious capacity. Looks like Congress has always had trouble making up its mind.

✳ ✳ ✳ ✳

Philadelphia, Pennsylvania
(1774–76, 1777, 1778–83, 1790–1800)

Carpenters' Hall served as the original meeting spot for America's First Continental Congress. Now-famous delegates such as George Washington, John Adams, and Patrick Henry hammered out policy in an attempt to assuage difficulties cast upon them by England, their ruling country. But a war for independence was on the horizon.

Baltimore, Maryland
(December 20, 1776–February 27, 1777)

In an attempt to evade hostile British forces that were then massing in Philadelphia, revolution leaders tapped Baltimore's Henry Fite House as America's next headquarters. However, expensive living costs and distasteful conditions would quickly drive Congress back to Philadelphia's Independence Hall.

Lancaster, Pennsylvania
(September 27, 1777)

Sensing a continued British threat, Congress resolved to meet in Lancaster, Pennsylvania. Only one session was convened at Lancaster's Court House before Congress issued another headquarters resolution. The group would next meet across the Susquehanna River in the town of York.

York, Pennsylvania
(September 30, 1777–June 27, 1778)

The York Court House served as America's next capital. Nine months later, Congress learned that the British Army had evacuated Philadelphia, so they decided to return to Independence Hall. But Congress' stay in York had been a productive one: While there, the assembled throng passed the Articles of Confederation and signed a treaty of alliance with France.

Princeton, New Jersey
(June 30, 1783–November 4, 1783)

During the final days of the Revolutionary War, peace negotiations with England were under way, but many battle-weary soldiers thought that an end to the hostilities would never come. As a result, some turned mutinous. On June 20, 1783, a group of these disgruntled soldiers surrounded Philadelphia's Independence Hall. While violence was ultimately averted, Congress felt that Pennsylvania had failed to provide them with proper protection. The assembly passed another resolution naming Nassau Hall in Princeton, New Jersey, as its command center. In well-practiced fashion, Congress would only stay at its new home for a few short months before relocating to Annapolis, Maryland.

Annapolis, Maryland
(November 26, 1783–August 19, 1784)

For the brief period that the Annapolis State House acted as America's capital, two key incidents took place: A definitive peace treaty with Great Britain was ratified, and General George Washington resigned as commander in chief of the Continental Army. Next, it was on to Trenton.

Trenton, New Jersey
(November 1, 1784–December 24, 1784)

If the French Arms Tavern sounds like an unlikely spot to house Congress, a closer look suggests otherwise. Formerly a personal residence, the ornate building was considered Trenton's most beautiful home. The dwelling was leased by the New Jersey legislature for use by Congress, but little action was taken under its roof during the assembly's time there. Congress would spend just two months at its New Jersey home before relocating to New York.

Manhattan, New York
(January 11, 1785–Autumn 1788;
March 4, 1789–August 12, 1790)

Taking a cue from the French Arms Tavern before it, New York's Fraunces Tavern became the second of four Manhattan venues to host Congress (City Hall and two locations on Broadway round out the roster). The delegates would remain at these transitory sites for five years before making their final move to Washington, D.C.

Washington, D.C.
(1800–present)

The Congressional Act of July 16, 1790, provided that Washington, D.C., would serve as America's permanent seat of government beginning in December 1800. Since then, the president, Congress, and other branches of government have been based in this agreeable location beside the Potomac River.

"If people see the Capitol going on, it is a sign we intend the Union shall go on."

—ABRAHAM LINCOLN

A Gay Time in the Oval Office?

Was James Buchanan—the nation's only bachelor chief executive—also its first homosexual president? Historians still aren't sure.

✳ ✳ ✳ ✳

THE YEAR 1834 WAS a momentous one for 42-year-old James Buchanan. Already a veteran political leader and diplomat, Buchanan won a seat in the U.S. Senate and formed a friendship with the man who would be his dearest companion for the next two decades.

Buchanan and his chum and housemate, William Rufus de Vane King, a U.S. senator from Alabama, became virtually inseparable. They shared quarters in Washington, D.C., for 15 years. Capitol wits referred to the men—who attended social events together—as "the Siamese twins."

Buchanan's bond with Senator King was so close that the future president described it as a "communion." In praising his friend as "among the best, purest, and most consistent public men I have ever known," Buchanan added that King was a "very gay, elegant-looking fellow." The adjective "gay," however, didn't mean "homosexual" back then; It commonly meant "merry."

It's also useful to understand that it was not unusual for educated men to wax rhapsodic about other men during the 19th century. Admiring rather than sexual, this sort of language signified shared values and deep respect.

Historians rightly point out a lack of evidence that either of the bachelors found men sexually attractive. They note that when Buchanan was younger, he asked a Pennsylvania heiress to marry him. (She broke off the engagement.) Later, he was known to flirt with fashionable women.

Buchanan's "Wife"

Whatever the nature of his relationship with Buchanan, King seemed to consider it to be something more than casual. After the Alabaman became U.S. minister to France in 1844, he wrote home from Paris, expressing his worry that Buchanan would "procure an associate who will cause you to feel no regret at our separation."

Buchanan did not find such a replacement, but it was apparently not for want of trying. He wrote to another friend of his attempts to ease the loneliness caused by King's absence: "I have gone a wooing to several gentlemen, but have not succeeded with any one of them..."

Sometimes the pair drew derisive jibes from their peers. The jokes often targeted King, who was a bit of a dandy with a fondness for silk scarves. In a private letter, Tennessee Congressman Aaron V. Brown used the pronoun "she" to refer to the senator and called him Buchanan's "wife." President Andrew Jackson mocked King as "Miss Nancy" and "Aunt Fancy."

High-Flying Careers Derailed

Despite the childish jokes, both Buchanan and King advanced to ever-more-important federal posts. President James K. Polk selected Buchanan as his secretary of state in 1845. King won the office of U.S. vice president (on a ticket with Franklin Pierce) in 1852. Voters elected Buchanan to the White House four years later.

Unfortunately, neither man distinguished himself in the highest office he reached. King fell ill and died less than a month after taking the oath as vice president. Erupting conflicts over slavery and states' rights marred Buchanan's single term in the Oval Office. Historians give him failing marks for his lack of leadership as the Civil War loomed. The pro-slavery chief executive (he was a Pennsylvania Democrat) opposed the secession of the Southern states but argued that the federal government had no authority to use force to stop it. As a result, Buchanan made no effort to save the Union, leaving that task to his successor, Abraham Lincoln.

What's Sex Got to Do with It?

Would Buchanan have risen to the highest office in the land if his peers honestly believed he was homosexual? It's hard to say. Today's perception is that 19th-century Americans were more homophobic than their 21st-century descendents, yet in an era when sexuality stayed tucked beneath Victorian wraps, there was a de facto "don't ask, don't tell" policy for virtually every profession.

Whatever their private proclivities, Buchanan and King clearly excelled in their public lives—at least until Buchanan got into the White House. Based on what little evidence history provides, neither man's sexual orientation had much—if any—bearing on what he accomplished, or failed to accomplish, in his career.

"What is right and what is predictable are two different things."

"I like the noise of democracy."

"The test of leadership is not to put greatness into humanity, but to elicit it, for the greatness is already there."

"The ballot box is the surest arbiter of disputes among free men."

—QUOTES FROM JAMES BUCHANAN

Stealing the President

A group of men decided that people would be willing to pay a lot of money to see the 16th president of the United States—even if he was dead.

✳ ✳ ✳ ✳

Breaking Out Boyd

THE PLOT WAS hatched in 1876, more than a decade after President Lincoln's assassination by John Wilkes Booth. Illinois engraver Benjamin Boyd had been arrested on charges of creating plates to make counterfeit bills. Boyd's boss, James "Big Jim" Kinealy, a man known around Chicago as "the King of the Counterfeiters," was determined to get Boyd out of prison in order to continue his operation.

Kinealy's plan was to kidnap Lincoln's corpse from his mausoleum at the Oak Ridge Cemetery in Springfield, Illinois, and hold it for ransom—$200,000 in cash and a full pardon for Boyd. Not wanting to do the dirty work himself, Kinealy turned to two small-time crooks: John "Jack" Hughes and Terrence Mullen, a bartender at The Hub—a Madison Street bar frequented by Kinealy and his associates.

Kinealy told Hughes and Mullen that they were to steal Lincoln's body on Election Night (November 7), load it onto a cart, and take it roughly 200 miles north to the shores of Lake Michigan. They were to bury the body in the sand, to stow it until the ransom was paid. The plan seemed foolproof until Hughes and Mullen decided they needed a third person to help steal the body—a fellow named Lewis Swegles. It was a decision Hughes and Mullen would come to regret.

The Plan Backfires

The man directly responsible for bringing Boyd in was Patrick D. Tyrrell, a member of the Secret Service in Chicago. Long before their current role of protecting the president was established, one of the main jobs of members of the Secret Service was to track down and arrest counterfeiters. One of Tyrrell's informants was a small-time crook by the name of Lewis Swegles. Yes, the same guy who agreed to help Hughes and Mullen steal the president's body. Thanks to the stool pigeon, everything the men were planning was being reported back to the Secret Service.

On the evening of November 7, 1876, Hughes, Mullen, and Swegles entered the Lincoln Mausoleum, unaware of the Secret Service agents lying in wait. The hoods broke Lincoln's sarcophagus open and removed the casket, and Swegles was sent to get the wagon. Swegles gave the signal to make the arrest, but once the Secret Service men reached the mausoleum, they found it to be empty. In all the confusion, Hughes and Mullen had slipped away, leaving Lincoln's body behind.

Unsure what to do next, Tyrrell ordered Swegles back to Chicago to see if he could pick up the kidnappers' trail. Swegles eventually found them in a tavern, and on November 16 or 17 (sources vary), Hughes and Mullen were arrested without incident.

Lincoln Is Laid to Rest (Again)

With no laws at the time pertaining to the stealing of a body, Hughes and Mullen were only charged with attempted larceny of Lincoln's coffin and a count each of conspiracy. After a brief trial, both men were found guilty. Their sentence for attempting to steal the body of President Abraham Lincoln: One year in the Illinois state penitentiary in Joliet.

As for Lincoln's coffin, it remains at its home in Oak Ridge Cemetery; it has been moved an estimated 17 times and opened 6 times. On September 26, 1901, the Lincoln family took steps to ensure that Abe's body could never be stolen again: It was buried 10 feet under the floor of the mausoleum, inside a metal cage and under thousands of pounds of concrete.

Presidential Nicknames

✳   ✳   ✳   ✳

✳ George Washington: His Excellency; Old Sink or Swim

✳ James Madison: Little Jemmy; His Little Majesty
(Madison was 5'4" tall)

✳ Andrew Jackson: Old Hickory; The Hero of New
Orleans; King Andrew the First

✳ Martin Van Buren: The Red Fox of Kinderhook; The
Little Magician

✳ John Tyler: The President Without a Party

✳ Franklin Pierce: Handsome Frank

✳ Abraham Lincoln: Honest Abe; The Rail-Splitter

✳ Ulysses S. Grant: The American Caesar

✳ Chester A. Arthur: The Dude President

✳ Grover Cleveland: Grover the Good; Uncle Jumbo

✳ Theodore Roosevelt: Theodore the Meddler

✳ Harry S. Truman: Haberdasher Harry

✳ John Fitzgerald Kennedy: The King of Camelot

✳ Richard Nixon: Richard the Chicken-Hearted; Iron Butt

✳ Gerald Ford: Mr. Nice Guy; Jerry the Jerk

✳ Ronald Reagan: Dutch; The Great Prevaricator

✳ Bill Clinton: Slick Willie; Teflon Bill

✳ George W. Bush: Bush 43, Bush the Younger, or Bush II;
Uncurious George or Incurious George; The Decider-in-
Chief; The Velcro President

Fast Facts

* The gag rule was instituted in the Senate in 1836 so that the senators would not have to accept, debate, or vote on anti-slavery petitions.

* The eagle on the United States Great Seal faces the olive branch in its right talons. However, until 1945 the eagle on the Presidential Seal faced the arrows gripped in its left talons.

* Both the GOP and the Democratic Party owe the popularization of their respective mascots—the elephant and the donkey—to a political cartoonist of the 1870s, Thomas Nast of *Harper's Weekly*.

* George Washington was a distant relation of King Edward I, Queen Elizabeth II, Sir Winston Churchill, and General Robert E. Lee.

* One of the U.S. presidents was not a U.S. citizen at his time of death: John Tyler, a Virginia native, died on January 18, 1862, as a citizen of the Southern Confederacy.

* West Virginia is the only state to have formed by seceding from a preexisting state.

* In 1814, the original Library of Congress was burned down by the British along with the Capitol. To replace it, Congress bought Thomas Jefferson's personal book collection, which consisted of approximately 6,500 volumes.

* The middle initial "S" in Harry S. Truman's name does not actually stand for anything—his parents could not agree on a middle name.

* The first time John F. Kennedy and Richard Nixon engaged in public debate was not as presidential candidates in 1960 but as young congressmen in 1947, when a political club invited them to discuss the pros and cons of a labor bill.

Who Is the Real First President?

You've had it driven into your brain since early childhood that George Washington was the first president of the United States. Care to make a little wager on that?

✳ ✳ ✳ ✳

Y OU'RE SITTING IN a bar minding your own business when some guy comes up to you and says, "Did you know that John Hanson was the real first president of the United States?"

You've never even heard of John Hanson. You look at the guy and say, "Are you nuts? Everyone knows George Washington was the first president." Then he pulls out a copy of *John Hanson, Our First President* by Seymour Wemyss Smith.

Suddenly, your sure thing doesn't seem so sure anymore.

His Case, Part 1

Your antagonist tells you that from 1776 to 1777, the representatives from the 13 colonies gathered in the Continental Congress to hammer out the details of the new nation's first governing charter, the Articles of Confederation.

He adds that by 1779, all the states except for Maryland had ratified the Articles of Confederation. Maryland was holding out until New York and Virginia ceded their western lands to reduce their territorial sizes and, consequently, their clout in the new union. Unanimous ratification was required to bring the Articles into force.

His Case, Part 2

Then he begins to enlighten you about John Hanson.

Hanson, he says, was a highly respected scholar and politician who had served in the Maryland assembly since 1757. He had been an early and raucous proponent of independence and was recognized by his peers as one of the movement's leading lights.

Hanson, he continues, was elected to the Contiental Congress in 1779, and once there, he did more than just fill a seat. He worked diligently to get his home state to accept the Articles of Confederation, arguing that the diplomatic recognition the new country desperately needed would only be given if the Articles were ratified. His efforts were rewarded in March 1781 when Maryland finally signed on, thus making the Articles of Confederation a done deal.

His Case, Part 3

Now, he says, comes the clincher.

He points out that Congress members, grateful for Hanson's efforts, unanimously elected him as president of the United States in Congress Assembled in November 1781. Hanson, he says, was the first person to hold that office under the ratified Articles of Confederation, and as such, was the first president of the United States.

To bolster his case, he offers a list of precedent-setting initiatives over which Hanson presided during his one-year term in office. Among these were the founding of a national census and a postal service, expanding America's foreign relations, negotiating foreign loans, and establishing the Thanksgiving holiday— all of which were continued under the presidency of George Washington.

Finally, he adds one more little tidbit. Washington, he contends, was actually the eighth president of the United States—there were six successors to Hanson before Washington took office.

"Pay up," he says.

How You Win

Before you start crying in your beer, here's a counterargument that will win you the bet.

Yes, Hanson was the first president of the United States in Congress Assembled. But in this capacity, Hanson's primary

function was to preside over the Continental Congress, which was a legislative body responsible for establishing the structure of the nascent American nation in accordance with the Articles of Confederation. Hanson's role was similar to what today might be the House chair or speaker of the Parliament.

The Articles emphasized state sovereignty over a strong national or central government. They made no provision for a federal government as it exists today, an executive branch, or a chief executive—in other words, there was no office of president of the United States.

The 13 colonies did empower the Congress to carry out several specific functions on their behalf, such as territorial defense, foreign relations, coinage and currency, and a postal service. As president of the United States in Congress Assembled, Hanson did preside over these "national" operations.

But when the flaws of the Articles of Confederation became apparent (namely, they offered no effective way for the Congress to raise an army for defense or taxes to fund its assigned tasks), Congress members set about writing a more balanced governing charter. The result was the creation of the U.S. Constitution in 1787–1788, which established a robust federal government and a chief executive office formally called the "president of the United States of America."

And the first person to hold that office was none other than George Washington. John Hanson, though a hero in his time, is mostly forgotten today—a victim of both bad historical timing and constitutional semantics.

Enjoy your beer.

What's in a Name: Presidents

✳ ✳ ✳ ✳

✳ There have been six presidents named James, five Johns, four Williams, three Georges, and two Andrews, Franklins, and Thomases.

✳ There have been two Adamses, two Bushes, two Harrisons, two Johnsons, and two Roosevelts. All except the Johnsons were related to each other.

✳ Rutherford is the longest presidential first name at 10 letters.

✳ Seventeen of the 44 presidents to date have no known middle name.

✳ Several presidential middle names were originally surnames: Baines, Birchard, Delano, Fitzgerald, Walker, Knox, Milhous, Quincy, and Simpson. Most of these were the president's mother's maiden name.

✳ George Herbert Walker Bush is the only president with two middle names.

✳ There are no duplicate presidential middle names, with the partial exception of Herbert Walker and Walker.

✳ Three presidents used their middle names as their given names: Calvin Coolidge, Grover Cleveland, and Woodrow Wilson.

✳ Only three of the ten most common surnames in the United States (Smith, Johnson, Williams, Jones, Brown, Davis, Miller, Wilson, Moore, and Taylor) have been the surnames of presidents (Andrew and Lyndon Johnson, Woodrow Wilson, and Zachary Taylor).

Chapter 18

Crime

Lana Turner and the Death of a Gangster

On the evening of April 4, 1958, Beverly Hills police arrived at the home of actress Lana Turner to discover the dead body of her one-time boyfriend Johnny Stompanato, a violent gangster.

✳   ✳   ✳   ✳

Lana Turner's first credited film role came in 1937 with *They Won't Forget*, which earned her the moniker "Sweater Girl," thanks to the tight-fitting sweater that her character wore.

Offscreen, Turner was renowned for her love affairs, including eight marriages. It was shortly after the breakup of her fifth marriage that Turner met Johnny Stompanato. When she learned that he had ties to underworld figures such as Mickey Cohen, she realized the negative publicity that those ties could bring to her career, so she tried to end the relationship. But Stompanato incessantly pursued her, and the pair were involved in a number of violent incidents, which came to a head on the night of April 4.

Coming to the Rescue

After hearing Stompanato threaten to "cut" Turner, her 14-year-old daughter, Cheryl Crane, rushed to her defense. Fearing for her mother's life, the girl grabbed a kitchen knife and ran to Turner's bedroom, where she saw Stompanato with his arms raised in the air in a fury. Cheryl then rushed past Turner and stabbed Stompanato. Turner called her mother, who brought their personal physician to the house, but it was too late. By the time the police were called, a good deal of time had passed and evidence had been moved around. According to the Beverly Hills police chief, who was the first officer to arrive, Turner immediately asked if she could take the rap for her daughter.

At the crime scene, the body appeared to have been moved and the fingerprints on the murder weapon were so smudged that they could not be identified. The case sparked a media sensation, and the tabloids turned against Turner, essentially accusing the actress of killing Stompanato and asking her daughter to cover for her. Mickey Cohen, who paid for Stompanato's funeral, publicly called for the arrests of Turner and Crane.

"The Performance of a Lifetime"?

During the inquest, the press described Turner's testimony as "the performance of a lifetime." But police and authorities knew that Turner did not commit the crime. At the inquest, it took just 20 minutes for the jury to  return a verdict of justifiable homicide, and the case was not brought to trial. However, Turner was convicted of being an unfit mother, and Crane was remanded to her grandmother's care until she turned 18, which further tainted Turner's image. There was an aura of guilt around Turner for years, though she was never seriously considered a suspect in the actual murder.

As fate would have it, Turner's film *Peyton Place*, which features a courtroom scene about a murder committed by a teenager, was still in theaters at the time of the inquest. Ticket sales skyrocketed as a result of the sensational publicity, and Turner parlayed the success of the film into better screen roles, including a part in a remake of *Imitation of Life* (1959), which would become one of her most successful pictures.

The Real Dracula

*The origins of Dracula, the blood-sucking vampire of Bram
Stoker's novel (and countless movies, comics, and costumes) are
commonly traced to a 15th-century Romanian noble, whose
brutal and bloody exploits made him both a national folk hero
and an object of horror.*

✳ ✳ ✳ ✳

VLAD III TEPES (1431–1476) was born into the ruling
family of Wallachia, a principality precariously balanced
between the Ottoman (Turkish and Muslim) and Holy Roman
(Germanic and Catholic) empires. His father, Vlad II, became
known as Dracul ("Dragon") due to his initiation into the
knightly Order of the Dragon. As a part of Dracul's attempt
to maintain Wallachia's independence, he sent Vlad III and his
brother Radu to the Turks as hostages. When Dracul was assassinated in 1447 and a rival branch of the family took the throne,
the Turks helped Vlad III to recapture it briefly as a puppet ruler
before he was driven into exile in Muldovia. In 1456, Vlad III—
known as Dracula, or "son of Dracul"—regained the throne with
Hungary's backing and ruled for six years.

Dracula faced several obstacles to independence and control.
The Ottoman Empire and Hungary were the largest external
threats. There was also internal resistance from the boyars
(regional nobles), who kept Wallachia destabilized for their
own interests, as well as from the powerful Saxon merchants of
Transylvania, who resisted economic control. Lawlessness and
disorder were also prevalent.

Dracula Im*pales* in Comparison

Dracula's successes against these obstacles are the source of the
admiration for him that still pervades local folklore, in which
there are many tales of his exploits. Locally, he is portrayed as
a strong, cunning, and courageous leader who enforced order,
suppressed disloyalty, and defended Wallachia.

His methods, however, are the source of his portrayal elsewhere as a sadistic despot who merited an alternate meaning of the word *Dracula*—namely "devil." He eliminated poverty and hunger by feasting the poor and sick and, after dinner, burning them alive in the hall. He eliminated disloyal boyars by inviting them to an Easter feast, after which he put them in chains and worked them to death building his fortress at Poenari. Ambassadors who refused to doff their caps had them nailed to their heads. Dishonest merchants and bands of gypsies suffered similarly grisly ends.

Vlad's favorite method of capital punishment was impaling (a gruesome and purposefully drawn-out form of execution in which victims were pierced by a long wooden stake and hoisted aloft), though he also found creative ways of flaying, boiling, and hacking people to death. When the Turks cornered him at Tirgoviste in 1461, he created a "forest" of 20,000 impaled captive men, women, and children, which so horrified the Turks that they withdrew. They called him *kaziklu bey* ("the Impaler Prince"), and though Vlad himself did not use it, the epithet Tepes ("Impaler") stuck.

Dracula Deposed

The Turks then backed Radu and besieged Vlad at Poenari. Dracula escaped to Hungary with the help of a secret tunnel and local villagers. Dracula gradually ingratiated himself and, upon Radu's death, took back Wallachia's throne in 1476. Two months later, Dracula was killed in a forest battle against the Turks near Bucharest. Tradition has it that his head was put on display by the Turks and his body buried in the Snagov Monastery, but no grave has ever been found there.

Four centuries later, while researching a story about "Count Wampyr," Bram Stoker learned that *dracula* could mean "devil" and changed his character's name. Connecting Stoker's Count to Vlad III has been popular in the press and in the tourist industry, but Stoker's character probably has little basis in Vlad Dracula.

Life on the Body Farm

One won't find cows, chickens, or pigs that go "oink" at the Body Farm—just scores of rotting human bodies.

✳   ✳   ✳   ✳

E-I-E-I-Oh, Gross

THE BODY FARM (officially known as the University of Tennessee Forensic Anthropology Facility) was the brain-child of Dr. William Bass, a forensic anthropologist from Kansas. Its purpose is to help law enforcement agencies learn to estimate how long a person has been dead. After all, determining times of death is crucial in confirming alibis and establishing time lines for violent crimes.

After 11 years of learning about human decomposition, Bass realized how little was actually known about what happens to the human body after death. With this in mind, in 1981, he approached the University of Tennessee Medical Center and asked for a small plot of land where he could control what happens to a post-mortem body and study the results.

Bass's Body Farm drew the attention of readers when popular crime novelist Patricia Cornwell featured it in her 1994 book, *The Body Farm*. In it, Cornwell describes a research facility that stages human corpses in various states of decay and in a variety of locations—wooded areas, the trunk of a car, underwater, or beneath a pile of leaves—all to determine how bodies decompose under different circumstances.

Reading the Body

According to Bass, two things occur when a person dies. At the time of death, digestive enzymes begin to feed on the body, thus "liquefying" the tissues. If flies have access to it, they lay eggs in the corpse. Eventually, the eggs hatch into larvae that feast on the remaining tissues. By monitoring and noting how much time it takes for maggots to consume the tissues, authorities can estimate how long a person has been dead. Scientists can also compare the types of flies that are indigenous to the area with the types that have invaded the body to determine whether the corpse has been moved.

But the farm isn't all about tissue decomposition—scientists also learn about the normal wear-and-tear that a human body goes through. For instance, anthropologists look at the teeth of victims to try to determine their ages at the time of death. The skull and pelvic girdles are helpful in determining a person's gender, and scientists can also estimate how tall the person was by measuring the long bones of the legs or even a single finger. Other researchers watch what happens to the five types of fatty acids that leak from the body into the ground. By analyzing the profiles of the acids, scientists can determine the time of death and how long the body has been at its current location.

Unfortunately, the perps are catching on. Some criminals try to confuse investigators by tampering with bodies and burial sites, spraying victims with pesticides that prevent insects (such as maggots) from doing their job.

Further Afield

At the University of New Mexico, scientists have collected more than 500 human skeletons and stored them as "skeletal archives" to create biological profiles based on what happens to bones over time. And in Germany, the Max Planck Institute for Computer Science has been working on a 3-D graphics program based on forensic data to produce more accurate likenesses of victims.

Many other proposed farms never got off the ground due to community protest, but since the inception of Bass's original Body Farm, another farm has been established at Western Carolina University. Ideally, Bass would like to see body farms all over the nation. Since decaying bodies react differently depending on climate and surroundings, Bass says, "It's important to gather information from other research facilities across the United States."

The Clairvoyant Crime-Buster

Before there were TV shows such as Ghost Whisperer *and* Medium, *which make the idea of solving crimes through ESP seem almost commonplace, there was psychic detective Arthur Price Roberts.*

✳ ✳ ✳ ✳

He Saw Dead People

A MODEST MAN BORN in Wales in 1866, Roberts deliberately avoided a formal education because he believed too much learning could stifle his unusual abilities. As a young man, he moved to Milwaukee; it was there that the man who never learned to read was nicknamed "Doc."

One of his earliest well-known cases involved a baffling incident in Peshtigo, Wisconsin. A man named Duncan McGregor suddenly went missing in July 1905, leaving no clue as to his whereabouts. Police searched for him for months; finally his desperate wife decided to visit the psychic detective, who had already made a name for himself in Milwaukee. She didn't even have to explain the situation to Roberts; he knew immediately upon meeting her who she was.

Roberts meditated on the vanished man, then sadly had to tell Mrs. McGregor that her husband had been murdered and that his body was in the Peshtigo River, caught near the bottom in a pile of timber. Roberts proved correct in every detail.

Mystery of the Mad Bombers

Roberts solved numerous documented cases. He helped a Chicago man find his brother who had traveled to Albuquerque and had not been heard from for months; Roberts predicted that the brother's body would be found in a certain spot in Devil's Canyon, and it was.

After coming up with new evidence for an eleventh-hour pardon, Roberts saved a Chicago man named Ignatz Potz, who had been condemned to die for a murder he didn't commit. But his biggest and most famous coup came in 1935 when he correctly predicted that Milwaukee would be hit by six large dynamite explosions and lose a town hall, banks, and police stations. People snickered; such destruction was unheard of in Milwaukee. Roberts made his prediction on October 18 of that year; in little more than a week, the Milwaukee area entered a time of terror.

First, a town hall in the outlying community of Shorewood was blasted, killing two children and wounding many other people. A few weeks later, the mad bombers hit two banks and two police stations. Federal agents descended upon the city, and several local officers were assigned to work solely on solving the bombings. Finally, the police went to Roberts to learn what was coming next. Roberts told them one more blast was in the works, that it would be south of the Menomonee River, and that it would be the final bomb. Police took him at his word and blanketed the area with officers and sharpshooters.

And sure enough, on November 4, a garage in the predicted area was blown to smithereens in an explosion that could be heard as far away as eight miles. The terrorists, two young men ages 18 and 21, had been hard at work in the building assembling 50 pounds of dynamite when their plan literally backfired. Few people argued with Roberts's abilities after that.

His Final Fortune

Roberts's eeriest prediction, however, may have been that of his own death. In November 1939, he told a group of friends that he would be leaving this world on January 2, 1940. And he did, passing quietly in his own home on that exact date. Many of his amazing accomplishments will probably never be known because a lot of his work was done secretly for various law enforcement agencies. But "Doc" Roberts had an undeniable gift, and he died secure in the knowledge that he had used it to help others as best he could.

Crime and Punishment

A tale of greed and murder ushered in a new era of forensic science.

✳ ✳ ✳ ✳

JACK GRAHAM'S MOTHER, Daisie King, knew her only son was no angel. Barely into his twenties, Graham had little patience for lawful employment, and he'd already been convicted of check forgery and running illegal booze.

By 1953, however, it seemed that Graham was settling down. He'd married and had two children. His mother, a successful businesswoman, bought a house in Colorado for the young couple, built a drive-in restaurant, and made Graham its manager.

But the drive-in lost money. Graham blamed his mother's meddling in the management, but he later admitted he had skimmed money. He also confessed to vandalizing the place twice—once by smashing the front window and the second time by rigging a gas explosion to destroy equipment that he'd used as security for a personal loan. A new pickup truck Graham bought mysteriously stalled on a railway track, with predictable results. This too proved to be an attempt at insurance fraud.

Flight to Doom

In the fall of 1955, King wanted to see her daughter in Alaska, so she prepared for her trip there via Portland and Seattle. On November 1, Graham saw her off on United Airlines Flight 629. Eleven minutes after takeoff, the plane exploded in the sky. Forty-four people died, including Daisie King.

FBI agents were soon at the crash site to help identify bodies. The painstaking task of gathering wreckage from a three-mile trail of scraps began. By November 7, Civil Aeronautics investigators concluded that sabotage was likely the cause of the disaster.

Criminal investigators joined the FBI technical teams. Families of passengers and crew members were interviewed while technicians reassembled the plane's midsection where the explosion likely occurred. In the wreckage, bomb fragments and explosives residue were identified.

Avalanche of Evidence

Inevitably, investigators took an interest in Graham. Not only would he receive a substantial inheritance from his mother's estate, but he had also taken out a $37,500 travel insurance policy on her. Moreover, he had a criminal record, and according to witnesses, a history of heated arguments with his mother.

Graham was first interviewed on November 10. In a search of his property on November 12, agents discovered a roll of primer cord in a shirt pocket and a copy of the travel insurance policy secreted in a small box. Circumstantial evidence—including that provided by his wife, half-sister, and acquaintances—contradicted his statements.

Finally, Graham admitted that he'd built a bomb and placed it in his mother's luggage. On November 14, he was arraigned on charges of sabotage. At the time, the charge did not carry the death penalty, so he was brought back into court on November 17 and charged with first-degree murder.

A Case of Firsts

Despite the confession, investigators continued to gather evidence, putting together what may have been the most scientifically detailed case in U.S. history up to that date. The case had other firsts as well. It was the first case of mass murder in the United States via airplane explosion. Graham's trial, which began on April 16, 1956, also marked the first time TV cameras were permitted to air a live broadcast of a courtroom trial.

On May 5, 1956, the jury needed only 69 minutes to find Graham guilty. On January 11, 1957, he was executed at the Colorado State Penitentiary, remorseless to the end.

A Tangled Web

So-called "black widows"—women who marry and then kill their spouses and sometimes families for profit—stand out for their sheer unlikelihood as perpetrators. The following black widows got caught up in their own webs.

✳   ✳   ✳   ✳

Artiste of Arsenic

NORWEGIAN IMMIGRANT Belle Poulsdatter, known as the Black Widow of the Heartland, was married for 17 years before her husband died and she collected $8,000 in life insurance. Belle moved her family to LaPorte, Indiana. There she married wealthy widower Peter Gunness, who later died when a meat grinder tumbled from a high shelf and landed on his head. His death was ruled accidental, and Belle collected Peter's insurance money and his estate.

Belle advertised for farmhands in a newspaper that catered to immigrants. Of those who responded, Belle hired the ones who came with a sturdy bank account as well as a sturdy back. Laborers came and went—and some simply disappeared.

After the Gunness farmhouse burned to the ground in 1908, the bodies of Belle's children and an unidentified headless female were found in the cellar. A search of the property revealed the bodies of Belle's suitors and laborers buried in the hog pen— some killed by arsenic, others by skull trauma. The widow was nowhere to be found. Belle's remaining beau-cum-farmhand was convicted of murdering Belle and her family. However, the identity of the headless corpse was never determined, leading some to believe that Belle staged the entire thing and escaped.

The Giggling Grandmother

From the mid-1920s to the mid-'50s, Nannie Doss left a trail of corpses: her mother, two sisters, a nephew, and a grandson. A mother of four trapped in an unhappy marriage, Nannie also murdered two of her children with rat poison before her first husband left her. She collected on the children's life insurance policies. Nannie married three more times, but each husband contracted a mysterious stomach ailment and died, leaving his widow his insurance settlement, home, and estate.

The doctor of Nannie's fifth husband ordered an autopsy, which showed rat poison in his system. After her arrest, the bodies of her former spouses were exhumed; all showed traces of poison. Nannie giggled as she confessed her crimes to the police, earning her the nickname "The Giggling Grandmother."

Nobody Buys the Doppelgänger Bit

Frank Hilley had been married to Marie for more than 20 years when he was admitted to the hospital with stomach pain and diagnosed with acute infectious hepatitis in 1975. He died within a month, and Marie collected on his life insurance policy. Three years later, she took out a life insurance policy on her daughter Carol, who then developed a strange illness with symptoms of nausea and numbness in her extremities. Physicians detected an abnormal level of arsenic in Carol's system and suspected foul play. Frank's body was exhumed and tests revealed that he had died of arsenic poisoning.

Marie was arrested in October 1979 for the attempted murder of her daughter. But when she was released on bond a month later, she promptly disappeared. Despite her indictment for murder, Marie remained a fugitive for more than three years before marrying John Homan in Florida under the alias Robbi Hannon.

In a bizarre turn of events, Marie invented a twin sister, "Teri," staged Robbi's death, and then returned to her husband pretending to be her grief-stricken twin. The ruse was discovered, and Marie was sent to Alabama, where she was wanted on other charges. Her scheme fell apart, and she was convicted of murder and attempted murder and sentenced to life in prison. Marie served four years of her sentence before she escaped during a furlough; she was captured and died of hypothermia.

Literary Superstar/Murderer

Norman Mailer is widely considered one of the 20th century's greatest (and most controversial) writers. But his involvement with criminal Jack Henry Abbott wound up being one of the more contentious episodes of his career.

✳ ✳ ✳ ✳

Author/Prisoner

BORN ON A U.S. Army base in Michigan in 1944, Jack Henry Abbott bounced around foster homes as a young child. He became a regular resident of juvenile detention centers, until he finally entered a reformatory at age 16. Five years later, while in prison, he stabbed a fellow inmate to death. In 1971, he escaped lockup and robbed a bank in Colorado. Abbott was recaptured and had more time added to his sentence.

In 1977, Abbott heard that famed author Norman Mailer was writing a book called *The Executioner's Song* about convict Gary Gilmore, who was scheduled to be executed—the first instance of capital punishment to occur in the United States in years.

Abbott began writing to Mailer, offering to help the writer understand the mind-set of a convict. In particular, Abbott—who had been in jail for most of his adult life—offered insight into the mind of a "state-reared" long-term convict. "The model we emulate is a fanatically defiant and alienated individual," Abbott wrote, "who cannot imagine what forgiveness is, or mercy or tolerance, because he has no experience of such values."

Mailer, who admitted that he knew little about prison violence, began corresponding with Abbott. As the men communicated, Mailer realized that the convict was a powerful writer.

The Beast on Broadway

The Executioner's Song won the 1980 Pulitzer Prize for fiction. Mailer continued to correspond with Abbott. In June 1980, Mailer's friend Robert Silvers, editor of *The New York Review of*

Books, published some of Abbott's letters. The missives created a sensation, and Abbott was offered a $12,500 advance for a book. Titled *In the Belly of the Beast*, it was to be published in the summer of 1981.

In late spring 1981, Abbott came up for parole. Despite his own admission that "I cannot imagine how I can be happy in American society," and even though he had spent virtually his entire adult life locked up, Abbott was indeed paroled. The parole board may have been influenced by a letter from Mailer, who offered Abbott a job as a research assistant at $150 per week. On June 5, Abbott flew to New York City, where he met Mailer and moved into a halfway house.

"Exceptional Man"

In early July, *In the Belly of the Beast* was published to over-whelming acclaim. *The New York Times Book Review* found Abbott to be "an exceptional man." The convict had become a celebrity.

This lasted for about three weeks. Little did anyone know that Abbott was rapidly spinning out of control. He hated the city, and he grew increasingly paranoid. Mailer tried to get Abbott to hang on until August, when he could accompany his family to Maine. But it was not to be.

On July 18, Abbott went to a small café where he argued with a 22-year-old waiter named Richard Adan. Abbott stabbed Adan to death with a knife. The next day, the *Times* unwittingly ran a glowing review of *In the Belly of the Beast*.

Afterword

Abbott fled the city, but he was eventually caught and brought back to stand trial for murder. The trial was stormy, with the press demonizing Mailer for his part in getting Abbott released, in particular Mailer's claim that "Culture is worth a little risk."

The jury didn't agree and convicted Abbott of manslaughter. In February 2002, Abbott hung himself in his prison cell.

Long Live the Crime-Fighting King

If a chubby, jumpsuit-wearing Elvis came karate-kicking out of a limo, what would you do?

✳ ✳ ✳ ✳

IF IT HAPPENED in 1977, shortly before Elvis ended his reign as the King of Rock and Roll, you would run, which is just what two troublemaking teens did when Presley gave them the scare of their lives.

On the night of June 23, "The King" was en route to his hotel in Madison, Wisconsin, when his 1964 Cadillac limo stopped at a red light. Peering out the window to the right, Presley was watching a young man reading the gas meters at the Skyland Service Station when two teenage misfits charged at the employee and began beating him up.

Presley, wearing his aviator sunglasses and a dark blue Drug Enforcement Agency jumpsuit over his sequined outfit, busted out of the limo and ran toward the scene, kicking karate-style and saying, "I'll take you on!"

The hoodlums ran off. As it turned out, the employee's father owned the service station, and one of the attackers had recently been fired from there. Getting back in the limo, Presley reportedly laughed and said, "Did you see the looks on their faces?"

Having witnessed Presley's crime-prevention skills, the King's entourage and a caravan of fans continued on to the hotel. The next day, Presley gave a lackluster concert, prompting one local critic to write: "So, long live the King. His reign is over. But that is no reason for us not to remember him fondly."

Presley died 51 days later, on August 16, 1977. Skyland closed years ago, and the site was most recently home to a used car lot, which paid homage to Elvis and Madison's famed gas station altercation with a marble plaque.

The Mystery of the Missing Comma

Legend suggests that a punctuation error sparked one of history's greatest unsolved mysteries.

✳ ✳ ✳ ✳

KING EDWARD II of England is primarily remembered for his weakness for certain men and the way he died. He spent most of his life in thrall to his alleged lovers, Piers Gaveston and, later, Hugh le Despenser, granting their every wish. When Edward married 12-year-old Princess Isabella of France in 1308, he politely greeted her upon her arrival in England—and then gave her wedding jewelry to Gaveston.

Isabella became accustomed to being pushed aside in favor of her husband's preferred companions. Even after Gaveston was murdered for being a bad influence on the king, Edward did not change, turning his affections to the greedy Despenser, whom the queen loathed and feared. When the opportunity arose for her to negotiate a treaty with her brother, the king of France, she took it, traveling to Paris and refusing to return.

After nearly 20 years in an unhappy marriage, Isabella had had enough. Along with her lover, Roger Mortimer, she raised an army and led it to England to depose her husband. Once the king was in custody, the queen forced him to abdicate the throne to their 14-year-old son, Edward III. She then sent a letter detailing how her deposed husband should be treated in captivity.

However, something very important was missing from Isabella's orders. In the letter, she wrote, *"Edwardum occidere nolite timere bonum est."* Many historians think she intended this to mean, "Do not kill Edward, it is good to fear." However, she neglected to add a necessary comma. If the comma is inserted in a different place, the letter means "Do not be afraid to kill Edward; it is good." It's clear how Edward's jailers construed the message: Soon Edward was murdered in his jail cell.

Northern Justice

Here are some weird ways to break the law in Canada.

✳ ✳ ✳ ✳

The Cow: Queen of Beasts

Canada's animal cruelty laws impose a penalty of two years in jail on anyone who tortures or mistreats any animal. But woe to the soul who abuses a cow— inflicting harm on Bessie can result in up to five years behind bars.

Trick, Treat, or Be Eaten

Polar bears may be an endangered species, but they're not uncommon in the northern town of Churchill, Manitoba. Visits from the snowy bruins present so much of a problem in that community (people must huddle in their houses as the large predators rumble through) that the municipal government has banned furry Halloween costumes for fear of attracting the bears.

Crime Comics

Canada has notoriously vague obscenity laws—one problematic prohibition takes a stand against images that show the "commission of crimes, real or fictitious." At least that stops super villains in Canadian comic books from dressing as polar bears or bothering cows.

Carousing in Calgary

Calgary, Alberta, is cracking down on horseplay after closing time, as shown in its new system of bylaws. Spitting can cost you $115 out of pocket, and relieving yourself in public carries a price tag of $300. Curiously enough, brawling and beating the wazoo out of one other will result in only a $250 penalty. Better to punch than to pee!

No Yellow Margarine

While the rest of Canada can enjoy yellow margarine with impunity, Quebec's dairy industry has successfully lobbied the provincial government to pass a law that all margarine must be a color other than yellow (so that the consumer doesn't mistake it for butter). Plans for blue and pink spreads were quickly abandoned in favor of the blander-hued off-white.

Simple Skill-Testing Questions

No one can simply be given a prize in a Canadian raffle, drawing, or sweepstakes; instead, Canadians are required to "earn" the prize, usually by answering a "skill-testing question," which often takes the form of a simple arithmetic problem. So if you're not up on your basic math, there's a chance that Jet Ski's not coming home with you.

Just Say "Boo!"

Canadian law specifically states that it is a criminal act to frighten a child or a sick person to death. Apparently, it's just peachy to pop out of a laundry hamper while wearing a hockey mask—as long as your victim is a healthy adult.

One in Five Songs

Non-Canadians have a difficult time understanding Canadian Content laws, or "Can-Con," which require that a certain portion of television and radio programming be filled with homegrown Canadian talent. Basically, it's an attempt to keep Canadian culture from being overwhelmed by Hollywood and Nashville.

No Fake Witches

Canada has complete freedom of religion, so if you're claiming to be a witch, you better be the real deal. Genuine witches, or wiccans, are considered fine, but should you grab a broom and a pointy hat and pretend to be a witch, you are violating the Criminal Code of Canada. Those who "pretend to exercise or use any kind of witchcraft, sorcery, enchantment, or conjuration" can be punished. Makes you wonder whether the jails are filled to capacity on Halloween.

Crockefeller

What began as a search for a missing girl uncovered 30 years of fraud, fake identities, and possible foul play. Before Christian Karl Gerhartstreiter was a convict, he was a con artist.

✳ ✳ ✳ ✳

IT STARTED AS a case of parental kidnapping not uncommon in custody battles: In July 2008, Clark Rockefeller, a descendant of the moneyed oil family, absconded with his seven-year-old daughter during a court-supervised visitation in Boston.

But oddly, FBI databases had no record of a Clark Rockefeller, and the wealthy family denied any connection. His ex-wife, millionaire consultant Sandra Boss, confessed that he had no identification, no social security number, and no driver's license. *So who was this guy?* The FBI released his picture, hoping for information. That's when the stories—and aliases—began pouring in.

Fake Foreign Exchange

His real name was Christian Karl Gerhartstreiter; he came to the United States in 1978 at age 17, claiming to be a foreign-exchange student from Germany. In truth, he showed up unannounced on the doorstep of a Connecticut family he'd met on a train in Europe, who'd suggested he look them up if he ever visited the States.

After living with them briefly, he posted an ad describing himself as an exchange student in search of a host and was taken in by the Stavio family. They threw him out after it became clear that he expected to be treated like royalty. During this time, Gerhartstreiter adopted the mannerisms and the snobbish accent of the *Gilligan's Island* character Thurston Howell III.

Bogus Brit

In 1980, "Chris Gerhart" enrolled as a film major at the University of Wisconsin-Milwaukee and persuaded another student to marry him so that he could get his green card (they divorced as soon as he got it). Shortly after the wedding, he headed to Los Angeles to pursue a film career—this time posing as the dapper British blue blood Christopher Chichester.

He settled in the swanky town of San Marino, living in a building with newlyweds John and Linda Sohus. The couple went missing in 1985, around the same time Chichester moved away; allegedly, he went back to England following a death in the family.

Chichester resurfaced in Greenwich, Connecticut, as former Hollywood producer and business tycoon Christopher Crowe. It was under this name that, in 1988, he tried to sell a truck that had belonged to the Sohuses. The police traced the missing truck to Connecticut, and they soon realized that Chichester and Crowe were the same person. But by then, he'd already vanished.

Mock Rock

He became Manhattan's Clark Rockefeller, the new darling of the elite. It was there that he met and married the Ivy League-educated business whiz Sandra Boss. For most of their 12-year marriage, Sandra believed his elaborate stories. She even believed he'd filed the paperwork for their marriage to be legal (he hadn't).

But eventually Sandra grew suspicious. She divorced Clark and won full custody of their daughter, Reigh; the two moved to London. Clark was limited to three court-supervised visits per year. It was on the first of these visits that he kidnapped Reigh.

Conclusion

In August 2008, the con man was arrested in Baltimore, and Reigh was returned to her mother. In June 2009, a judge sentenced him to four to five years in prison.

In 1994, new owners found skeletal remains in the backyard of the San Marino home. Neighbors remember Chichester borrowing a chainsaw shortly before the Sohuses disappeared and complaining of "plumbing problems" when asked why the backyard was dug up. Gerhartstreiter was charged with the murder of John Sohus in March 2011.

The Sex-sational Trial of Errol Flynn

After the actor was charged with rape, his status in Hollywood was in doubt—but his popularity soared ever higher.

✳ ✳ ✳ ✳

I N THE 1940S, the saying "in like Flynn" was used to compliment one who was doing exceedingly well. Moreover, it usually referred to sexual conquests made by a lucky "man about town."

The saying originated with Hollywood heartthrob Errol Flynn (1909–1959), the swashbuckling star of *Captain Blood* (1935) and *The Adventures of Robin Hood* (1938). A "man's man" in the most cocksure sense, Flynn was known for his barroom brawls and trysts both on-camera and off, which became part of his star image. He was proud of his conquests, but he hit a rough patch when he was charged with two counts of statutory rape.

Swashbuckling Seductor

The alleged crimes took place during the summer of 1942. In one instance, 17-year-old Betty Hansen claimed that Flynn seduced her after she became ill from overimbibing at a Hollywood party; in the other, 15-year-old Peggy Satterlee insisted that Flynn took advantage of her on his yacht. Both women claimed that he referred to them by the nicknames "S.Q.Q." ("San Quentin Quail") and "J.B." ("Jail Bait"), thereby suggesting that Flynn knew that they were underage. Flynn was arrested that fall and charged with two counts of rape. Proclaiming his innocence, he hired high-powered attorney Jerry Giesler, who called the women's motives and pasts into question and stacked the jury with nine females in the hope that Flynn's considerable charm might win them over. The move would prove prophetic.

Women of Questionable Character

When the defense presented its case, Giesler went directly for the jugular. His cross-examination revealed that both women had engaged in sexual relations *before* the alleged incidents with

Flynn and that Satterlee had even had an abortion. Even more damning, Satterlee admitted to frequently lying about her age and was inconsistent in a number of her answers. Suddenly, the women's veneer of innocence was stripped away.

There was also Satterlee's claim that Flynn had taken her below deck to gaze at the moon through a porthole. Giesler challenged the testimony of an astronomer hired by the prosecution, who admitted that, given the boat's apparent course, such a view was physically impossible through the porthole in Flynn's cabin.

The Verdict

By the time Flynn took the stand, the members of the all-female jury were nearly won over by his charm. The effect was not at all surprising for a man whom actress and costar Olivia de Havilland once described as "... the handsomest, most charming, most magnetic, most virile young man in the entire world."

When a verdict of "not guilty" was read, women in the courtroom applauded and wept. Afterward, the jury forewoman noted, "We felt there had been other men in the girls' lives. Frankly, the cards were on the table and we couldn't believe the girls' stories."

Continued Fortune in a Man's World

Cleared of all charges, Errol Flynn resumed his carousing ways and grew even more popular in the public eye. Many felt that, despite the verdict, Flynn had indeed had sexual relations with the young women, but most were willing to forgive the transgression because the liaisons seemed consensual and the allegations of rape looked like little more than a frame-up.

Young men would continue to regard the amorous actor as an ideal to emulate, and women would continue to swoon. But years of hard living eventually took their toll on Flynn: By the time he reached middle age, his looks had all but vanished. At the premature age of 50, Flynn suffered a massive heart attack and died. His passing only served to cement his legendary status amongst Hollywood's actors and its rapscallions alike.

"Fifteen Women on a Dead Woman's Chest..."

If Calico Jack had heard the famous pirate song sung this way, would he have declared it blasphemous? Don't bet your booty.

✳ ✳ ✳ ✳

Women of the Sea

THROUGHOUT HISTORY, WOMEN have received more than their share of omissions, and this certainly was the case during the Golden Age of Piracy. Although it's true that men were the predominant players in this high-seas melodrama, women also had important roles. Most people have heard of Captain Kidd, Blackbeard, and Calico Jack, but those same folks might scratch their heads while trying to recall Anne Bonny, Lady Mary Killigrew, and Mary Read.

Despite their relative anonymity, female swashbucklers were as much a part of the pirate experience as garish outfits and handheld telescopes. Female pirates date back at least as far as the fifth century, but the most notable figures appeared long after that. Mary Killigrew, a lady under Queen Elizabeth I, operated in the late 16th century. In her most celebrated outing, Killigrew and her shipmates boarded a German vessel off of Falmouth, Cornwall. Once on deck, they killed the crew and stole their cargo. When she was later brought to trial for the murders, Killigrew was sentenced to death. With some well-placed bribes and a queen who was sympathetic to her plight, however, Killigrew was eventually acquitted.

The Story of Mary and Anne

The exploits of pirates Mary Read and Anne Bonny rank among those of their male counterparts. Read was born in London in the late 17th century and spent her entire childhood disguised as a boy. The reasons for her unusual dress are unknown, but Read's thirst for adventure has never been in question.

Working as a "footboy" for a wealthy French woman, "Mark" Read eventually grew tired of such drudgery and signed on for sea duty aboard a man-o'-war. From there, the cross-dressed woman joined the Flemish army, where she served two stints. Eventually, Read booked passage on a ship bound for the West Indies. While on this fateful journey, her vessel was attacked and captured by none other than Captain (Calico) Jack Rackham.

A dashing figure in her male persona, Read drew the amorous gaze of Bonny, who was Calico Jack's mistress and a pirate in her own right. Upon the discovery of Read's gender, the two became friends, and they struck a deal to continue the ruse. The game wouldn't last long: A jealous Calico Jack confronted the pair, and he too learned the truth. Finding appeal in the prospect of having two female pirates on his crew, the captain let things stand.

The adventure-loving Read took well to her life of piracy and soon fell in love with a young sailor. This upset a veteran crewmate, who challenged the would-be lothario to a duel. Fearful that her man would be killed by the strapping seaman, Read demanded her own showdown. She was granted her wish. After the combatants discharged their pistols, both stood unscathed. When they reached for their swords, Read cunningly ripped her shirt open and exposed her breasts. The stunned seaman hesitated, and Read drove her cutlass home.

Courageous Buccaneers

Read's victory would be short-lived. Charged with piracy after their ship was seized by Jamaican authorities in 1720, Calico Jack, Read, and Bonny were tried and sentenced to hang. When asked in court why a woman might wish to become a pirate and face such a sentence, Read replied, "As to hanging, it is no great hardship, for were it not for that, every cowardly fellow would turn pirate and so unfit the seas, that men of courage must starve."

Animals on Trial

During the Middle Ages, people believed that animals were legally responsible for their crimes and misdeeds. But punishment was not administered without fair trial.

✳   ✳   ✳   ✳

T HE YEAR WAS 1386. In the French city of Falaise, a child was killed and partially devoured by a sow and her six piglets. Locals refused to let such a heinous crime go unpunished. However, rather than killing the sow, they brought it to trial... dressed in men's clothing. The pig was tried for murder, convicted, and hanged from the gallows in the public square.

Porkers weren't the only animals to face trial during medieval times. Bees, snakes, horses, and bulls were also charged with murder. Foxes were charged with theft. Rats were charged with damaging barley.

In the early 1700s, Franciscan friars in Brazil brought "white ants" (probably termites) to trial because they "did feloniously burrow beneath the foundation of the monastery and undermine the cellars...threatening its total ruin." The monks wanted to excommunicate the termites for devouring their food, furniture, and foundation. But the termites' appointed lawyer defended them on several grounds, including their work ethic. It seems the "white ants" were more industrious than the "gray friars."

History

The first recorded animal trials occurred in Athens, Greece. More than 2,000 years ago, the Athenians instituted a special court to try objects (such as stones and beams) and animals that caused human deaths. They believed that in order to protect moral equilibrium and prevent the wrath of the Furies, these murders had to be avenged.

Animal trials peaked in the Middle Ages and were still being held as late as the 18th century. During this time, people believed that animals committed crimes against humans and that, like humans, animals were morally and legally responsible for their actions. As a result, animals received the same punishments as humans, ranging from a knock on the head to excommunication or death.

Legal Rights

Animals accused of crimes in Europe's Middle Ages received the same rights under the law as humans, including the right to a fair trial. Domestic animals were often tried in civil courts and punished individually. Animals that existed in groups (such as weevils, eels, horseflies, locusts, caterpillars, and worms) were usually tried in ecclesiastical courts. They weren't stretched on the rack to extract confessions, nor were they hanged with individual nooses; instead, they received a group anathema.

The accused animals were also entitled to legal representation. When weevils in the French village of St. Julien were accused of threatening its vineyards in 1587, Pierre Rembaud argued in their defense. The innocent weevils should not be blamed, said Rembaud; rather, the villagers should recognize God's wrath and don sackcloth. The court ruled in favor of the weevils and gave them their own parcel of land.

As for the six little piglets in Falaise? They also must have had good counsel—they were acquitted on the grounds of their youth and their mother's poor example.

Capital Punishment

Murder wasn't the only crime to carry a death sentence. Often, animals accused of witchcraft or other heinous crimes received the ultimate punishment. In 1474, a cock was burned at the stake in Basel, Switzerland, for the crime of laying an egg. As was widely understood, this could result in the birth of a basilisk, a monster that could wreak havoc in a person's home.

Pigs were often brought to the gallows for infanticide. A 900-pound sow smothering her piglets was usually an accident, but in those times, people saw it as a sign of evil thanks to the biblical account of the demon-possessed herd at Gadarenes.

Animals had slim hopes for survival when they were accused of severe crimes. However, there is the amazing account of a jenny that was saved when the parish priest and the citizens signed a certificate that proclaimed her innocence. It stated that they had known the "she-ass" for four years and that "she had always shown herself to be virtuous and well-behaved both at home and abroad and had never given occasion of scandal to anyone."

Contemporary Courtrooms

Although animals are not tried as humans in the United States, they are not immune to the gavel. In April 2007, a 300-pound donkey named Buddy entered a courtroom at the North Dallas Government Center in Texas.

While it was Buddy's owner who was on trial, the donkey was accused of the "crime." His owner's neighbor had been complaining about Buddy's braying and foul odor.

When the defense attorney asked Buddy if he was the accused donkey, Buddy twitched his ears and remained silent. For the next few minutes, he was calm and polite—hardly the obnoxious beast that had been described in the accusations.

While the jury deliberated, the neighbors reached a settlement. The day ended peacefully. Buddy—like his ancestors—had his day in court.

＊ The Case for Animal Rights: In *The Criminal Prosecution and Capital Punishment of Animals* (1906), author E. P. Evans refers to the words of his contemporary Henry Salt: "If animals may be rendered liable to judicial punishment for injuries done to man, one would naturally infer that they should also enjoy legal protection against human cruelty."

Not on My Ship!

Sometimes a hunch is just a hunch—but sometimes amateur sleuthing pays off.

❋ ❋ ❋ ❋

Four years before the RMS *Empress of Ireland* sank, Captain Henry Kendall was working as a skipper aboard the SS *Montrose*. Fancying himself an amateur detective, the captain would often "size up" his passengers, deciding which ones were on the straight and narrow and which ones weren't.

During a voyage from Antwerp, Belgium, to Quebec City, Kendall read a *Daily Mail* article that outlined the tale of Dr. Hawley Harvey Crippen and his girlfriend Ethel Le Neve. Scotland Yard was seeking the fugitive Crippen in connection with the murder of his wife, and Le Neve was believed to be traveling with him.

With the intriguing story in the back of his mind, Kendall noticed that one of his passengers bore a strong resemblance to Dr. Crippen. His suspicions mounted when he saw that the passenger, a Mr. Robinson, was traveling with a peculiar man of slight build to whom he referred as his son.

While observing the pair from a stealthy distance, Kendall saw the two "men" embrace in a romantic manner. Convinced that he had identified Crippen and his girlfriend, Kendall contacted Scotland Yard by wireless.

When authorities boarded the *Montrose* in Canada, a positive identification was made. Kendall's sleuthing had proven correct. Ethel Le Neve was acquitted of any wrongdoing, but Crippen wasn't quite so fortunate: He was convicted of murdering his wife and was hanged on November 28, 1910. The incident marked the first time that a fugitive was tracked down via wireless telegraph.

Chapter 19

Health and the Human Body

What Was the Black Death?

The Black Death—also called the Great or Black Plague—first swept through Europe in the 1340s, killing nearly 60 percent of the population. It returned periodically, spreading panic and death, and then disappeared into history. Was it bubonic plague, as many people believe, or was something else to blame?

✳ ✳ ✳ ✳

"The Great Mortality"

BROUGHT TO ITALY in 1347 by Genoese trading ships, the Black Death spread through Europe like wildfire. Contemporaries described scenes of fear and decay—the sick were abandoned and the dead were piled in the streets because no one would bury them. Anyone who touched the infected, the dead, or their belongings could also get the disease.

It could take as long as a month before symptoms showed, which was plenty of time for the infection to spread. But after the dreaded blackened spots began to appear on a victim's body, death came quickly—usually within three days. At least one autopsy recorded that the person's internal organs had almost liquified and that the blood within the body had congealed.

Transmission

One problem with the bubonic plague theory is that the plague doesn't transmit from person to person—it can only be transmitted through the bites of fleas that have left an infected rat after its death. The signature symptom—black swellings, or *buboe*— begin showing within two or three days of contraction. Accounts of the Great Plague often mention people who became infected merely by touching an infected person. Some writings describe the infection as spreading via droplets of body fluid (whether sweat, saliva, or blood), which isn't possible for bubonic plague but is a defining characteristic of hemorrhagic fever.

Incubation and Quarantine

It must be mentioned that it is possible for a bubonic plague infection to spread to the lungs and become pneumonic plague, and this kind of infection is transmittable from person to person. However, pneumonic plague is extremely rare—it occurs in only 5 percent of bubonic plague cases. It also isn't easily transmitted, and it certainly isn't virulent enough to have been responsible for the widespread person-to-person infection rates during the Black Death. Both have a very short incubation period, taking only a few days from infection to death. However, historical records show that the plague took as long as a month to manifest. It was no coincidence that cities started to mandate a strict 40-day quarantine. Officials had observed from multiple cases that that much time was needed to determine whether someone was infected. If it had been pneumonic plague, such a long period of quarantine would not have been needed—those infected would have been dead within a single week.

Another anomaly that casts doubt on the bubonic plague theory is that the plague spread in Iceland. There were no rats in Iceland, and there wouldn't be until 300 years later. However, the plague still ravaged the island, killing nearly 60 percent of its population.

Rapid Spread

The Black Death spread through Europe faster than any disease people had ever seen. It made the trip from Italy to the Arctic Circle in less than three years and is recorded as having traveled 150 miles in England within six weeks. Rats can't travel that quickly, but people can. Frightened citizens fled from cities where the epidemic raged, not knowing that they were infected, and they spread the disease as they went. Many parish records indicate that after strangers arrived in town, the plague emerged there within a few weeks.

In contrast, studies of confirmed bubonic plague outbreaks show that that disease spreads very slowly. One outbreak in India in 1907 took six weeks to travel only 100 yards, and another in South Africa from 1899 to 1925 moved only eight miles per year.

An Odd Connection

A strange piece of evidence that backs up the hemorrhagic fever theory is the connection between the Black Death and Human immunodeficiency virus (HIV). A large percentage of Europe's native population has a genetic mutation known as CCR5-delta 32 (which is generally shortened to "32"). This mutation prevents CCR5 receptors in the body's white blood cells from acting as entry ports. Viruses such as HIV cannot enter the white blood cells through those receptors (and aren't able to use the cells to replicate), so people with the 32 mutation are partially protected from viruses that work in this manner.

Molecular biologists have traced the increased frequency of the mutation to approximately 700 years ago—around the time of the Great Plague. They conclude that the plague must have been a disease that used the CCR5 receptors in the same way that HIV does, which points to a viral culprit.

Interestingly, the 32 mutation is found only in people of European descent. It is not found in people from eastern Asia or Africa—places the plague didn't touch.

Luna Ticks

Does the gleaming globe really have magical powers, or is it just our state of mind?

✳   ✳   ✳   ✳

FOR CENTURIES, THERE have been reports of abnormal human behavior under the whole of the moon. Full moons have been linked to fluctuating rates of birth, death, crime, suicide, mental illness, natural and spiritual disasters, accidents of every description, fertility, and all kinds of indiscriminate howling. People with too much spare cash and not enough common sense have been known to buy and sell stocks according to the phases of the moon. The word *lunatic* was coined to describe irrational and maniacal individuals whose conduct is seemingly influenced by the moon; their desolate domicile is dubbed the "loony bin."

Don't Blame It on the Moon

So is there a scientific relationship between the moon and human behavior? In 1996, researchers examined more than 100 studies that looked into the effects of the moon—full or otherwise—on an assortment of everyday events and anomalies, including births and deaths, kidnappings and carjackings, casino payouts and lottery paydays, aggression exhibited by athletes, assaults and assassinations, suicides and murders, traffic accidents, and aircraft crashes.

Dr. Ivan Kelly, a professor of educational psychology and human behavior, found that the "phases of the moon accounted for no more than $3/100$ of 1 percent of the variability in activities usually termed lunacy." This represents a percentage so close to zero that it can't be considered to have any theoretical, practical, or statistical interest. Because there was no significant correlation between the aforementioned occurrences and the periods and phases of the moon, it's safe to assume that the only moonshine that's causing trouble is the kind that's brewed in the Ozarks.

15 Unusual Facts About the Human Body

1. Don't stick out your tongue if you want to hide your identity. As with fingerprints, everyone has a unique tongue print!

2. Your pet isn't the only one in the house with a shedding problem. Humans shed about 600,000 particles of skin every hour. That works out to about 1.5 pounds each year, so the average person will lose around 105 pounds of skin by age 70.

3. An adult has fewer bones than a baby. We start life with 350 bones, but because bones fuse together during growth, we end up with only 206 as adults.

4. Did you know that you get a new stomach lining every three to four days? If you didn't, the strong acids your stomach uses to digest food would also digest your stomach.

5. Your nose is not as sensitive as a dog's, but it can detect 50,000 different scents.

6. The small intestine is about four times as long as the average adult is tall. If it weren't looped back and forth upon itself, its length of 18 to 23 feet wouldn't fit into the abdominal cavity.

7. This will really make your skin crawl: Every square inch of skin on the human body has about 32 million bacteria on it, but fortunately, the vast majority of them are harmless.

8. The source of smelly feet, like smelly armpits, is sweat. And people sweat buckets from their feet. A pair of feet has 500,000 sweat glands and can produce more than a pint of sweat a day.

9. The air from a human sneeze can travel at speeds of 100 miles per hour or more—another good reason to cover your nose and mouth when you sneeze (or duck when you hear one coming your way).

10. Blood has a long road to travel: Laid end to end, there are about 60,000 miles of blood vessels in the human body. And the hard-working heart pumps about 2,000 gallons of blood through those vessels every day.

11. You may not want to swim in your spit, but if you saved it all up, you could. In a lifetime, the average person produces about 25,000 quarts of saliva—enough to fill two swimming pools.

12. By age 60, 40 percent of women and 60 percent of men will snore. The sound of a snore can seem deafening: While snores average around 60 decibels, the noise level of normal speech, they can reach more than 80 decibels. Eighty decibels is as loud as the sound of a pneumatic drill breaking up concrete. Noise levels over 85 decibels are considered hazardous to the human ear.

13. If you're clipping your fingernails more often than your toenails, that's only natural: The nails that get the most exposure and are used most frequently grow the fastest. Fingernails grow most quickly on the hand that you write with and on the longest fingers. On average, nails grow about one-tenth of an inch each month.

14. No wonder babies have such a hard time holding up their heads: The human head is ¼ of our total length at birth but only ⅛ of our total length by the time we reach adulthood.

15. If you say that you're dying to get a good night's sleep, you could mean that literally. You can go without eating for weeks without succumbing, but ten days is tops for going without sleep. After that, you'll be asleep—forever!

7 Medical Myths

If you trust the source, you're most likely going to trust the information. That's what makes the following medical myths so hard to discredit—you usually hear them first from Mom, Dad, or someone else you trust. But it is nice to know the truth.

✳ ✳ ✳ ✳

1. **Chocolate and fried foods give you acne:** Some speculate that this myth originated with the baby boomer generation, who had worse acne than their parents and also more access to chocolate and fried foods. But wherever this idea came from, it's wrong. Pimples form when oil glands under the skin produce too much of a waxy oil called sebum, which the body uses to keep skin lubricated. When excess sebum and dead skin cells block pores, that area of the skin gets irritated, swells, and turns red—the telltale signs of a pimple. It is unknown why sebaceous glands produce excess sebum, but hormones are the prime suspects, which explains why teenagers are affected by acne more than others. Stress and heredity may also be factors, but chocolate bars and onion rings are off the hook.

2. **Coffee will sober you up:** If you've had too much to drink, no amount of coffee, soda, water, or anything else is going to sober you up. The only thing that will do the trick is time. The liver can metabolize only about one standard drink (12 ounces of beer, 6 ounces of wine, or 1.5 ounces of hard liquor) per hour, so if you're drinking more than that every 60 minutes, you'll have alcohol in your system for some time. The idea of coffee's sobering effect may have started because caffeine acts as a stimulant, counteracting the sedative effect of alcohol to a small degree. However, it has no effect on the amount of alcohol in the blood. So if you've been drinking, spend your money on a cab rather than a cappuccino.

3. **Cold weather can give you a cold:** "Put your jacket on or you'll catch a cold!" How times have you heard that? You may not want to tell her this, but dear old Mom was wrong about this. Viruses (more than 200 different kinds), not cold weather, cause colds. In order for you to catch a cold, a virus must travel from a sick person's body to yours. This usually happens via airborne droplets that you inhale when an infected person coughs or sneezes. You can also get a cold virus by shaking hands with an infected person or by touching an item on which the virus has found a temporary home, such as a phone or door handle. Colds are more prevalent during the colder months because people tend to spend more time inside, making it much easier for viruses to jump from person to person.

4. **Cracking your knuckles causes arthritis:** The knuckles are the joints between the fingers and hand, and these joints contain a lubricant called synovial fluid. When you crack your knuckles, you are pulling apart two bones at the joint, which means that the synovial fluid has to fill more space. This decreases the pressure of the fluid, and dissolved gases that are present, such as nitrogen, float out of the area in tiny bubbles. The bursting of these bubbles is the familiar sound we hear when someone "cracks" his or her knuckles. This bubble-bursting is not the same as arthritis, which is when the body's immune system attacks joints. However, constant knuckle-cracking can injure joints and weaken fingers.

5. **Don't swallow gum—it takes seven years to digest:** People have been chewing on this one for years. This myth has probably been around since chewing gum became popular in the late 19th century and most likely originated thanks to a single word: indigestible. Gum is comprised of flavor, sweeteners, softeners, and gum base; the body is able to break down the first three ingredients, but gum base is indigestible. This simply means that your body can't dissolve it and extract nutrients from it. In the end, gum base works its way

through your digestive system much like fiber—in two or three days, it goes out in basically the same shape in which it went in.

6. **Feed a cold and starve a fever:** This bit of folk wisdom has been bouncing around for centuries. This advice may have evolved from the idea that illnesses could be classified as either low temperature (those that give chills, such as a cold) or high temperature (those that give fever). With chills, it sounds reasonable to feed a person's internal fireplace with food. Likewise, the logic follows that when an illness raises the body's temperature, cutting back on the "fuel" should help. However, scientific evidence does not endorse this advice—many illnesses must simply run their course. Nevertheless, if you are stuck in bed with a cold and a loved one brings you your favorite healthful foods, it is still okay to chow down. Alternatively, you may lose your appetite while fighting a fever-based sickness. When you're sick, it's okay to miss a meal or two as long as you are keeping up with fluid intake.

7. **You can get the flu from a flu shot:** Vaccinations are misunderstood because they are created from the offending viruses themselves. But when you get a flu shot, you're not being injected with a whole virus—you're receiving an inactive—or dead—virus. That means that the part of the virus that can infect you and make you sick is turned off, but the part of the virus that stimulates your body to create antibodies is still on. The body's antibodies will kill the flu virus should you come into contact with it later. Even pregnant women are advised to get flu vaccinations, so you know they're safe. The only people who should avoid them are those who have severe allergies to eggs, because eggs are used to create the vaccines. No vaccine is 100 percent effective, so there is still a chance you can get the flu after receiving the shot, but that doesn't mean that the vaccination gave it to you.

To Pee or Not to Pee

Some alternative medicine proponents claim urine therapy can cure a long list of ailments.

✳ ✳ ✳ ✳

URINE IS SOMETHING even moderately healthy people see on a daily basis. Urine has an ammonialike smell due to the nitrogenous wastes that make up a small percent of it (the remaining 95 percent is water.) The chief constituent of the nitrogenous wastes in urine is *urea*, which is a product of protein decomposition. When your body doesn't need urea or can't use it, it's processed through your system and comes out in your urine.

There are accounts from ancient Rome of people to the west who brushed their teeth with urine for a whitening effect. A Sanskrit text from India promotes the benefits of "pure water," or urine, for treating afflictions of the skin. The prophet Muhammad is said to have advocated drinking camel urine when sick, and some point to a line in the Bible as proof that urine is supposed to be used medicinally: "Drink waters from thy own cistern, flowing water from thy own well." (Proverbs 5:15)

Urine *is* sterile (though only fresh urine qualifies) and does contain various elements that are found in medicines, but scientists and doctors have found that it doesn't contain enough to really be an effective treatment. It's said urine can be used to treat athlete's foot, but there's no scientific evidence that this is true.

However, urine therapy *is* helpful on a battlefield, when water is unavailable to clean a wound. In those cases, using urine in place of water is better than nothing. Research shows that while you're unlikely to die if you drink your own pee, there have been some cases in which folks became ill after doing so. As always, it's always a good idea to consult your doctor before experimenting with any home health remedies.

Facts of Sleep

* On average, humans sleep three hours less per night than other primates. Chimps, rhesus monkeys, and baboons sleep ten hours per night. Maybe that's because they don't have to get up and go to work!

* When we sleep, we drift between rapid-eye-movement (REM) sleep and non-REM sleep in alternating 90-minute cycles. Non-REM sleep starts with drowsiness and proceeds to deeper sleep, during which it's harder to be awakened. During REM sleep, our heart rates increase, our breathing becomes irregular, our muscles relax, and our eyes move rapidly beneath our eyelids.

* Elephants stand while they're dozing in non-REM sleep, but once REM sleep kicks in, they lie down. Ducks are constantly at risk of being attacked by predators, so they keep half their brain awake while they sleep.

* Studies have shown that our bodies experience diminished capacity when we've been awake for longer than 17 hours, and we behave as if we were legally drunk. After five consecutive nights with too little sleep, we actually get intoxicated twice as fast.

* The longest recorded period that a person has gone without sleep is 18 days, 21 hours, and 40 minutes; it took place in a rocking-chair marathon. By the end, the winner was experiencing paranoia, hallucinations, blurry vision, slurred words, and an inability to concentrate.

* During their first year, babies cause between 400 and 750 hours of lost sleep for parents.

* At least 10 percent of all people sleepwalk at least once in their lives. Men are more likely to sleepwalk than women.

* Sleepwalking occurs most commonly in middle childhood and preadolescence, and it often lasts into adulthood.

* It is a myth that snoring is harmless. It can be a sign of sleep apnea, a life-threatening sleep disorder. As many as 10 percent of people who snore have sleep apnea, which causes sufferers to stop breathing as many as 300 times every night and can lead to a stroke or heart attack.

* REM sleep was initially discovered years before the first studies that monitored brain waves overnight were conducted in 1953, though scientists didn't understand its significance at first.

* When we sleep, our bodies cool down. Body temperature and sleep are closely related—that's why it's hard to sleep on a hot summer night. We sleep best in moderate temperatures.

* Caffeine can overcome drowsiness, but it actually takes about 30 minutes before its effects kick in, and they are only temporary. It is better to get adequate sleep before driving all night than to rely on caffeine to keep you alert.

"A ruffled mind makes a restless pillow."

—CHARLOTTE BRONTË

"It appears that every man's insomnia is as different from his neighbour's as are their daytime hopes and aspirations."

—F. SCOTT FITZGERALD

"Sleep, that deplorable curtailment of the joy of life."

—VIRGINIA WOOLF

Buried Under Memories

Wouldn't it be nice if you could remember everything you've experienced throughout your entire life? For one California woman, that capability is both a gift and a curse.

✳ ✳ ✳ ✳

Life on Automatic Rewind

JILL PRICE, A woman in her forties who lives in Los Angeles can remember every detail of her life from when she was 14 years old to the present. And not just hazy, familiar memories but detailed minutia from much larger events, such as the date, time, and day of the Mount St. Helens eruption or the time of day that the Rodney King beating took place. "My life is like a split-screen," says Price. "Though I'll be living in the present, a dozen times or more a day I'll be pulled back into reliving specific memories of the past." All it takes is a song, a familiar smell, or seeing something on television to send Price back in time. She can tell you what she was doing, what her friends said, and even what her mother ordered for lunch at a restaurant.

Price traces the beginning of her unique "gift" to a traumatic time as an eight-year-old when her family moved from their East Coast home to California. The move triggered a series of traumatic recurrences of anxiety and depression—bouts that she has learned to control over time. After marrying at age 37, she suffered a miscarriage and her husband died after suffering a stroke. The onslaught of memories, both good and bad, was relentless. For years, Price tried to suppress her unpleasant memories, but they always returned with a vengeance.

Don't forget! (Ever)

A New Diagnosis

In 2000, after enduring these flashbacks for more than 20 years, Price finally wrote to Dr. James McGaugh, a neuroscientist and a leading memory researcher affiliated with the University of California, Irvine. She was looking to understand why she held onto such vivid memories. McGaugh was astounded by Price's dilemma. Relying on historical almanacs and a diary that Price had kept since she was ten, McGaugh and his research team tested her recollection of events going back to her childhood.

McGaugh performed a series of evaluations on Price, including CAT scans and MRIs, which determined that a part of her brain was three times the size of those of other women her age. The medical team published a research paper about Price and proposed a new name for her condition—"hyperthymestic syndrome," or quite simply, having a superior memory. According to McGaugh, the enlarged parts of Price's brain are consistent with those of people who are diagnosed with obsessive-compulsive disorder (OCD). For example, just as many OCD patients have a tendency toward compulsive collecting, Price has a doll collection totaling in the hundreds.

Company in Numbers

After Price was diagnosed with hyperthymestic syndrome, two other people with similar symptoms surfaced. However, unlike Price, who has struggled with reliving vivid and traumatic memories, others seem to have more control over their memories and the impact they have on their current lives. Perhaps this is because the others didn't have the trauma that plagued Price as a child. In 2008, Price published *The Woman Who Can't Forget*, a memoir about her condition.

"The constant onslaught of memories is both a blessing and a curse," says Price, noting that sometimes her pleasant memories serve as a form of comfort. "I have this warm, safe feeling that helps me to get through anything." Unfortunately, she also has to deal with reliving unpleasant memories that replay in full detail. "Over the years, it has paralyzed my life. It has eaten me up."

War (Medicine) Is Hell

Civil War–era medicine was certainly crude, but doctors in battle learned and developed new procedures at an astonishing rate.

✳ ✳ ✳ ✳

A CIVIL WAR BATTLEFIELD was not a good place to get wounded. If your injury didn't kill you immediately, the inexperienced surgeon who was hacking off your damaged limb probably would. And if you survived that, the infection you'd get from the filthy field hospital would likely do you in. However, the war was a vast training ground and laboratory for physicians, nurses, and other medical personnel.

Not So Spick-and-Span

When thousands of young men from around the country gathered in training camps, their bodies were assaulted by pathogens of all types. Most soldiers had never been away from their family farms and had no immunity to dysentery, cholera, malaria, and other diseases. There was also no plumbing, so troops relieved themselves in pits dug for that purpose or randomly out in the woods. This bred hordes of flies that spread disease. Eventually the medical establishment tried to remedy the situation—proper sanitation was probably the first medical contribution of the war.

Surgery in the Fields

Camps were not the only places that needed better sanitation. Field operating stations were crude places where doctors regularly operated on soldiers without cleaning neither their hands nor instruments. Infections in wounds after surgery were so common that doctors considered them part of the healing process.

While never fully curbing the spread of infection, surgeons did start to understand how to use antiseptics to clean wounds and equipment. Doctors today still fight infection after surgery, and they can look back to their Civil War counterparts as their forebears in infection prevention.

A Smooth Ride

Another vital advancement in medicine concerned how the wounded were moved. Many wounded died on the battlefield due to the lack of quick transport to field hospitals. Eventually, medical authorities established designated ambulance teams with attendants who were trained to do the job correctly and special wagons that were designed to hold the stretchers. Farther down the line, specially equipped trains transported the wounded to field hospitals behind the lines.

Recovery Room

Possibly the most important medical advancement in the Civil War was the establishment of better hospitals. Because of the vast number of wounded pouring off the battlefields, small civilian hospitals were practically worthless. At first, doctors commandeered schools, warehouses, and other buildings as substitutes. Eventually, though, doctors learned that disease quickly spread through these cramped, dank places, and they began building "pavilion hospitals," which featured long, wide, airy bays for patients. Ventilation kept a steady stream of fresh air moving through. Toilets with running water kept human waste—and the resulting flies—away from patient areas. Efficient design ensured that supply rooms, operating rooms, and kitchens were conveniently situated.

Nursing Back to Health

Another essential medical development during this time was the growth of the role of female nurses. Nursing was already a profession, but during the war it became an important way for women to help the war effort. Nurses weren't prevalent in the war's early months, but news of wounded soldiers' suffering elicited a call for volunteers. About 2,000 women served as nurses during the war. They attended to the wounded, assisted doctors, maintained sanitary conditions, and generally helped keep order. The most famous Civil War nurse was Clara Barton, who organized volunteers and traveled to the front lines of many battles. She formed the American Red Cross in 1881.

I Vant to Transfer Your Blood!

A blood transfusion is the process of transferring blood from one person to another, and it's a life-saving procedure that is performed about once every two seconds somewhere in the world. Read on for some interesting facts about blood and the history of transfusions.

✳   ✳   ✳   ✳

✳ One of the earliest recorded blood transfusions took place in the 1490s. Pope Innocent VIII suffered a massive stroke and his physician advised that the pontiff undergo a blood transfusion. Unfortunately, the methods used were crude and unsuccessful; the Pope died within the year and so did the three young boys whose blood was used.

✳ Fifty-three percent of all transfusions go to women, while 47 percent go to men.

✳ Richard Lower, an Oxford physician in the mid-1600s, performed blood transfusions between dogs and eventually (and successfully) between a dog and a human. Several years later, cross-species transfusions was deemed unsafe.

✳ Donating blood seems to appeal to folks with a sense of civic duty: 94 percent of blood donors are registered voters.

✳ In 73 countries, many of which are undeveloped and have a great need for blood, donation rates are less than 1 percent.

✳ One unit of blood can be separated into several components: red blood cells, plasma, platelets, and cryoprecipitate, which is a substance that is helpful in the clotting process.

✳ If you're older than 17 and weigh at least 110 pounds, you may donate a pint of blood (the most common amount of donation) about every two months in the United States and Canada.

* If only 1 additional percent of Americans would donate blood, shortages would disappear.

* In 1818, British obstetrician James Blundell performed the first successful transfusion of human blood. His patient, a woman suffering postpartum hemorrhaging, was given a syringeful of her husband's blood. The woman lived, and Blundell became a pioneer in the study of blood transfusions.

* More than 85 million units of blood donations are collected globally every year. About 35 percent of these are donated in developing and transitional countries, which make up about 75 percent of the world's population.

* Those with type O blood are considered "universal donors," which means that their blood can be cross-matched with all blood types successfully.

* Only 38 percent of the U.S. population is eligible to donate blood and less than 10 percent actually do so on an annual basis. There are 43,000 pints of blood used each day in the United States and Canada.

* Thirteen tests (11 of which screen for infectious diseases) are performed on every unit of donated blood.

* In 1916, scientists introduced a citrate-glucose solution that allowed blood to be stored for several days. Due to this discovery, the first "blood depot" was established during World War I in Britain.

* Try not to need a blood transfusion in the summer or around the holidays. Shortages of all blood types happen during these times.

* The first test to detect the probable presence of HIV was licensed and implemented by blood banks in the United States in 1985. Since then, only two people have contracted HIV from a blood transfusion.

Eat Worms, Lose Weight!

No matter what anyone tells you, the only way to successfully lose weight is to eat less and exercise more. Yet this common-sense knowledge hasn't stopped millions from trying anything to make the road to weight loss smoother—including purposely ingesting parasites. Wouldn't it be easier to just go for a walk and skip dessert?

✳ ✳ ✳ ✳

Those Wacky Early 1900s

THERE WAS A time when cocaine was the cure for a sore throat and smoking was considered a healthy habit. So not many feathers were ruffled when ads showed up advertising a tapeworm pill for ladies who were looking to slim down. The ads, which first appeared between 1900 and 1920, claimed that by ingesting a pill containing tapeworm larvae, you could give a hungry worm a happy home and lose that pesky weight. You could eat all you wanted, content in the knowledge that your new friend would be devouring most of the calories you consumed, thus allowing you to lose weight without thinking twice about it.

No one can prove that the pills advertised back then actually contained worm larvae. They could've been placebos, and for the foolish folks who tried the diet fad, we can only hope that was the case.

The Worm Is Back!

The weight-loss-via-tapeworm idea subsequently died down for many years (obesity was not much of an issue during the Great Depression and both World Wars), but talk of it resurfaced in the 1960s. Rumors that a new appetite-suppressant candy introduced to the market contained worm eggs started getting around, though, of course, this was entirely false.

After a remarkable weight loss of an estimated 65 pounds, acclaimed opera star Maria Callas endured gossip that she had purposely acquired a tapeworm to do it. Though the singer was indeed diagnosed with a tapeworm, her doctor suggested that it was due to her fondness for eating beef tartare. Other celebrities are rumored to have swallowed tapeworm pills to whittle down their figures, including model Claudia Schiffer, though this was never confirmed.

An Internet search reveals companies that advertise "sterile tapeworms" for a variety of medicinal uses (whether they're selling a real product or scamming the public is another story). The fine print is lengthy, however, as using tapeworms to treat any condition has not been approved by the USDA. To get your worms, you'll likely have to go to Mexico. But these stowaways will get you in big trouble if you try to bring them back across the border.

Tapeworms: Not Good Pets

Around the world, a lot of time and attention is spent trying to keep worms from getting into the human body via water, food, or skin. Simply put: Having a tapeworm is not a good thing. In the case of the fish tapeworm especially, the essential vitamin B_{12}— a vital ingredient for making red blood cells—is sucked out of the host's body.

Adult tapeworms can grow up to 50 feet long and live up to 20 years. Depending on the worm, a host's symptoms range from epileptic seizures, diarrhea, nausea, fatigue, a swollen belly (oh, the irony), and even death. While it's likely that a person with a worm will lose weight, he or she will also suffer from malnutrition— B_{12} isn't the only nutrient eaten by the parasite. And tapeworm eggs are inevitable byproducts of a tapeworm. The fish tapeworm can produce a million eggs in a single day, and the larvae tend to burrow out of the intestines and find homes elsewhere in the body—like the brain, for example. Worms also have the habit of popping out of various orifices without warning. Still interested in tapeworms as a method of weight loss? Then perhaps it's your head, and not your pants size, that's the issue!

A Pain in the Brain

Neurologist and psychiatrist Walter Freeman believed that lobotomy was a cure-all for a wide variety of mental illnesses. Over a period of more than three decades, he scrambled the frontal lobes of more than 3,400 patients—sometimes with disastrous results.

✳   ✳   ✳   ✳

WALTER FREEMAN BECAME interested in the curative properties of lobotomy after attending a conference at which a colleague discussed how chimpanzees that suffered damage to their frontal lobes became docile and inactive. Freeman became particularly intrigued when neurologist Egas Moniz started performing a procedure that replicated the frontal lobe damage found in the chimps on human patients with psychiatric illnesses.

Working with neurosurgeon James Watts at George Washington University Hospital, Freeman developed a technique that used an instrument called a *leucotome* to scramble the frontal lobes of the brain. In 1936, he performed his first lobotomy on a Kansas woman who was suffering from agitated depression. By all accounts, the procedure was successful.

Honing His Craft

Initially, Freeman and Watts cored the frontal lobes through six holes that were drilled into the top of the skull. Later, the doctors replaced the leucotome with a device that resembled a butter knife, and they moved the entry holes from the top of the skull to the sides. Freeman eventually developed a procedure called transorbital lobotomy that allowed him to quickly and easily reach a patient's brain by driving an ice-picklike device through the bone at the back of the eye socket.

He performed the procedure on patients young and old who were suffering from depression, manic depression, schizophrenia, obsessive-compulsive disorder, and a variety of undiagnosed conditions. More than a few of Freeman's colleagues questioned lobotomy's effectiveness, and many became incensed by his publicity-mongering.

Not All's Well that Ends Well

However, there was often a damaging downside to the procedure: Three percent of Freeman's patients died, and many of those who survived saw their emotions, inhibitions, and personalities subjugated. It wasn't uncommon for lobotomy patients to forget how to perform basic life functions such as dressing themselves and using the toilet; many would simply stare at the wall for hours. This suited officials at some mental institutions just fine—a gentle, dazed patient was certainly preferable to an aggressive one. One of the more famous of Freeman's patients—John F. Kennedy's sister Rosemary—was among the unfortunate. After she received a lobotomy for her mood swings, she became an invalid and needed full-time care for the next 64 years.

By the mid-'50s, tranquilizers had come into vogue as the treatment for many of the same disorders. Freeman moved to California and continued to perform lobotomies. He became known for his showmanship. His records show that he lobotomized 3,439 patients during his career.

Leaving Lobotomy Behind

Freeman's practice came to an end in 1967 after he killed a patient by accidentally severing a blood vessel in her brain. His surgical privileges were revoked by the hospital where the procedure had been attempted, and Freeman performed no more lobotomies.

Today, lobotomy is essentially a thing of the past. In the United States, fewer than 20 such surgeries are performed annually for the treatment of psychiatric illness. However, researchers are investigating a futuristic approach to Freeman's work: an implantable device that uses electrodes to stimulate parts of the brain to help control conditions such as obsessive-compulsive disorder and Parkinson's disease. Any way you look at it, it's better than an ice pick through the eye.

Feeling Stressed?

* As many as four out of ten adults are so stressed out that it affects their health. It is reported that 75 to 90 percent of all visits to doctors are for symptoms caused by stress.

* In the United States, tranquilizers, antidepressants, and medications for anxiety account for one-fourth of all prescriptions written.

* Studies show that 60 percent of all employee absences are due to stress. The number of employees who report that they are "highly stressed" is greater now than at any other time in history. Stress that is strong enough to cause disability has doubled over the past decade.

* In a 14-year study of stress among 12,500 Swedes, it was found that those who had little control over their work were twice as likely to develop heart disease. Those who had little support at work were nearly three times as likely to develop heart disease.

* In the United States, about a million people each day are absent from work due to stress-related disorders.

* Stress is a significant contributing factor to cardiovascular disease, gastrointestinal disorders, skin problems, neurological disease, and emotional disorders. It also attacks the immune system, causing an increase in colds, herpes, arthritis, and even cancer and AIDS.

* Post-traumatic stress disorder (PTSD) is delayed stress that can occur after people are exposed to a disturbing or frightening experience. Any extremely stressful situation can result in PTSD, whether someone is involved or just an observer.

* Stress can be good for you! *Eustress* is the stress you feel when you play a fun game, fall in love, have a baby, ride a roller coaster, or take a vacation.

Lincoln and the Blue Mass

In the 19th century, the cures for many common ailments were often worse than the sicknesses.

✳ ✳ ✳ ✳

ABRAHAM LINCOLN WAS known for the composure, patience, and calmness he displayed while under the stress of the presidency. But in his earlier years, the gangly prairie lawyer was known to suffer from bouts of anxiety, depression, tremors, and insomnia—and to make others suffer from his outbursts of rage.

During an 1858 debate, Lincoln became so infuriated that "his voice thrilled and his whole frame shook." He grabbed a man by the collar and shook the poor fellow "until his teeth chattered."

Why was his personality so different in the period prior to his becoming president? A recent study by the University of Chicago Medical Center may have unveiled the answer.

Blue Mass for a Blue Mood

According to the study, Lincoln had been using a common medication for depression known as "blue mass." Taken as a syrup or in pill form, blue mass was used for a number of ailments, such as birthing pains, constipation, parasitic infestations, toothache, and tuberculosis. Although it differed from pharmacist to pharmacist, the concoction could include chalk, sugar, honey, licorice root, rose water, rose petals—and even mercury. A daily dosage of blue mass would have likely contained somewhere between 100 and 9,000 times more mercury than modern medicine would find acceptable.

Symptoms of mercury poisoning can include severe mood swings and aggression. Lincoln stopped taking the medicine in 1861, a few months after his inauguration, noting that it made him "cross." His mood swings disappeared after that.

Heterochromia: The Eyes Have It

The vast majority of human beings have two eyes that are the same color. But some people just can't commit.

✳ ✳ ✳ ✳

PUT PLAINLY, HETEROCHROMIA is the presence of different-colored eyes in the same person. The condition is a result of the relative excess or lack of pigment within an iris or part of an iris. The causes for having mismatched eyes vary, and some are more worrisome than others.

Heterochromia is largely hereditary; many people are born with eyes that don't quite match, and nothing seems to be wrong with them. Females experience it far more than males, and most cases of genetic heterochromia have been found to occur between ages 2 and 19. Some folks who had different-colored eyes as a child report their eyes matching when they reached adulthood.

Diseases such as glaucoma, neurofibromatosis, and Waardenberg syndrome can also cause the condition. Waardenberg syndrome is a rare disorder that affects skin pigmentation and is responsible for varying degrees of hearing loss. Most people with Waardenberg syndrome have two different-colored eyes, making the presence of, say, one green eye and one blue eye cause for a trip to the doctor.

Another way that two eyes might be different colors is due to eye injury. A foreign object in the eye (we're talking about more than an eyelash), ocular inflammation, and other injuries can cause trauma to the eye and alter its color.

Now, heterochromia doesn't just mean that a person has one brown eye and one green eye: Many people with this trait will have one eye that is two colors—part green and part blue, for example. And whatever the color of the iris happens to be, it doesn't have any bearing on the ability of the eye to see. Having heterochromia doesn't mean that a person has poor eyesight.

What (Color) Is It, Lassie?

Animals—including cats and horses—can also show signs of heterochromia. However, much more common than heterochromia in humans is heterochromia in dogs. Many breeds—including Siberian huskies, Australian sheep dogs, Great Danes, dalmatians, and Alaskan malamutes—exhibit striking cases of the condition; perhaps you've seen a Siberian husky with one ice-blue eye and one dark brown eye. Just as is the case with humans, the vision of a dog with this condition is completely normal; unlike humans, however, heterochromia in canines is rarely considered a cause for medical concern.

Got Heterochromia? Cool!

If you see multicolored eyes when you look in the mirror and you have a clean bill of health, then celebrate! Chances are good that you'll never have to think up something interesting to talk about at a cocktail party.

Many people find folks with different-colored eyes fascinating. Some say that they find it mysterious (and more than a little sexy) because it's such a rare characteristic. Largely, heterochromia is seen as really cool, the prevailing notion being that people with mismatched eyes are smarter and more intriguing, possess greater depth, and are generally cooler than everyone else.

If you have heterochromia, you're in good company. Check out this list of celebrities and historical figures that have boasted dual-tone eyes:

* actress Kate Bosworth

* actor Christopher Walken

* king and conqueror Alexander the Great

* actress Mila Kunis

* comedian and actor Dan Aykroyd

* singer Carly Simon

Hair, There, and Everywhere

Amaze your friends with these follicle facts.

✳ ✳ ✳ ✳

✳ If you're an average nonbalding person, you have approximately 100,000 hairs on your head.

✳ The amount of hair you have varies based on natural color: Blondes have the most, with about 140,000 hairs; dark-haired folks have about 110,000; and redheads have a meager 80,000–90,000 strands attached to their noggins.

✳ People normally shed about 50 to 100 hairs a day.

✳ More than half of all men start going bald by the time they're 50. But they'll have lost at least 50 percent of the hair on their heads before it even becomes noticeable.

✳ The Guinness World Record holder for longest hair is a Chinese woman named Xie Qiuping, whose hair measures a whopping 18.5 feet long. That's a lot of brushing!

✳ Most hair grows about six inches per year, and men's hair tends to grow faster than women's.

✳ Hair is actually made of keratin, a kind of dead protein. It's the same stuff that makes up your fingernails.

✳ Hair follicles determine whether locks are straight, curly, or wavy: Flat follicles produce curls, while oval-shaped follicles lead to waves. Round follicles make for straight strands.

✳ A single strand of hair can support up to 100 grams of weight (just under a quarter of a pound).

✳ Using that calculation, your entire head of hair could support more than 22,000 pounds—or 10–15 tons—of weight. That'd be the weight equivalent of 85 Arnold Schwarzeneggers (during his bodybuilding heyday).

6 Modern Health Problems

Modern life—with its emphasis on information, automation, computerization, and globalization—has made work easier and given us more leisure options, but we now have a whole host of new health problems.

✻ ✻ ✻ ✻

1. **Computer vision syndrome:** If you spend all day staring at a computer screen, you may be at risk for computer vision syndrome (CVS), also called occupational asthenopia. CVS encompasses all eye or vision-related problems suffered by people who spend a lot of time on computers. According to the American Optometric Association, symptoms of CVS include headaches; dry, red, or burning eyes; blurred or double vision; trouble focusing; difficulty distinguishing colors; sensitivity to light; and even pain in the neck or back. As many as 75 percent of computer users have symptoms of CVS due to glare, poor lighting, and improper workstation setup. To overcome CVS, keep your monitor about two feet away from you and six inches below eye level, and be sure that it's directly in front of you to minimize eye movement. Adjust lighting to remove any glare or reflections. You can also adjust the brightness on your monitor to ease eyestrain. Even simple steps can help, such as looking away from your monitor every 20 or 30 minutes and focusing on something farther away. And you can always use eyedrops to perk up your peepers!

2. **Earbud-related hearing loss:** Earbud headphones fit inside the ear but don't cancel out background noise, which causes users to turn up the volume, often to 110 to 120 decibels— loud enough to cause hearing loss after only 75 minutes. As a result, young people are developing the types of hearing loss normally seen in much older adults. Experts recommend turning down the volume and limiting the amount of time

spent listening to music players to about an hour a day. Headphones that fit outside the ear canal also help, as can noise-canceling headphones that reduce background noise so listeners don't have to crank up the volume.

3. **Generalized anxiety disorder:** We all have worries, uncertainties, and fears, but generalized anxiety disorder (GAD) is excessive or unrealistic unease or concern about life's problems. Although the disorder often manifests without any specific cause, large issues of modern life (such as terrorism, the economy, and crime) can bring it about, as can individual circumstances. GAD affects about 6.8 million people in the United States, and symptoms include restlessness, fatigue, irritability, impatience, difficulty concentrating, headaches, upset stomach, and shortness of breath. Anxiety disorders like GAD are treated with antianxiety drugs, antidepressants, psychotherapy, or a combination of these.

4. **Orthorexia nervosa:** It seems like every day there's a new report about something you shouldn't eat. People who have the eating disorder orthorexia nervosa are obsessed with eating healthful food and have constructed strict diets that they follow religiously. Although many people who have orthorexia nervosa become underweight, nutritional purity is actually their goal. Among the signs of orthorexia nervosa are: spending more than three hours a day thinking about healthful food; planning meals days in advance; feeling virtuous from following a strict healthful diet, but not enjoying eating; feeling socially isolated (such strict diets make it hard to eat anywhere but at home); and feeling highly critical of those who do not follow a similar diet.

5. **Sick building syndrome:** Rising energy costs aren't just harmful to your wallet: If you work in an office building, they could be making you physically ill. Businesses have found that by packing buildings with insulation and then adding caulking and weather stripping, they can seal buildings tight,

keep indoor temperatures constant, and cut energy costs in the process. Such measures require the heating, ventilation, and air conditioning (HVAC) systems to work hard to recycle air. After all, when the building is sealed, you can't open a window to let fresh air circulate. The result is sick building syndrome, which the Environmental Protection Agency (EPA) classifies as a situation in which building occupants experience discomforting health effects, even though no specific cause can be found. Symptoms include headache; eye, nose, or throat irritation; dry cough; dry or itchy skin; dizziness; nausea; fatigue; and sensitivity to odors. The EPA estimates that 30 percent of all U.S. office buildings could be "sick," so they recommend routine maintenance of HVAC systems, including cleaning or replacing filters; replacing water-stained ceiling tiles and carpeting; restricting smoking in and around buildings; and ventilating areas where paints, adhesives, or solvents are used.

6. **Social anxiety disorder:** Despite all the ways to interact with others in our technologically savvy world, those with social anxiety disorder (SAD) feel boxed in by the shrinking globe. According to the National Institutes of Health (NIH), people with social anxiety disorder have an "intense, persistent, and chronic fear of being watched and judged by others and of doing things that will embarrass them," and that fear can be so intense that it interferes with work, school, and other ordinary activities and can make it hard to make and keep friends. But the condition also has physical manifestations, including trembling, upset stomach, heart palpitations, confusion, and diarrhea. Its cause hasn't been nailed down, but SAD probably manifests due to a combination of environmental and hereditary factors. About 15 million people in the United States are affected by social anxiety disorder, which usually begins during childhood. Like other anxiety disorders, treatment of this disorder often involves medication and psychotherapy.

Medical Oddities at the Mütter Museum

Located at the College of Physicians of Philadelphia, the Mütter Museum is perhaps the most grotesque and shockingly fascinating museum in the United States. Its collection of human skulls, preserved brains (eyes included), and freaks of nature will surely entertain those with even the most morbid curiosities.

✳ ✳ ✳ ✳

"Disturbingly Informative"

IT'S ALSO ONE of the most elegant museums that's open to the public, with red carpet, brass railings, and redwood-lined display cases. It might even appear a bit highbrow if the curators themselves didn't acknowledge what a uniquely abnormal exhibit they are pushing—a refreshing attitude that's evident in their motto, "Disturbingly Informative."

The museum was founded in 1859 when Dr. Thomas Mütter donated several thousand dollars and his personal collection of 1,700 medical specimens to the College of Physicians. Merging it with their own meager collection, the institution used Mütter's money to build new quarters to house it all and opened it to both students and the public.

Dem Bones, Dem Bones

Further acquisitions expanded the museum's collection tremendously, as doctors contributed specimens they'd acquired through their own private practices and studies. A large number of them are skeletal, such as a woman's rib cage that became cartoonishly compressed by years of wearing tight corsets and the 19th-century Peruvian skulls that show primitive trephinations (holes cut or drilled in the head).

There are also the combined skeletons of infants born with a shared skull and the bones of a man who suffered a condition in which superfluous bone grows in patches, eventually fusing the skeleton together and immobilizing its owner. Most popular, though, is the skeleton of a man measuring 7'6", the tallest of its kind on display in North America, which stands next to that of a 3'6" female dwarf.

Other items include heads that are sliced like loaves of bread—both front to back and side to side—and outdated medical instruments, many of which look torturous. Curators also have acquired more than 2,000 objects removed from people's throats and airways, a vintage iron lung, and photographs of some of medicine's most bizarre human deformities.

Where's the Gift Shop?

The museum even has its own celebrities of sorts. For example, there's Madame Dimanche, an 82-year-old Parisian whose face and the drooping ten-inch horn growing from her forehead have been preserved in lifelike wax. There's also the unidentified corpse of a woman known simply as the Soap Lady, whose body was unearthed in 1874. The particular composition of the soil in which she was buried transformed the fatty tissues in her body, essentially preserving her as a human-shaped bar of soap.

And who can forget Eng and Chang, the conjoined brothers who toured the world with P. T. Barnum and inspired the term "Siamese twins"? The Mütter Museum not only has a plaster cast of their torsos, but also their actual connected livers.

23 Phobias and Their Definitions

1.	Ablutophobia	Fear of bathing
2.	Acrophobia	Fear of heights
3.	Pupaphobia	Fear of puppets
4.	Ailurophobia	Fear of cats
5.	Alektorophobia	Fear of chickens
6.	Anthropophobia	Fear of people
7.	Anuptaphobia	Fear of staying single
8.	Philophobia	Fear of being in love
9.	Atychiphobia	Fear of failure
10.	Autophobia	Fear of oneself or of being alone
11.	Aviophobia	Fear of flying
12.	Caligynephobia	Fear of beautiful women
13.	Coulrophobia	Fear of clowns
14.	Cynophobia	Fear of dogs
15.	Gamophobia	Fear of marriage
16.	Ichthyophobia	Fear of fish
17.	Melanophobia	Fear of the color black
18.	Mysophobia	Fear of germs or dirt
19.	Nyctophobia	Fear of the dark or of night
20.	Ophidiophobia/Herpetophobia	Fear of snakes
21.	Ornithophobia	Fear of birds
22.	Phasmophobia/Spectrophobia	Fear of ghosts
23.	Thanatophobia or Thantophobia	Fear of death or dying

The Blue People of Kentucky

Skin, the largest organ of the body, comes in many colors. But who ever heard of people with blue skin?

✳ ✳ ✳ ✳

THERE EXISTS A rare blood disorder called *methemoglobinemia* that affects a very small percentage of the population. The disorder causes a Caucasian person's blood to carry a higher than normal level of *methemoglobin*, a form of hemoglobin that does not bind oxygen. Too much methemoglobin can make the blood dark brown and give the skin a distinctly bluish tint. A person can get this disorder via exposure to certain drugs in the antibiotic and bromate family, but the most famous occurrences of methemoglobinemia were caused by genetics.

In the early 1800s, a man named Martin Fugate lived in the Appalachian Mountains with his wife, who was said to have carried the recessive gene that causes methemoglobinemia (metHB for short). One carrier of metHB won't make a blue person, but two will: The Fugates married into the Smith family, who also carried the gene. In 1832, a blue baby was born. Now, a blue baby or two might not make news, but because of serious inbreeding between the clans, eventually there was a concentration of these "Blue Fugates" in the hills near Hazard, Kentucky.

In the 1960s, the Fugates were diagnosed with metHB and treated with an injection of the chemical methylene blue, effectively replacing the missing enzyme in their blood. Those who were treated changed to a pinkish hue that lasted as long as they took regular doses of the chemical. The results were temporary.

Now that people get around more easily and inbreeding is more of a social taboo, the gene that carries metHB is becoming even rarer than before. The "blue people of Kentucky" are increasingly harder to find, though thanks to nature's infinite combinations, it's still possible to encounter a blue man or woman.

Burn, Baby, Burn

Here are 12 activities and the approximate number of calories they burn. Before launching into any of these activities, be sure to consult your doctor.

✳ ✳ ✳ ✳

1. **Running:** Burning about 450 calories every 30 minutes (based on an 8-minute mile), running also gives a fantastic cardiorespiratory workout. Leg strength and endurance are maximized, but few benefits accrue to the upper body.

2. **Rock Climbing:** Rock climbing relies on quick bursts of energy to get from one rock to the next. Your strength, endurance, and flexibility will greatly benefit, and you'll burn about 371 calories every half hour.

3. **Swimming:** Swimming provides an excellent overall body workout, burning up to 360 calories in a half hour (depending on the stroke used). The best swim workout is based on interval training: Swim two lengths, catch your breath, and then repeat.

4. **Cycling:** Cycling is an excellent non-weight-bearing (your weight is not being supported by your body) exercise, and depending on your speed, it burns anywhere from 300 to 400 calories in a half hour. It provides a great cardio workout and builds up thighs and calves. However, it doesn't provide much in the way of an upper-body workout.

5. **Boxing:** If you're game enough to step into the ring, you'll be rewarded with a 324-calorie deficit for every half hour of slugging it out. In addition, your cardiorespiratory fitness and muscular endurance will go through the roof.

6. **Racquetball:** Churning through about 300 calories in 30 minutes, racquetball gives you a fantastic cardiorespiratory workout, builds lower body strength and endurance,

and, with all that twisting and pivoting, develops great flexibility around the core (back and abs).

7. **Basketball:** The nonstop action of b-ball will see you dropping around 288 calories every half hour, while at the same time developing flexibility, endurance, and cardiorespiratory health.

8. **Rowing:** Burning about 280 calories per half hour, rowing is a very effective way to rid yourself of extra energy. It also builds up endurance, strength, and muscle in your shoulders, thighs, and biceps. The key to rowing is technique—coordinate the legs, back, and arms to work as one.

9. **Tennis:** Tennis is a fun game that demands speed, agility, strength, and reaction time. It consumes about 250–300 calories in a half-hour session, thus providing a great opportunity to burn excess calories while developing cardiorespiratory fitness.

10. **Cross-Country Skiing:** As soon as you start mushing through snow, you'll be churning through those calories at the rate of 270 every half hour. The varied terrain will provide a great interval training workout too!

11. **Ice-Skating:** Ice-skating gives you all the benefits of running without the joint stress. A half hour on the ice consumes about 252 calories. Skating provides an excellent workout for your thighs, calves, hamstrings, and buttocks. The twists and turns also tighten and tone your abs. Holding out your arms helps you balance and also works the deltoids, biceps, and triceps.

12. **Swing Dancing:** You can burn about 180 calories in a half hour by swing dancing. You'll be developing flexibility, core strength, and endurance—and you won't even feel as if you're exercising.

Calories based on a 150-pound person. (A heavier person will burn more calories.)

Cigarette Trivia

Cigarettes—we think we know everything there is to know about them. But cigarettes have been around for more than 1,000 years, and quite a lot of weird and surprising things have happened around them in that time.

✳   ✳   ✳   ✳

✳ The earliest forms of cigarettes were used in religious ceremonies by Mayan priests in Central America and Aztec priests in South America in the ninth century.

✳ In 1854, Philip Morris, a London tobacco seller, was the first to manufacture cigarettes. Forty cigarettes could be hand-rolled in one minute; by 1880, the first cigarette-rolling machine could produce 120,000 in ten hours.

✳ Although most people associate sports cards with gum, they were first found in cigarette packs (they helped protect cigarettes from being crushed). The most valuable baseball card—a 1909 Honus Wagner—was found in American Tobacco Company packs. In 2007, one such Wagner card sold for $2.8 million because only around 50 still exist. Wagner worried that children would begin smoking after buying the cigarettes to get his card so he had it pulled from circulation.

✳ Smoking was initially considered "unladylike," but tobacco companies defeated the ban on public smoking by women by appealing to women's rights groups. They invited debutantes to light up their "torches of freedom" at the 1929 Easter Parade in New York City. And they did!

✳ Joe Camel and the Marlboro Man weren't the only mascots touting cigarettes. Ads have featured dentists, doctors, babies, and even Santa enjoying a puff. In the 1940s, Madison Avenue highlighted the health benefits of smoking: calming nerves, soothing throats, boosting energy, and maintaining a healthy weight.

* According to the Department of Health and Human Services, the 1994 tobacco blend of cigarettes lists some 600 ingredients. When lit, they create more than 4,000 chemical compounds, of which 43 cause cancer. For a short time in the 1950s, Kent's "Micronite" filter was even made of asbestos!

* Cigarette consumption peaked in the United States in 1965. At the time, 50 percent of men and 33 percent of women were smokers.

* President Richard Nixon signed the Public Health Cigarette Smoking Act of 1969, which banned cigarette ads on television and radio in the United States starting on January 2, 1971. Why not January 1? Advertisers were given a one-day reprieve to take advantage of ad time during the New Year's Day football games!

* Get around smoking bans with an e-cig, or electronic cigarette. A plastic e-cig looks like a regular cigarette (the tip even glows red when you're "smoking") but holds a cartridge of flavored liquid (with or without nicotine) and is run with a lithium battery. Smokers can inhale, but the exhaled vapor instantly disappears and has no odor.

* According the American Cancer Society, 6.3 trillion cigarettes were consumed in 2010.

* Check your entrée for ashes! A study by the U.S. Substance Abuse and Mental Health Services Administration found that the occupations with the highest rates of smoking (45 percent) were food-preparation and serving-related.

* State cigarette taxes range from $1.30 per pack in Virginia to $4.35 in New York. Globally, taxes range from 0 percent of retail price to 80 percent. Because of these wild fluctuations, cigarette smuggling is widespread. In Delaware, for example, as many as 82 percent of all cigarettes purchased are smuggled.

On the Double

Ever heard someone say, "There's another person inside of me, just waiting to get out?" For parasitic twins, that's pretty much true.

✳ ✳ ✳ ✳

O Brother, Where Art Thou?

PARASITIC TWINS ARE formed by the same biological defect that causes conjoined twins: Both begin promisingly, like any other set of identical twins: A single egg is fertilized and begins dividing into individual babies. With conjoined twins, both embryos continue to fully develop and both are typically born alive—albeit fused together at the chest or abdomen or, occasionally, the head. But in the case of parasitic twins, only one embryo continues growing normally; the other stops at some point and begins feeding off the blood supply of its twin like a parasite. Though conjoined twins occasionally share a heart, they usually have their own brains. With parasitic twins, however, the undeveloped embryo lacks both heart and brain, and so it is never actually alive, although it can grow hair and limbs and even fully functioning genitalia.

Conjoined twins are rare, but parasitic twins are even rarer. In fact, there have been only 90 documented cases of parasitic twins throughout medical history.

Stuck on You

Parasitic twins can be either internal or external. In the case of external twins, the parasitic twin appears as an extra set of limbs or a faceless, malformed head growing out of the host twin's abdomen. Usually the limbs just hang there uselessly. But sometimes the nervous systems are attached, so the host twin can actually feel the parasitic twin being touched.

Internal parasites, which are known as *fetus-in-fetu,* occur when the host embryo envelops the parasitic embryo early in pregnancy. In this case, the parasitic twin continues to grow inside of its host's abdomen. As the parasite grows, the host appears to be pregnant. Usually doctors mistake this strange growth for a tumor—and are shocked when they go in to remove it and find that they are actually performing a C-section to remove the patient's stillborn twin.

Sideshow Stars

The earliest known case of parasitic twins occurred in Genoa, Italy, in 1617 with the birth of Lazarus and Joannes Baptista Colloredo; Joannes was a fully formed torso growing out of Lazarus's stomach. Lazarus, aka "The Boy with Two-Heads," traveled Europe exhibiting himself.

One of the more fascinating carnival cases is that of Myrtle Corbin, a Ringling Brothers star in the 1880s who was known as a *dipygus* parasitic twin. Everything from the waist down was double: two sets of legs, two backsides, and yes, two sets of reproductive organs. Though normally the extra set of limbs is useless, that wasn't so for Myrtle: Her "sister" delivered two of her five children, making Myrtle's the only case of dipygus twins to give birth.

Modern Marvels

An eight-limbed Indian toddler named Lakshmi Tatma made the news in 2007 when her village decided that she was a reincarnation of her namesake, the Hindu goddess of wealth. Her parents turned down offers to sell her to a circus, opting instead for surgery to remove her extra arms and legs.

Modern medicine has made it possible to uncover more cases of fetus-in-fetu. In 2006, the world learned of a 36-year-old Indian farmer—who'd been plagued all his life by a very distended belly and a chronic shortness of breath—who went in to have a stomach tumor removed. During surgery, doctors discovered a hand (with fingernails) inside the man, and then another, followed by a hair-covered head, teeth, and genitalia.

Famous Hypochondriacs

If you've ever had a cold and thought you were dying of pneumonia or had a sore back and were sure you had meningitis, you might be a hypochondriac. Read on for some famous folks who were also hypochondriacs.

✳ ✳ ✳ ✳

Hypochondriacs are individuals who are abnormally anxious about their health. What is a normal ache or pain for one person is a catastrophe for a hypochondriac, and often, hypochondriacs will imagine symptoms that were never there in the first place. Chronic hypochondriacs struggle their whole lives in fear and doubt, and ironically, they are often in poor health.

Howard Hughes

As a boy, Howard Hughes—movie mogul, captain of industry, pilot, and ladies man—suffered from an undiagnosed affliction that caused minor hearing loss. His mother doted on him as a result, and Hughes came to associate being sick with being loved and cared for, so he was "sick" quite a bit. As an adult, Hughes's hypochondria was made worse by his legendary obsessive-compulsive disorder and his generally poor mental health. In the mid-1970s, Hughes died a recluse in terrible physical condition.

Adolf Hitler

The neurotic dictator was known for his aversion to germs and illness. He would examine his own feces on a regular basis to check for consistency and eventually took on Dr. Theodor Morell, a quack doctor who put Hitler on a steady regimen of amphetamines. The drugs did nothing but make the already jumpy leader all the more unwell—and unpredictable.

Hans Christian Andersen

The Danish writer of beloved fairytales such as "The Little Mermaid," Andersen was said to be so terrified of being buried alive that, during his travels through Europe, he slept with a note that read, "I only seem dead," so that no one might mistake him for a corpse.

Charles Darwin

Author of *The Origin of the Species*, Charles Darwin was kind of a mess health-wise. It's still unknown what exactly he suffered from, but he was ill for most of his life. Many believe today that while Darwin was probably suffering from at least one affliction, his hypochondria likely made him sicker and prevented him from getting well.

Florence Nightingale

For 53 years, starting when she was 37 years old, Red Cross founder Florence Nightingale lived the life of a bedridden, almost agoraphobic invalid. But was she actually sick? Some have diagnosed Nightingale as suffering from an organic disease—likely chronic brucellosis, which causes joint and muscle pain. Others believe she was a hypochondriac who used physical symptoms to manipulate people. Recent evidence suggests that she suffered from bipolar disorder.

Sara Teasdale

Poet Sara Teasdale was born to a severely overprotective mother who would send her daughter to bed with the slightest cough or sniffle. By the time Sara was nine years old, she was thoroughly convinced that she was simply a sickly person. Nearly every year, she would take a "rest" and convalesce at home, surrounded by medicine, blankets, and, if she was feeling up to it, her writing materials.

Tennessee Williams

Playwright Tennessee Williams was a serious hypochondriac. Convinced that he was slowly losing his sight, he had four eye operations for cataracts. Worried that his heart would stop beating, he drank and took pills to both head off impending disease and calm his nerves, neither of which worked. Williams died in 1983 as a result of choking on the cap of an eye-drop bottle, though drugs and alcohol may have been involved as well.

No Hoopskirts, Please

An unusual woman with a passion for health care emerged as both a "dragon" and an "angel."

✳   ✳   ✳   ✳

ANYONE WHO THINKS that women's lives during the Civil War revolved around nothing more than tea and crinolines—or thinks that life ends at 50—should discover Dorothea Dix. A tireless crusader for the humane treatment of the mentally ill before the war, by age 59 in 1861, she'd written five books, founded 32 mental hospitals, and visited Western Europe, Russia, and Turkey. At the war's outbreak, she obtained the post of superintendent of women nurses for the Union army—until then, men had primarily served as army nurses. Dix's many responsibilities included nurse recruitment, organizing first-aid stations, purchasing supplies, and helping establish training facilities and field hospitals. She performed this work without pay for the duration of the war.

Female nurses were a brand-new concept, and U.S. military officials were skeptical about the ability of women to serve effectively. As a result, Dix—a rather severe woman herself—established rules for hiring and conduct that are quite amusing by modern standards. To dissuade the "wrong" kinds of ladies from rushing to her cause, she announced that she would hire only plain-looking women over the age of 30. The dress code she imposed allowed only black or brown dresses, no bows, no jewelry, no curls . . . and certainly no hoopskirts! This code helped to exclude young women who Dix saw as flighty or marriage-minded from the ranks. Although she may have gone a bit overboard with these restrictions, she did recruit more than 3,000 women, including Louisa May Alcott.

Dorothea Dix was greatly admired for her work, and she succeeded in giving her multitude of nurses a high reputation. A train she was riding was once stopped by Confederate troops but was allowed to proceed when a Southern officer recognized Dix and remembered her work with the mentally ill. On the flip side of the coin, however, she became equally well known in some circles for her prickliness. The daughter of a fire-and-brimstone preacher, Dix often quarreled with hospital bureaucracy and took male doctors to task for habits such as guzzling liquor on the job and dribbling tobacco juice on patients. After losing arguments with military administrators, she'd often ignore their deeply embedded protocols. Although soldiers called her an "angel of mercy," her detractors called her "Dragon Dix."

Inexplicably—and indeed, sadly—Dorothea Dix considered the Civil War chapter in her career to be a failure, as she did not force the acceptance of—and establish lasting rules for—female nurses, as she'd intended. Still, she tirelessly worked for human health causes and soldiers' rights well beyond the war's end, helping families track down missing soldiers, assisting wounded veterans in obtaining their pensions, and continuing her crusade for the mentally ill for another 20 years.

"In a world where there is so much to be done, I felt strongly impressed that there must be something for me to do."

—DOROTHEA DIX

Amazing Facts About the Human Body

✳ ✳ ✳ ✳

✳ The longest cells in the human body are the motor neurons. They can be up to four and a half feet long and run from the lower spinal cord to the big toe.

✳ The human eye blinks an average of 3.7 million times per year.

✳ The longest-living cells in the body are brain cells, which can live a human's entire lifetime.

✳ Fifteen million blood cells are produced and destroyed in the human body every second.

✳ The brain requires more than 25 percent of the oxygen that is used by the human body.

✳ If your mouth was completely dry, you would not be able to distinguish the taste of anything.

✳ The human body has enough fat to produce seven bars of soap.

✳ Food travels from the mouth, through the esophagus, and into the stomach in seven seconds.

✳ The pupil of the eye expands as much as 45 percent when a person looks at something pleasing.

✳ The human heart creates enough pressure while pumping to squirt blood up to 30 feet.

✳ The average adult body consists of approximately 71 pounds of potentially edible meat, not including organ tissue.

Chapter 20

Music

Big Band Leaders

✳ "It Don't Mean a Thing if It Ain't Got that Swing" was one of his most famous numbers, and swing Duke Ellington (1899–1974) certainly did.

✳ With mega-hits including "In the Mood," "Chattanooga Choo Choo," and "Moonlight Serenade," Iowan Glenn Miller (1904–1944) achieved big-band supremacy.

✳ Jimmy Dorsey (1904–1957) was content to remain with his band while younger brother Tommy branched out on his own. We owe hits such as "What a Difference a Day Made" and "I Believe in Miracles" to the fruitful pairing.

✳ William "Count" Basie's (1904–1984) band featured energetic ensemble work and generous soloing. "One O' Clock Jump" and "Jumpin' at the Woodside" kept toes a-tappin'.

✳ The younger half of the famed Dorsey Brothers, Tommy Dorsey (1905–1956) found shared success with such songs as "I Get a Kick out of You" and "Lullaby of Broadway." He later broke off on his own and scored again with "I'm Getting Sentimental over You" and "I'll Never Smile Again."

✳ Responsible for ditties such as "Hi-De-Ho" and "Minnie the Moocher," former law school student Cab Calloway (1907–1994) was known for his energetic "scat" singing and frequent appearances at the Cotton Club.

* Known as "King of the Vibes," percussionist/bandleader Lionel Hampton (1908–2002) elevated the vibraphone to first-class status. "On the Sunny Side of the Street" and "Hot Mallets" rank as two of his most popular songs.

* Gene Krupa (1909–1973) was one of the most influential drummers of the 20th century. He invented the arrangement of drums that would come to be known as the standard "kit" and raised drum solos to an art form.

* Benny "the King of Swing" Goodman (1909–1986) scored more than 100 hits during his career. "Sing, Sing, Sing," "Blue Moon," "Moonglow," and "Jersey Bounce" were among them.

* Born Arthur Arshawsky, Artie Shaw (1910–2004) was noted for hits such as "Begin the Beguine" and "Everything's Jumping." He was also noted for his eight marriages: Actresses Lana Turner and Ava Gardner took the plunge with Shaw. The last of the great big band leaders, Shaw died at age 94 on December 30, 2004.

* Once a contortionist in a traveling circus, trumpeter Harry James (1916–1983) twisted himself into a successful bandleader. His swinging version of "You Made Me Love You" turned fanatic teens into quivering masses.

* Considered by many to be the greatest drummer to ever pick up sticks, Buddy Rich (1917–1987) had a career that spanned seven decades. His caustic humor, which was as finely honed as his precision drum licks, made him a talk-show favorite.

* John Birks "Dizzy" Gillespie (1917–1993) was known for his comical antics—hence his nickname—in addition to his world-class trumpeting. With his trademark upturned horn, Gillespie ushered in the bebop era.

Maria Callas: Operatic Superstar

In the history of modern opera, there have been a handful of legendary performers whom even non-opera fans can name-drop. But perhaps the most famous of them all is Maria Callas.

✳ ✳ ✳ ✳

IN 1923, MARIA CALLAS was born in New York City to unhappily married Greek immigrant parents. Maria's mother, Evangelia, was a domineering and ambitious woman. When she learned that her chubby youngest daughter, Maria, could sing—and we mean *really* sing—she packed up her two daughters and returned to Athens (sans husband). There, Evangelia hoped to enroll Maria at the famous Athens Conservatoire. Maria wasn't accepted, but she continued to train. She eventually landed a spot at the Conservatoire, wowing the admissions committee with her powerhouse style.

Hardest-Working Diva in the Business

At the Conservatoire, Callas devoured librettos and scores for ten hours a day, learning her parts as well as those of the other singers. But this wasn't entirely because of her insatiable interest in music: She was near-sighted, so seeing the conductor was difficult (and glasses were not considered an option). In order to be able to follow along and not miss a beat, Callas had to know all the parts of every opera in which she took part.

In 1940, Callas signed on at the Greek National Opera. Two years later, the rising star landed a principal role in Eugen d'Albert's *Tiefland*. Reviewers and audiences were unanimous: Maria Callas was unlike anything they had ever seen. The emotion she brought to the roles she played was raw and real—a powerful combo of virtuoso voice and extraordinary acting.

A Rise to Fame, Fortune—and Drama

By the time she left Greece in 1945 at age 21, Callas had given 56 performances in seven operas and had appeared in 20 recitals.

Her teachers advised her to make a name for herself in Italy, the center of the opera world. After a short detour to the Metropolitan Opera in New York, she came under the guidance of Tullio Serafin, a maestro whom Callas credited with launching her career. Perhaps he did, but it was Callas's abilities that redefined the very concept of vocal range.

Her big break came when she stepped in as a replacement for a singer who fell ill before the opening of *I Puritani* in Venice. Never mind that she only had six days to learn the part and that she was already singing the large role of Brunnhilde in *Die Walküre*: It was a challenge she couldn't refuse.

Diets, Rumors, Heartbreak

Throughout the '50s, Callas dominated the Italian opera world, essentially launched the Chicago Lyric Opera with an inaugural performance of the lead in *Norma,* gave star turns in London, and was on the cover of *Time* magazine. But with increased exposure came scrutiny. Always a robust woman, she lost about 80 pounds mid-career. Rumors circulated that she had taken a tapeworm pill. There were stories of divalike behavior after a string of cancelled performances, lawsuits, and contract troubles. True, Callas was a force to be reckoned with, but those close to her knew that many of the stories were either embellished or totally fabricated.

Whether it was caused by the stress of being the world's most famous opera singer, the weight loss (some believed the diet contributed to her vocal decline), or simply the march of time, Callas's voice began to lose some of its luster during the '60s. Romantically, she had long been involved with shipping magnate Aristotle Onassis. But when Onassis moved on to Jackie Kennedy, Maria was devastated.

In 1977, at age 53, Maria Callas died in Paris of a heart attack. In 2007, the BBC named her the greatest soprano of all time and she posthumously won a Grammy Lifetime Achievement Award. And she's still one of opera's greatest legends.

Welcome to the Jungle: Animal-Led Bands

They may be hairy, smelly, and totally wild, but that's not all these singers have in common with most rock stars. In fact, a handful of musicians have employed actual animals to front their bands.

✳ ✳ ✳ ✳

Caninus

CLAIMING TO BE the world's first-ever animal-fronted band, Caninus features two pit bull terriers named Budgie and Basil performing with a bunch of death-metal dudes. The dogs, as one might expect, basically bark over loud, distorted guitars and lots of double-bass drumming.

Hatebeak

Another metal act, Hatebeak decided to go with a parrot vocalist named Waldo. The human members say they knew Waldo loved metal because he would stand on one leg any time they blasted their music. Fittingly enough, this band has worked on a joint project with the dogs from Caninus. Squawk 'n' roll!

K9 Fusion

Looking to make your neighbors good and angry? K9 Fusion is another excellent option to blare from your stereo. A dog and his canine pals supposedly play all the instruments on this animalistic project (with the exception of programmed drums added by their owner).

Thai Elephant Orchestra

Perhaps the most inventive of the animal bands, the Thai Elephant Orchestra gives a group of 16 elephants from the Thai Elephant Conversation Center the chance to play specially designed instruments. An American composer and a worker from the center came up with the idea. We suggest that they be forced to listen to eight straight hours of Hatebeak as punishment.

Behind the Music of Our Time

* The inspiration for Duran Duran's "Hungry like the Wolf" came from Little Red Riding Hood.

* Blondie singer Deborah Harry worked as a Playboy bunny before becoming a full-fledged pop star.

* Elvis scored an unimpressive grade of C in his junior high music class.

* Both pre-fame Elvis Presley and Buddy Holly didn't make the cut when they tried out for *Arthur Godfrey's Talent Scouts,* a 1950s-era talent show.

* One of the background singers on The Righteous Brothers song "You've Lost That Lovin' Feeling" was none other than a then-unknown Cher.

* The Dire Straits album *Brothers in Arms* has the honor of being the first CD to sell a million copies.

* Swedish pop group ABBA reportedly turned down a $1 billion offer to reunite.

* The '80s Toto hit "Rosanna" was written for actress Rosanna Arquette (sister of Patricia and David), who was rumored to have had a relationship with keyboardist Steve Porcaro.

* Singer and pianist Tori Amos was expelled from the prestigious Peabody Conservatory music school in Baltimore when she was 11 years old. Apparently she hated to read sheet music.

* The beach hit "Wipe Out" was written on a lark. The Surfaris needed a quick ditty to fill the B-side of their single "Surfer Joe." "Joe" didn't make much of a dent, but "Wipe Out" became a surfer standard.

* The Beach Boys' original band name was Carl and the Passions.

KISS and Makeup

Why were protective parents around the globe convinced that the name of the rock band KISS was really an acronym for "Knights in Satan's Service?"

✳ ✳ ✳ ✳

Branding a Band

ONSTAGE, THEY LOOKED like they'd come straight from the gates of hell, dressed head-to-toe in black, their faces adorned with macabre makeup. When KISS hit the road in 1973—playing to all of three people at their first gig—rock 'n' roll was undergoing an image transformation. The emergence of androgynous rockers such as David Bowie, along with the popularity of glam groups such as the New York Dolls, forced bands to find new, exciting, and controversial methods to market their product. When four young rockers from New York City combined comic-book characters and colorful costumes with a morbid mentality, they needed an appropriate handle to describe themselves—one that was easy to spell and mysterious to keep their fans confused. Drummer Peter Criss had been in a group called Lips, which prompted the crew to dub themselves KISS.

What's in a Name?

According to the boys in the band, the name was spelled in capital letters to make it stand out and was never meant to be an acronym for anything. But that revelation didn't stop members of religious flocks—who considered rock to be the sound of Satan—from claiming that the group's moniker was a devilish derivation. In fact, the KISS name has spawned several acronymic identities, including "Keep It Simple, Stupid," "Kids In Satan's Service," and "Korean Intelligence Support System." Judging from the millions of records they've sold in their nearly 40 years in the business, as well as their relentless licensing of KISS-related merchandise, a more appropriate name for the band might be CASH.

Former Names of 17 Famous Bands

See if you can match the former names to these famous bands.

❋ ❋ ❋ ❋

1. Cheap Trick	A. Angel and the Snake
2. U2	B. Atomic Mass
3. The Beatles	C. Tea Set
4. Styx	D. The Detours
5. Queen	E. Tom and Jerry
6. Led Zeppelin	F. Feedback, The Hype
7. The Beach Boys	G. Golden Gate Rhythm Section
8. Green Day	H. Smile
9. KISS	I. Unique Attraction
10. The Who	J. Fuse
11. Def Leppard	K. The Tradewinds
12. Pink Floyd	L. Sweet Children
13. Boyz II Men	M. Mookie Blaylock
14. Blondie	N. The New Yardbirds
15. Simon and Garfunkel	O. Wicked Lester
16. Journey	P. The Pendletones
17. Pearl Jam	Q. The Quarrymen

Answer Key: 1. J; 2. F; 3. Q; 4. K; 5. H; 6. N; 7. P; 8. L; 9. O; 10. D; 11. B; 12. C; 13. I; 14. A; 15. E; 16. G; 17. M

"I Will Hear in Heaven"

Beethoven's musical accomplishments should be enough to secure his place in history. However, his legacy has been embellished by the misconception that he was totally deaf when he composed his major works.

✳   ✳   ✳   ✳

LUDWIG VAN BEETHOVEN, the most famous composer of his era, was born in Bonn, Germany, in 1770. When he was in his early twenties, he moved to Vienna to study with Joseph Haydn, just after the death of his idol, Wolfgang Amadeus Mozart. In a letter to a friend in 1801, Beethoven wrote that his hearing had been deteriorating for at least three years and he was beginning to show signs of deafness.

Beethoven wasn't completely deaf until 1817. At that point, he had already composed his most celebrated works, including the second through eighth symphonies, the *Appassionata* Sonata, the *Emperor* Concerto, and his only opera, *Fidelio*. His poor hearing inevitably ended his career as a virtuoso pianist, but with the use of rudimentary hearing aids, he continued composing.

Beethoven died in 1827, having been totally deaf for ten years. His final words were allegedly, "I will hear in heaven."

✳ Once he was completely deaf, Beethoven's primary work was limited to the completion of the Ninth Symphony. Despite his deafness, when the work debuted in concert, he insisted on conducting the orchestra. Unbeknownst to him, a second conductor was also employed to beat out time. At the end of the performance, Beethoven was unable to hear any response, and he began to weep. It wasn't until someone took his arm and turned him around that he was able to see the audience applauding wildly.

Nobody Puts Josie in the Corner

It's hard to imagine Beyoncé being denied a table at a fancy club or restaurant. Nowadays, a major recording artist—regardless of race—can go into pretty much any place he or she wants and receive serious VIP treatment. But it wasn't always that way.

✳   ✳   ✳   ✳

One Busy Lady

BY THE TIME St. Louis native Josephine Baker was 19 years old, her legendary career was well under way. The exceedingly talented, exceptionally beautiful African American singer had a way with song, dance, and stage banter—and as a stripper, her sex appeal couldn't be denied. The total package catapulted her to fame first across Europe and then the United States, where she performed her act to sold-out crowds.

During World War II, Baker spent much of her time performing for troops in Africa, the Middle East, and her home country while still finding time to gather intelligence for the French Resistance. She was a tireless volunteer for the Red Cross, and after the war, she and her husband adopted 12 children from around the world—just under a Baker's dozen. Baker's wish was to make her family "a World Village." After touring the world, Josephine and her family returned to America in 1950 with the hope of creating a kind of multicultural utopia.

Sorry, Ma'am, We're Full

You'd think that with her significant fame and tireless war efforts, people would be falling all over themselves to offer the great Josephine Baker pretty much anything she wanted. Unfortunately, this was America in the 1950s, which meant that if you were black you were considered a second-class citizen.

In 1951, Baker wanted to dine at the famous Stork Club in New York City. The Stork Club was a "see-and-be-seen" establishment, full of the city's elite in entertainment, politics, and culture. Classy joint or not, the Stork Club refused to serve Baker simply on the basis of her skin color.

Baker didn't take the news well: She caused quite a row and was asked to leave, but not before being accused of being a Communist sympathizer (after all, this was America in the 1950s). Grace Kelly was at the club at the time, and she came to the aid of the embarrassed and furious Baker. Kelly vowed never to return to the club, and she kept her promise. After Baker left, with Kelly by her side, she vowed to get her revenge on the Stork Club by dedicating her life to the fight for equality for all people of color.

Igniting the Fight

Baker started by suing the club, thus forcing its discrimination policy into the open. She also began refusing gigs that didn't take place in integrated venues. This had a major effect on many establishments that couldn't afford *not* to change the policy. In particular, Las Vegas nightclubs were forced to look at their admittance policies, which set the ball rolling for much of the rest of the entertainment industry.

In 1963, Martin Luther King Jr. asked Baker to speak at his legendary March on Washington. Baker was the only woman speaker featured that day. As she surveyed the audience, she was pleased. "Salt and pepper. Just what it should be," she said. However, when King's widow, Coretta Scott King, asked her to step into her husband's shoes as leader of the American Civil Rights Movement, Baker declined. She wanted to raise her family and lead a quiet life. Indeed, Baker and her "Rainbow Tribe" of children lived in a castle in France in relative peace.

By the time of her death in 1975, Baker had made a career comeback. More than 20,000 people came to pay their respects at her funeral. To this day, Baker's name is invoked among legendary musicians, performers, and civil rights activists.

Fab Four Fast Facts

The Beatles—the best-selling band in history—formed from humble beginnings in England, in the late 1950s. Think you know everything about these lads from Liverpool? Read on for some lesser-known Fab Four facts.

* The Beatles' "I Want to Hold Your Hand" was the top-selling single of the 1960s.

* John Lennon's "(Just Like) Starting Over" is considered the best-selling posthumous hit of all time.

* Legend has it that Ringo Starr nearly missed his cue during the recording of "Hey Jude." The drummer left the studio to go to the loo right before his part was about to begin.

* The Beatles' classic "Yesterday" initially had the words "scrambled eggs" in its chorus until Paul McCartney came up with the magic word.

* The Beatles' "Lovely Rita Meter Maid" was inspired by a parking ticket that Paul McCartney received from a female officer—fittingly enough, while on London's Abbey Road.

* "A Day in the Life" ends with a high-pitched whistle that no human could possibly hear (although a dog could).

* Three different drummers played on recordings of The Beatles song "Love Me Do." Original drummer Pete Best played on the first recording, but when Ringo Starr replaced him in the band, the song was rerecorded. The story goes, though, that the producer didn't like Ringo's version, which led session drummer Andy White to step in. All three versions have been released on various albums over the years.

* On January 1, 1962, The Beatles auditioned at Decca Records, performing 15 songs in just under an hour. Producer Mike Smith flatly rejected them saying, "We don't like their sound. Groups of guitars are on their way out." Oops!

9 Fictional Bands with Hit Songs

Check out these fictional hit makers, but be prepared to get at least one song stuck in your head.

✳ ✳ ✳ ✳

1. **The Chipmunks:** The Chipmunks, a fictitious music group created by Ross Bagdasarian in 1958, featured singing "chipmunks" Alvin, Simon, and Theodore. The trio was managed by their human "father," Dave Seville. In reality, Dave Seville was the stage name of Bagdasarian, who electronically sped up his own voice to create the higher-pitched squeaky voices of the chipmunks. This process was so new and innovative that it earned a Grammy for engineering in 1959. The Chipmunks released a number of albums and singles, with "The Chipmunk Song (Christmas Don't Be Late)" spending four weeks atop the charts in the late 1950s.

2. **The Monkees:** Hey, hey...were the Monkees a real band? In 1965, auditions were held for "folk & roll musicians" to play band members on a new TV show called *The Monkees*. Actors Davy Jones and Micky Dolenz and musicians Mike Nesmith and Peter Tork were chosen to be The Monkees. The show won two Emmy Awards in 1967, and the band was so successful that it went on tour— with The Jimi Hendrix Experience as their opening act! The Monkees reached the top of the charts with the hits "I'm a Believer," "Last Train to Clarksville," and "Daydream Believer." Although the show was canceled in 1968 and the band officially broke up in 1970, they have continued to record and tour with some or all of the original members.

3. **The Archies:** Stars of the comic strip *Archie* and the Saturday morning cartoon *The Archie Show*, The Archies were a garage band "founded" in 1968. Band members included fictional characters Archie, Reggie, Jughead, Betty,

and Veronica. Producer Don Kirshner gathered a group of studio musicians to perform the group's songs. They scored with "Sugar, Sugar," which topped the pop charts in 1969 and was named *Billboard* magazine's song of the year, the only time a fictional band has ever claimed that honor. The Archies also reached the top 40 with "Who's Your Baby?," "Bang-Shang-a-Lang," and "Jingle Jangle."

4. **The Kids from *The Brady Bunch*:** Marcia, Marcia, Marcia! The music group known as The Kids from *The Brady Bunch* was made up of—who else—the young cast members from the popular sitcom that originally aired from 1969 to 1974. The cast recorded several albums, including *Christmas with the Brady Bunch* and *Meet the Brady Bunch*. None of the songs topped the charts, but some fan favorites include "Sunshine Day," "Time to Change," and "Keep On."

5. **The Partridge Family:** *The Partridge Family*—a popular TV show that aired in the early 1970s—focused on Shirley Partridge (played by Shirley Jones) and her brood of five children, who performed as a band. To promote the show, producers released a series of albums by The Partridge Family. Although the music was originally created by studio musicians with Jones singing backup, David Cassidy, who played eldest son Keith, quickly convinced producers to let him sing lead vocals. The show and the band became overnight sensations, making Cassidy a teen idol. The group's most popular hits include the show's theme song "C'Mon, Get Happy," "I Woke Up in Love This Morning," and "I Think I Love You," which spent three weeks at number one in late 1970.

6. **The Blues Brothers:** In April 1978, *Saturday Night Live* cast members John Belushi and Dan Aykroyd appeared on the show as The Blues Brothers. Dressed in black suits, fedoras, and sunglasses, Belushi sang lead vocals as "Joliet" Jake Blues while Aykroyd portrayed Elwood Blues, singing

backup and playing harmonica. Their first album, *Briefcase Full of Blues*, went double platinum and reached number one on *Billboard*'s album chart. The record produced two top 40 hits with covers of Sam and Dave's "Soul Man" and The Chips'"Rubber Biscuit." The band went on tour, even opening for The Grateful Dead. In 1980, Belushi and Aykroyd starred in *The Blues Brothers*, a feature film that chronicled the life of the fictional duo. Belushi died in 1982, but The Blues Brothers live on with Jim Belushi (John's brother), John Goodman, and other guests stepping in to fill Jake's shoes.

7. **The Heights:** *The Heights*, a TV show about a rock 'n' roll band of the same name, aired for only one season in 1992. By day, the characters worked blue-collar jobs, but at night, they lived their musical dreams. The show's theme song, "How Do You Talk to an Angel," which featured actor Jamie Walters on lead vocals, topped the charts in mid-November. Ironically, the show was canceled a week later.

8. **The Wonders:** When your band's name is The Oneders (pronounced "The Wonders") and everybody calls you The Oh-nee-ders, it's time to change your name. That's exactly what happened in the 1996 hit movie *That Thing You Do!*, written and directed by Tom Hanks. The movie's theme song—also called "That Thing You Do!"—went as high as number 18 on *Billboard*'s chart and was nominated for an Oscar for Best Original Song. In addition, the sound track reached number 21 on *Billboard*'s album chart.

9. **Gorillaz:** *Guinness World Records* named Gorillaz the most successful virtual band after its 2001 debut album, *Gorillaz*, sold more than six million copies. Created in 1999 by musician Damon Albarn and graphic artist Jamie Hewlett, this alternative rock band is made up of animated characters 2D, Murdoc, Noodle, and Russel. In 2006, the band's second album, *Demon Days*, received five Grammy nominations and won for Best Pop Collaboration with Vocals.

International Opera House

A thick black line runs across the center of the Haskell Free Library's reading room and on the floor above, under the seats of the opera house upstairs. But this isn't any old line—it's the United States–Canada border, which happens to slice straight through this neoclassical, three-story building.

❋ ❋ ❋ ❋

WALK THROUGH THE library's front door, and you are officially in Derby Line, Vermont. But if you check out books—your choice of English or French—you must do so in Stanstead, Quebec. If you are upstairs watching a show in the opera house, you will probably be sitting in the United States, but you would be applauding performers onstage in Canada.

A Vision of Unity

In 1904, American sawmill owner Carlos Haskell and his Canadian wife, Martha Stewart Haskell, built the Haskell Free Library and Opera House. Martha, the wealthy granddaughter of one of Derby's founding families, was born in Canada, lived in the United States, and had business interests in both. The couple's idea was that by building this institution across the border, it could be used by people of both countries, and the profits from the opera house would support the operation of the library.

When the Haskells passed away, the building was donated to the towns in which it is located. It is run by a private international board of four Americans and three Canadians that governs both the building and its facilities. Registered in 1976 on the National Register of Historic Places for Orleans County in Vermont, the Haskell Free Library and Opera House is also on the list of Provincial Heritage Buildings in Quebec. Modeled after the former Boston Opera House, this unique structure boasts painted scenes of Venice on its drop curtains in the opera hall and two-foot-thick walls made with granite from Stanstead.

The 400-seat theater has its drawbacks, though. There's no orchestra pit, incendiary special effects are not allowed, and it has tiny dressing rooms. There's also no air-conditioning, and the wooden chairs are hard—unless, of course, you pay 50 cents to rent one of the pink sofa cushions offered by the ushers.

Split Personality

Keith Beadle, one of the trustees, grew up using the library. He says that running a building straddling two countries can be a logistical problem, noting that maintenance issues are particularly difficult. Most work must be done by hiring companies from both sides of the border or the tradespeople will be working illegally at least part of the time. However, in a recent renovation, Canadian immigration agreed to let American workers make some improvements as long as they came in through the front door (on the American side). Still, when the American crews were up on the roof, they were on the Canadian side, and labor unions wanted to fine the contractor $8,000. They later resolved the dispute amicably.

The Sign of the Times

The turn-of-the-century building has always been a point of pride in the two communities. Much like Carlos and Martha Haskell, many townsfolk have intermarried, and some share relatives on both sides of the border.

But recent world events are pressing down hard on these two small towns. As part of the Department of Homeland Security, American border officials have stricter regulations. What worries the locals is that authorities may tighten the loose agreements that the two villages share, which allow them to come and go from the library without reporting to customs, even if they cross the border one or more times while inside.

But in the meantime, kids still hop back and forth over the black line, and tourists still take pictures of themselves straddling the border, with one foot inside Canada and one inside the United States.

Behind the Music of Our Time

❋ Buddy Holly's real name was Charles "Buddy" Holley. Decca Records misspelled his last name on his original recording contract, so he decided to stick with the e-free version.

❋ A printing mistake also led to Dionne Warwick's last name: The diva was Dionne Warrick until her first single, "Don't Make Me Over," came out with a typo. She also opted to stick with the altered alias.

❋ The main guitar riff from Guns N' Roses' only number-one hit, "Sweet Child O' Mine," actually was a result of guitarist Slash doing a musical exercise. Singer Axl Rose happened to hear him playing it and decided to turn it into a song. Slash always hated the idea.

❋ The rock 'n' roll classic "Johnny B. Goode," which was written by Chuck Berry, originally contained the controversial lyric "that little colored boy can play." It was later changed to "that little country boy can play."

❋ In the mid-1950s, before she became a famous author/poet and one of Oprah's best pals, Maya Angelou recorded an album called *Miss Calypso*.

❋ Sister Luc-Gabrielle, who was known as The Singing Nun and became internationally famous in 1963 with her hit "Dominique," committed double suicide with her lesbian lover in 1985.

❋ Dolly Parton and Sylvester Stallone teamed up in the film *Rhinestone*, which was an epic flop. But they also recorded a duet—"Stay Out of My Bedroom"—for the soundtrack.

❋ None of the Taylors in Duran Duran—Andy, John, and Roger—are related by blood.

❋ "I Will Survive" was originally slated to be a B-side for Gloria Gaynor's single "Substitute."

Blind Man's Bluff

Known for his soaring tenor, Roy Orbison was equally famous for his horn-rimmed dark glasses. Since he was rarely seen sans shades, many people believed the sweet-singing Roy was blind. But was he? Or was that just part of his persona?

❋ ❋ ❋ ❋

WHEN ORBISON'S FLASHY falsetto began to fly up the charts, he shared radio time and record sales with two other performers who had similar vocal talents and stage attire. Like Ray Charles and Stevie Wonder, "The Big O" wore dark glasses when he performed onstage. Unlike the aforementioned artists, though, Orbison was not sightless. Back in the day, it was rare for performers to wear any kind of eyewear onstage, and most people assumed that artists who wore dark glasses did so because they were blind. Although Orbison's eyesight was decidedly less than 20/20, he could see relatively well, and there was no devious dupe behind his choice of eye apparel. In 1963, he was asked to tour Europe with a potpourri of artists, including The Beatles. When Orbison arrived in England, he realized that he had left his regular glasses in the United States, so he was forced to wear his prescription sunglasses. He wore the shades onstage, which garnered a favorable reaction from both fans in the seats and beat writers reviewing the concerts. The rest is history.

❋ Orbison's personal life was marred by tragedy. In 1966, his wife died as a result of a motorcycle accident, and in 1968, two of his three sons died in a house fire. Offstage, he was a taciturn and humble man who tended to shun the spotlight. But contrary to rumor, Orbison did not hide behind his dark shades. Early in his career, he was often photographed without his trademark frames and was noticeably lensless when he starred in the 1967 film *The Fastest Guitar Alive.*

Marian Anderson at the Met

Even with all the post-production and computer-aided technology that modern pop stars have at their disposal, it seems improbable that they could move a crowd of thousands to tears with the sheer sound of their voices. But from 1915 to 1965, opera singer Marian Anderson did just that on a regular basis.

✳ ✳ ✳ ✳

MARIAN ANDERSON WAS born into a poor black family in Philadelphia in 1897. She was the first African American opera singer to perform at the New York Metropolitan Opera, and she even served as a U.S. goodwill ambassador because of the impressive and unifying quality of her voice and career. Singing at the opera may not seem like such a big deal, but Anderson did it during the time of Jim Crow laws, when arrangements were "separate but equal" for African Americans.

Anderson sang for sold-out audiences, but she often had to stay at the homes of fans because African Americans weren't allowed in many hotels; if she was allowed to stay at a hotel, she had to use the freight elevator to get to her room. She only received formal voice training to learn arias and *leiders* because kind musicians recognized her talent and tutored her for free (or, in the case of her church, donated money toward her education). "Reputable" music schools wouldn't look past the color of her skin.

International Success

Anderson was not afraid to travel, and going to Europe truly allowed her to shine. Opera fans adored her, the press raved about her, and she was treated like a celebrity while selling out shows. In 1936, she was even invited to perform in Nazi Germany—until her "hosts" confirmed that she wasn't 100-percent white.

After touring Europe, Anderson had a chance to be a superstar in the United States. However, she again faced racial adversity when the Daughters of the American Revolution (DAR) refused to allow her to perform in Constitution Hall, which the organization owned. Controversy ensued, which led First Lady Eleanor Roosevelt to resign from the DAR and help set up a dramatic and historic free concert performed to a mixed-race crowd at the base of the Lincoln Memorial.

Conquering the Met

Never one to comment on the injustice surrounding her career or the state of the country, Anderson continued to focus on her voice and touring. In 1954, at age 57, she was finally invited to perform as the gypsy Ulrica in the opera *Un Ballo in Maschera* at the New York Metropolitan Opera.

In a review of her performance, one reporter wrote, "There were those who wished the night had happened ten years earlier when the contralto was at her peak." He wasn't alone: Many reporters expressed their wonderment as to why it had taken so long for Anderson to finally get a chance to perform at the Met—some 20 years after she became a superstar.

Her popularity peaked at this time, but Anderson jumped at the chance to serve as a U.S. emissary with the United Nations when a big fan of hers, President Dwight Eisenhower, invited her to do so. Throughout Anderson's diplomatic tour of the world, America's racial policies were addressed by foreign citizens, who wondered how she could represent a country in which she herself could not be entirely free. Anderson simply explained in her calm way that she was very saddened by the situation.

Today, it's hard to imagine what life was actually like for Marian Anderson; we take for granted many things that she never got to experience. Anderson may not have been as visible as Martin Luther King Jr., but her faith, family support, talent, elegance, and silent contribution to civil rights causes made her an icon.

The Double Life of Billy Tipton

Not long ago, if a woman wanted to make a life for herself in the music business, she had to face quite a few obstacles. Read on for the incredible story of one person who didn't let gender stand in the way of a lifelong dream.

✳ ✳ ✳ ✳

It's a Girl!

DOROTHY LUCILLE TIPTON was born in 1914 in Oklahoma City to parents who were already on the brink of divorce. After the Tiptons split, Dorothy, or "Tippy," as her friends called her, was sent to live with an aunt in Kansas City. It was there that she first fell in love with music. Tippy tried to join the high school band but was told that girls weren't allowed.

Tippy moved back to Oklahoma, and for the majority of her young adulthood, she studied piano and saxophone on her own while working odd jobs to pay the bills. But her music career was going nowhere on account of her gender, so Tippy decided to take a risky step that would change the course of her life.

That's "Mr. Tipton" to You

In 1933, at age 19, "Tippy" started dressing as a man and adopted the name "Billy" so as to be taken seriously at jazz band auditions. The ruse was successful: Soon Tipton was cross-dressing for her professional career. By binding her breasts and getting creative with an athletic supporter and a sock, Tipton was now fully identifying (and identifiable) as a male.

Although it wasn't easy passing as a man, it began to pay off. Tipton quickly gained popularity as a talented musician, and the jobs started to roll in. Tipton scored radio performances, posh hotel gigs, national tours, and album deals. He also shared the stage with prestigious names such as The Ink Spots, the Delta Rhythm Boys, and big-band leaders such as Billy Eckstine. Tipton was officially a successful musician—with a big secret.

Love at First Sight

Initially, Tipton continued to dress as a woman in her private life but eventually did it full time. Incredibly, no one could tell that underneath the suit and tie was a female body—including more than a few women. After Tipton became "Billy," he had long-term relationships with women—including five wives—who never knew Tipton's true identity.

For seven years, Billy and a woman named Betty Cox lived together in what Cox has described as a rewarding, heterosexual relationship. Tipton told Cox that his genitals had been damaged in a car accident and that he would forever have to bind his chest as a result of upper-body injuries. This was enough for Cox, who said Tipton was "the most fantastic love of my life."

In 1962, Tipton married an exotic dancer named Kitty Oakes. Oakes said that they never had sex on account of her ill health and Tipton's "injuries." The couple eventually adopted three children. Tipton was a fully engaged parent, involving himself in the PTA, the Boy Scouts, and every charity event in town. The couple divorced in 1980.

A Secret Revealed

In 1989, when Tipton passed away at age 74 from hemorrhaging ulcers, coroners discovered his long-held secret. Oakes swore that she had no knowledge of Tipton's true gender, but her shocked sons believed differently. Oakes ran immediately to the Spokane newspapers and pleaded with them to keep the story under wraps, but she was too late—one of the Tipton boys had already agreed to speak to the press. The sensational story made it all the way to *The New York Times*. But Tipton's double life has served as an inspirational story to some: He's now regarded as a poster boy/girl for transgendered people around the world.

How It Happened: Band Names

These legendary rock groups threw around a couple of options before landing on their now-famous names.

✳ ✳ ✳ ✳

Deep Purple

Heavy metal pioneers Deep Purple went through nearly as many name changes in their early days as lineup changes in the years to follow. After taking names such as The Flower Pot Men and Their Garden, The Ivy League, and Roundabout, guitarist Ritchie Blackmore suggested the name Deep Purple, which had been his grandmother's favorite song from 1933.

Lynyrd Skynyrd

This Southern rock band has had its share of triumph and tragedy—including a plane crash that took the lives of three band members. The band was first known as The Noble Five and later My Backyard. Eventually, they settled on their moniker, a play on the name of Leonard Skinner, a gym teacher at Robert E. Lee High School in Jacksonville, Florida, who was known for a strict policy against boys having long hair.

U2

While it is sometimes rumored that the Irish quartet was named after the American U-2 spy plane, the fact is that the members—who had been playing under the name The Hype—chose their moniker because it was ambiguous and somewhat open-ended. In the end, the members actually just picked the name they disliked the least.

AC/DC

This group has a rather simple explanation for the origin of their name—that of the electrical "alternating current/direct current." Members say that they saw the abbreviation on the back of a sewing machine and felt that it described their band's raw energy.

Stage Names vs. Real Names

Study the list below, and see if you can match the entertainer with his or her birth name. No cheating!

❋ ❋ ❋ ❋

1. Bono

2. Busta Rhymes

3. C. C. DeVille

4. Chubby Checker

5. Dido

6. Elvis Costello

7. 50 Cent

8. Freddie Mercury

9. Gene Simmons

10. Liberace

11. MC Hammer

12. Moby

13. Pink

14. Seal

15. Shania Twain

16. Slash

17. Snoop Dogg

18. Sting

A. Alecia Moore

B. Bruce Anthony Johanssen

C. Calvin Broadus

D. Chaim Witz

E. Curtis Jackson

F. Declan Patrick McManus

G. Eileen Regina Edwards

H. Ernest Evans

I. Florian Cloud De Bounevialle Armstrong

J. Frederick Farookh Bulsara

K. Gordon Matthew Sumner

L. Henry Olusegun Olumide Samuel

M. Paul David Hewson

N. Richard Melville Hall

O. Saul Hudson

P. Stanley Kirk Burrell

Q. Trevor Tahiem Smith

R. Wladziu Lee Valentino

Answer Key: 1. M; 2. Q; 3. B; 4. H; 5. I; 6. F; 7. E; 8. J; 9. D; 10. R; 11. P; 12. N; 13. A; 14. L; 15. G; 16. O; 17. C; 18. K

Chapter 21

Food and Drink

World's Healthiest Foods

Four Best Brain Foods

1. **Blueberries**: Blueberries are rich in antioxidants, which may help fend off the degenerative effects of conditions such as Alzheimer's disease.

2. **Wild Salmon**: Salmon is loaded with omega-3 fatty acids—the kind of fat that helps create and strengthen brain tissue.

3. **Spinach**: Eating a can of spinach may not make your muscles bulge like Popeye's, but it will increase serotonin and dopamine levels in the brain.

4. **Flaxseed**: Sharpen your senses with a tablespoon of flaxseed oil per day. Its high levels of alpha-linolenic acid help improve cerebral cortex functioning.

Four Unexpectedly Healthy Foods

1. **Red Wine**: *Vive la France!* A substance found in red wine called resveratrol has anti-inflammatory properties and helps to protect against heart disease and cancer.

2. **Chocolate**: Chocolate has been shown to lower blood pressure, fight heart disease, and defend against cancer. Only dark chocolate and cocoa powder work, though— so put down that Snickers bar.

3. Blue M&M'S: Okay, so M&M'S aren't good for you, but the dye found in blue M&M'S has been shown to slow paralysis in rats with spinal cord injuries.

4. Coffee: Coffee has been proven to markedly improve memory and cognition.

The Spice of Life

Four spices have been shown to have powerful health properties.

1. Cinnamon: Recent studies indicate that this common spice helps control blood sugar levels, making it a good choice for type 2 diabetes sufferers.

2. Ginger: For centuries, ginger has been used to treat stomach ailments.

3. Cayenne Pepper: Here's good news for those who like spicy food: Cayenne pepper has been shown to alleviate circulatory problems and help the body fight infections.

4. Turmeric: Since ancient times, turmeric has been used to treat conditions ranging from diarrhea to skin disease.

Lowest Rates of Heart Disease Deaths, by Country

The typical American diet isn't particularly heart-healthy, as illustrated by the fact that Americans suffer more than 106 heart disease–related deaths annually per every 100,000 people. Folks at risk for heart disease might want to check out the diets of the people in the countries listed below, which have the fewest heart disease–related deaths per 100,000 people:

1. Japan—30

2. France—39.8

3. Spain—53.8

4. Portugal—55.9

5. Belgium—64.6

16 Things You Didn't Know About Popcorn

High in fiber, low in fat, and a tiny demon in every kernel. Here are 20 things you probably didn't know about popcorn.

✳   ✳   ✳   ✳

1. Popcorn's scientific name is *Zea mays everta*, and it is the only type of corn that will pop.

2. People have been enjoying popcorn for thousands of years. In 1948, popped kernels that were around 5,000 years old were discovered in caves in New Mexico.

3. It is believed that the Wampanoag Native American tribe brought popcorn to the colonists for the first Thanksgiving in Plymouth, Massachusetts.

4. Traditionally, Native American tribes flavored popcorn with dried herbs and spices, and possibly even chili. They also made popcorn into soup and beer, and crafted popcorn headdresses and corsages.

5. Christopher Columbus allegedly introduced popcorn to Europe in the late 15th century.

6. The first commercial popcorn machine was invented by Charles Cretors in Chicago in 1885.

7. American vendors began selling popcorn at carnivals in the late 19th century. When they began to sell the snack outside movie houses, theater owners were annoyed, fearing that popcorn would distract their patrons from the films. It took a few years for them to realize that popcorn could be a way to increase revenues. Popcorn has been served in movie theaters since 1912.

8. Nowadays, many movie theaters make a greater profit from popcorn than they do from ticket sales: For every dollar spent on popcorn, around 90 cents is pure profit. Popcorn also makes moviegoers thirsty, and thus more likely to buy sodas.

9. Native American folklore speaks of spirits that live inside each kernel of corn. When heated, or "angered," the spirits explode in a "puff," often considered an omen of bad luck.

10. What really makes popcorn pop? Each kernel contains a small amount of moisture. As the kernel is heated, this water turns into steam. Popcorn differs from other grains in that the kernel's shell is not water-permeable, so the steam cannot escape. Pressure builds up until the kernel finally explodes, turning inside out.

11. Unpopped kernels are called "old maids" or "spinsters."

12. There are two possible explanations for old maids. The first is that they didn't contain sufficient moisture to create an explosion; the second is that their outer coatings were damaged, so that steam escaped gradually, rather than with a pop. Good popcorn should produce less than 2 percent old maids.

13. Americans consume 17 billion quarts of popcorn each year. That's enough to fill the Empire State Building 18 times.

14. Nebraska produces more popcorn than any other state in the country—around 250 million pounds per year.

15. Popped popcorn comes in two basic shapes: snowflake and mushroom. Movie theaters prefer snowflake because it's bigger. Confections such as caramel corn use mushroom because it won't crumble.

16. According to *Guinness World Records*, the world's largest popcorn ball measured 12 feet in diameter and required 2,000 pounds of corn, 40,000 pounds of sugar, 280 gallons of corn syrup, and 400 gallons of water to create.

How Sweet It Is!

In the early 1900s, American entrepreneurs brought an exclusive treat of the wealthy to masses of children.

✳   ✳   ✳   ✳

THE ORIGINS OF cotton candy are a bit muddled, with sources pointing to four different possible inventors. Just before 1900, John Wharton and William Morrison received a patent for a machine that melted sugar and then used centrifugal force to stretch it into thin strands. They sold almost 70,000 boxes of their "fairy floss" at the 1904 St. Louis World's Fair for 25 cents a pop—a hefty sum for a candy treat in those days. Meanwhile, Thomas Patton received a patent for a similar machine; he supposedly peddled his fluffy treats for the Barnum & Bailey Circus. To further complicate the story, Josef Delarose Lascaux is said to have had a homemade cotton candy machine in his dentistry office. How convenient.

Who is the true inventor of the famous spun-sugar treat? Well, it's not Lascaux, Patton, or Wharton and Morrison—they simply automated a process for spinning sugar that had been used for at least 150 years. Recipes for creating wispy strands of solidified sugar by hand date to at least 1769. Around that time, confectioners provided meticulous instructions that detailed how to melt the sugar and then "take small portions and pass it quickly to and fro to form threads over an oiled rolling pin held in the left hand. A fork is best to use to take up the sugar." The process was labor-intensive and, therefore, expensive. But confectioners skilled in the art of hand-spinning sugar could create elaborate "nests" as Easter decorations and sparkling silvery-gold strands to top off desserts for their well-to-do clients.

Still, the accomplishments of these early 20th-century inventors were significant. They took a decadent candy, typically sold to few in small quantities at high cost, and turned it into an iconic treat enjoyed by millions of young people.

Does Processed Food=Junk Food?

Processed food hasn't always been synonymous with junk food.
In fact, in earlier eras, it was actually the safest kind of food to eat.

✳   ✳   ✳   ✳

THESE DAYS, "PROCESSED FOOD" usually means junk food—
something with little (if any) nutritional value because the
healthful parts have been removed or unhealthful ingredients
have been added. Fresh, unprocessed food is considered
superior—especially if it's organic. Despite its negative connotation, however, processed food is not always devoid of nutrition
or bad for you. Processed foods are not all alike, and fresh food
is not always safe. Myths about food abound.

Processing in the Good Old Days

Processed food isn't a contemporary phenomenon. Food
processing, in the form of salting and pickling, had been going
on for thousands of years to keep larders full between harvests.
It was the *unprocessed* foods that could be dangerous and even
deadly, because they were subject to spoiling and rotting. From
the ancient Egyptians, who used yeast to make bread and brew
beer, to the French confectioner who invented canning circa
1800 to keep Napoleon's troops supplied during their march
across Europe, civilization has invented ways to feed itself by
preserving (i.e., processing) food.

Modern Manipulation

In the 20th century, technology created a mass market for processed foods with previously unheard-of shelf lives. But this convenience came at a cost: Large amounts of trans fats, sodium, and
sugar were added, and manufacturing techniques often removed
nutritional components such as vitamins and fiber. Many snack
foods and fast foods are exceptionally high in calories and low in
nutritional value. And nitrites added as preservatives to cold cuts
and hot dogs may cause cancer. That's how processed food got a
bad rap and became synonymous with junk food.

Pretzel Facts

In the 1,400 years since the pretzel was invented, bakers have come up with a wide variety of shapes and flavors. The history of this contortionistic snack shows its versatility.

✳   ✳   ✳   ✳

✳ In A.D. 610, while baking bread, an Italian monk decided to create a treat to motivate his distracted catechism students. He rolled out ropes of dough, twisted them to resemble hands crossed on a chest in prayer, and baked them. The monk christened his snacks *pretiola*—Latin for "little reward." Parents who tasted their children's classroom treats referred to them as *brachiola*, or "little arms." When pretiola arrived in Germany, they were called *bretzels*.

✳ Perhaps because of its religious roots, the pretzel has long been considered a good-luck symbol. German children wear pretzels around their necks on New Year's Day. In Austria in the 16th century, pretzels adorned Christmas trees, and they were hidden along with hard-boiled eggs on Easter morning.

✳ The phrase "tying the knot" comes from the Swiss, who still incorporate the lucky pretzel in wedding ceremonies. Newlyweds traditionally make a wish and snap a pretzel in two.

✳ In Austria, signs outside many bakeries depict a lion holding a pretzel-shaped shield. According to a legend that dates back to 1510, pretzel bakers working before dawn heard Ottoman Turks tunneling under Vienna's city walls and then sounded an alarm. The city was saved, and the bakers were awarded their unique coat of arms by the Viennese king.

✳ Hard pretzels came about in the late 1600s, when a snoozing apprentice in a Pennsylvania bakery accidentally overbaked his pretzels.

* In 1861, Julius Sturgis opened the first commercial pretzel bakery in Lititz, Pennsylvania. He was given an original pretzel recipe as a thank-you from a down-on-his-luck job seeker after Sturgis gave the man dinner.

* Until the 1930s, pretzels were handmade, and the average worker could twist 40 per minute. In 1935, the Reading Pretzel Machinery Company introduced the first automated pretzel machine, which enabled large bakeries to make 245 pretzels per minute, or five tons in a day.

* More than $550 million worth of pretzels are sold in the United States annually; 80 percent of them are made in Pennsylvania.

* The average U.S. citizen consumes up to two pounds of pretzels per year, but Philadelphians snack on about 12 pounds of pretzels per person annually.

* Joey "Jaws" Chestnut ate 21 soft pretzels in 10 minutes during a pretzel-eating competition in 2007.

* In 2002, President George W. Bush was munching on a pretzel in the White House when he choked and lost consciousness while watching an NFL playoff game between Baltimore and Miami.

* Medieval street vendors carried pretzels on a stick and sold them to locals. Today, soft pretzels remain a popular push-cart item in Philadelphia and New York City.

* Joe Nacchio of Federal Baking in Philadelphia holds the record for baking the largest pretzel, which was 5 feet across and weighed 40 pounds.

Pepsi's Mad Scientists

In their quest to unseat Coca-Cola, the folks at Pepsi Co have created some pretty bizarre beverages.

✳ ✳ ✳ ✳

Crystal Pepsi

IN 1992, PEPSI released Crystal Pepsi, a colorless, caffeine-free cola that confused the heck out of people. The enormous flop lasted only a year on the market.

Pepsi Blue

In 2002, Coca-Cola released Vanilla Coke. Pepsi countered with their Pepsi Blue. This blue-tinted soda was said to contain "berry flavors," although just which berries was never specified. Despite massive advertising campaigns, Pepsi Blue tanked and left most markets by 2004.

Pepsi Fire and Pepsi Ice

In early 2005, fire and ice (in cola form) started appearing on grocery shelves all across Asia. Pepsi Fire had a flavor not unlike that of hot cinnamon candies, while Pepsi Ice had a bizarre "minty aftertaste." Not surprisingly, teens dared each other to mix the two together and drink "fire and ice" concoctions.

Pepsi Ice Cucumber

In Japan in 2007, Pepsi released a limited-edition flavor that spokespeople said was designed to make "people think of keeping cool in the summer heat." Apparently, that meant Pepsi Ice Cucumber. The beverage did not contain any real cucumbers but rather the "refreshing taste of a fresh cucumber."

Pepsi White

In 2008, a new product hit Japanese markets: Pepsi White. The drink's unusual flavor was supposed to be a combination of Pepsi and yogurt; however, the milky white concoction didn't taste like either one. Pepsi White was soon labeled a "limited edition" and was gone by year's end.

Sweet Lemons? It's a Miracle!

Get ready for a flavoriffic experience!

✳ ✳ ✳ ✳

MIRACLE FRUIT (SYNSEPALUM DULCIFICUM) is a small berry that can change the way common foods taste. After eating a miracle berry, stout beer tastes like a chocolate milkshake and cheese tastes like cake frosting. It can make hot sauce taste like a glazed donut, vinegar taste like apple juice, and oysters taste like chewing gum.

Miracle fruit is indigenous to West Africa. The berries can be freeze dried or refrigerated indefinitely. The fruit can also be ordered online in tablet or granulated form. The berry's glycoproteins bind to the tongue's taste buds, producing miraculin, which makes bitter and sour foods taste sweet. The effect lasts from 30 minutes to 2 hours.

Miracle fruit first gained popularity in the United States during the 1970s, when it was marketed as a diet aid. So why aren't we all eating the fruit? Allegedly, the Food and Drug Administration folded to pressure exerted by the sugar industry and stopped allowing its import. But if you would like to have your own miracle fruit, you can order it from online vendors.

There are four steps to enjoying miracle fruit:

1. Buy a selection of foods such as citrus fruits, rhubarb, bleu cheese, stout beer, and cheap tequila.

2. Wash the miracle fruit and put a berry into your mouth. Swirl it around for about a minute.

3. Bite into the berry, coating your tongue with its juice.

4. Taste, say, a lemon wedge—it should taste like sweetened lemonade.

Time Spent Wasting Time

Here's how much time is spent on everyday activities.

✳ ✳ ✳ ✳

Breathing: Not exactly "wasted" time, if you consider the alternative. Numerically speaking, if you live to be 75 (and assuming that you expire at the same exact time you were born, mind you), the time you'll spend breathing equates to 900 months or 3,900 weeks or 27,300 days or 655,200 hours or 39,312,000 minutes or 2,358,720,000 seconds.

Eating: Americans sure do love to eat. We stuff our faces for an average of 67 minutes per day. That comes out to roughly 24,455 minutes—about 407 hours—per year.

Housework: Ask the folks at the U.S. Department of Labor's Bureau of Labor Statistics, and they'll tell you that married moms spend about two hours per day on activities such as housework, preparing dinner, and gardening. Assuming a 50-year marriage, that's 36,500 hours. To the surprise of no wife anywhere in the United States, married men reported doing about half this amount of work.

Internet: Americans spend about 10½ hours per month, or 126 hours per year, browsing websites while at home.

Sleeping: By age 75, most people who sleep an average of 7 hours per night will have snoozed away 22 years— about a third of their lifetime. Incidentally, a 1900 news article put that total at 23 years and 4 months.

Stuck in Traffic: You might want to bring along a good book (like this one) the next time you head out on the open road. Every year, the Texas Transportation Institute conducts a study of 439 urban areas across the United States. In 2010,

Americans spent a grand total of nearly 4.8 billion hours stuck in gridlock. Residents in urban areas with populations of more than one million people average 42 hours per year in traffic. The delays are worst in the Washington, D.C., area, where people spend about 74 hours per year staring at the brake lights in front of them. Take heart, residents of Buffalo, New York... you spend a mere 17 hours per year cursing your fellow motorists.

Texting: We can thank technology for giving us something new to obsess over. According to CTIA: The Wireless Association, Americans sent and received approximately 2.12 trillion text messages in 2011—an average of almost 176.7 billion per month. That's about 5.8 billion texts per day. One girl in California reportedly sent and received 14,528 texts in a single month.

Watching Television: According to a 2009 A.C. Nielsen Research study, the average American spends about 153 hours every month at home watching the boob tube, up 1.5 percent from the year prior. A 2010 study found that the average American now spends 35 percent more time watching TV *while* using the Internet (way to multitask, America) than they did in 2009. Of course, since you're reading this book, these statistics probably don't apply to you.

Working: How does 89,784 hours sound? That's given an average eight-hour workday and a career that spans 43 years. That total may be skewed these days, however, given rising unemployment and the fact that technology has us connected to the office 24 hours a day, 7 days a week. For what it's worth, on average, American men work about an hour longer each day than their female coworkers.

The McDonald's Kingdom

In the restaurant biz, there's no greater success story than that of McDonald's. But there are plenty of people who believe the introduction of the "take-away meal" has ruined Americans' health. Regardless of which side of the bun you're on, there's no denying that the origins of McDonald's make for a great story.

❊ ❊ ❊ ❊

Setting the Record Straight

IF YOU THOUGHT Ray Kroc opened the first McDonald's, you're wrong. In 1940, brothers Dick and Maurice McDonald opened the first McDonald's on Route 66 in San Bernardino, California. As was common at the time, carhops served made-to-order food to hungry patrons. But in 1948, the McDonalds fired the carhops and implemented an innovative "Speedee Service System," a technique that streamlined the assembly process and became the benchmark for premade hamburgers. This process allowed profits to soar. The brothers charged 15 cents for hamburgers.

In 1953, the brothers decided to franchise their restaurant, and the second McDonald's opened in Phoenix, Arizona. This was the first Mickey D's to sport the famous golden arches. A year later, an entrepreneur and milkshake-mixer salesman named Ray Kroc visited. He was impressed with the enterprise, and he immediately joined the team. In 1955, he founded McDonald's Systems, Inc., and opened the ninth McDonald's restaurant in Des Plaines, Illinois. Six years later, Kroc bought the business from the McDonald brothers for $2.7 million. The poorly constructed deal stipulated that Dick and Maurice could keep the original restaurant but somehow overlooked their right to remain a franchise. Because of this error, Kroc opened a McDonald's restaurant down the block from the original store, and within a short time he drove the brothers out of the hamburger business.

Expansion

McDonald's restaurants spread like wildfire, and in 1967, the first McDonald's outside the United States opened in Richmond, British Columbia. Ten years later, McDonald's was operating on four continents. In 1992, a McDonald's was opened in Casablanca, Morocco.

In 1974, along with former Philadelphia Eagles player Fred Hill, McDonald's founded the Ronald McDonald House, an organization that aids families of critically ill children seeking medical treatment. As of 2008, there were 259 outlets.

Hold the Mayo

All has not gone smoothly for the Mc-empire over the past several years, however. In 2000, Eric Schlosser published *Fast Food Nation*, a critical commentary on fast food in general and McDonald's in particular. This was followed by several lawsuits: one regarding obesity (claiming McDonald's "lured" young children into their restaurants with playgrounds) and others asserting various claims of damaged health due to saturated fats.

A huge sensation followed the 2004 release of the film *Supersize Me*, a documentary that accused fast-food restaurants of ignoring America's escalating obesity crisis. Later that year, in response to the public's desire for healthier food, McDonald's did away with its super-size meals and implemented more chicken options. In 2005, it added a range of salads and low-sugar drinks and agreed to put nutritional information on all of its packaging.

In addition to its new, more health-conscious menu, McDonald's has also tightened its belt by closing restaurants to compensate for occasionally sagging financial reports. But not much deters McDonald's, which, for the foreseeable future, will remain the king of the world's fast-food empire.

That's Offal! 7 Dishes Made from Animal Guts

For anyone raised on the standard American hamburger, the idea of eating the brains, guts, or feet of a cow may seem gross and totally out of the question. But in many parts of the world, those are the parts of the animal that are considered delicacies.

✳   ✳   ✳   ✳

1. **Kokoretsi**: In Greece, it's all about lamb. Skewered, baked, roasted—if it's lamb, the Greeks are cooking it. But these folks would never waste good meat: Kokoretsi is a traditional Balkan dish often served for Easter that includes lamb intestines, hearts, lungs, or kidneys, or a combination of any of the above. Chunks of these organs are speared onto a skewer and then wrapped up in the small intestine, which forms a kind of sausage casing. The skewer is set over a fire and sprinkled with oregano and lemon juice. In a few hours, *opah!* Grilled guts for everyone!

2. **Haggis**: On paper, haggis just doesn't look appetizing: mix a sheep's heart, liver, and lung meat with oatmeal and fat, then stuff the mixture into the animal's stomach. Boil for three hours and enjoy. Traditionally served in a sauce with turnips and potatoes, this dish is also used in the popular Scottish pastime known as "haggis throwing."

3. **White pudding**: To make white pudding (which is common in Scotland, Ireland, Nova Scotia, and Iceland), you'll first need a big bowl of suet, or pork fat. It's the main ingredient in this dish, which also includes meat and oatmeal, and, in some older recipes, sheep brains. The pudding is similar to blood pudding, but without the blood. Sometimes formed into a sausage shape, white pudding can be cooked whole, fried, or battered and served in place of fish with chips.

4. Tripe: Served in many countries, the stomach, known as tripe, is the main ingredient in many regional dishes. Beef tripe is the most commonly used, but sheep, goat, and pig stomachs are often on the menu as well. Tripe is used often in soups and in French sausages, fried up in Filipino dishes, and used as a relish in Zimbabwe. In Ireland and Northern England, tripe is simply served with onions and a stiff drink.

5. Khash: If you're a cow, you'd better watch your step. Folks in Armenia are crazy about khash, a dish made primarily from cow's feet. First, the hooves are removed. Next, the feet are cleaned and cooked in plain boiling water overnight. By morning, the mixture is a thick broth and the meat has separated from the bone. Brain and stomach bits can be added for extra flavor. Armenians are careful about when they prepare this dish—it reportedly has strong healing properties and is usually served only on important holidays. Peppers, pickled veggies, and cheese go well with khash, but the most popular accompaniment is homemade vodka.

6. Yakitori: Who doesn't like a juicy grilled chicken skewer? Look closely if you purchase one from a Japanese street vendor, however. Yakitori are chicken skewers that contain more parts of the chicken than you may care to taste, including the heart, liver, gizzard, skin, tail, small intestine, and tongue.

7. Rocky Mountain Oysters: There aren't a lot of parts of a farm animal that you *can't* eat, as evidenced by Rocky Mountain Oysters, or bull testicles. After the testicles of a bull (or lamb or buffalo) are removed, they're peeled, dipped in flour, and deep-fried to a golden crunch. This dish is commonly found in the American West, where bulls are prevalent, and also in bull-populated Spain.

That's the Way the (Fortune) Cookie Crumbles

Perhaps it's fitting that a cookie that foretells the future has such a murky past. Who invented the fortune cookie? There's a debate over that. But here are some interesting facts about fortune cookies that can't be disputed. (Well, maybe just a little.) Confucius recommends that you read on; it will bring you health and prosperity.

✳ ✳ ✳ ✳

✳ According to journalist Jennifer 8. Lee, more than three billion fortune cookies are made each year, the vast majority of them in the United States. Lee should know: She's the author of *The Fortune Cookie Chronicles*, a 2008 book that traces the history of Chinese food in the United States.

✳ Fortune cookies are a mainstay in the United States, but they are also served in Britain, Italy, France, and Mexico. Surprisingly, it's extremely difficult to find fortune cookies in China. That's because the fortune cookie actually traces its origins back to Japan, not China.

✳ No one knows for sure who introduced the fortune cookie to the United States, but two entrepreneurs are given credit for it. One legend says that Japanese immigrant Makoto Hagiwara introduced fortune cookies to America in 1914 in San Francisco. A second story credits David Jung, founder of the Hong Kong Noodle Co. Legend suggests that Jung introduced the cookie in 1918 in Los Angeles. According to the story, he was concerned about the number of poor people living on the streets, so he passed out free fortune cookies to them. Each cookie contained an inspirational verse written by a Presbyterian minister.

✳ In 1983, San Francisco's Court of Historical Review held a mock trial to determine whether Hagiwara or Jung should get credit for bringing fortune cookies to U.S. diners. Not surprisingly, the judge ruled for San Francisco and Hagiwara. A piece of evidence that surfaced during the trial was a fortune saying, "S.F. judge who rules for L.A. not very smart cookie."

* Wonton Food, Inc., in Long Island City, New York, is the largest producer of fortune cookies in the United States. The factory churns out 4.5 million cookies every day. The company also boasts a database of 10,000 possible fortunes. Company officials say that only about 25 percent of these fortunes are used at any given time.

* The recipe for fortune cookies is surprisingly simple. The batter used to make them is a mixture of flour, sugar, and vanilla or citrus flavoring.

* Fortune cookies have proven to be fertile ground for jokes. One of the oldest and most popular? A diner cracks open a fortune cookie to find the following message inside: "Help! I'm trapped in a fortune cookie factory."

* Today, fortune cookies come in a wide variety of flavors. Diners can munch on fortune cookies that are covered in chocolate or caramel. Many bakeries also sell fortune cookies decorated for Christmas, Valentine's Day, and other holidays.

* Some fortune cookies don't contain fortunes at all. Crack one open, and you'll often find lucky lottery numbers or a philosophical message: "No one is richer than he who has many friends." Some fortune cookies even contain riddles or jokes.

* Unless you really love them, you won't gain too much weight eating fortune cookies. The average fortune cookie contains about 30 calories and no fat. So eat away!

Meet the Real Chef Boyardee

American consumers see a lot of familiar faces on store shelves when they go shopping. Among the most famous is the mustachioed Chef Boyardee, the king of canned pasta. Many people assume he's just another fictitious corporate spokesperson, but in fact, he was a real person.

Must Be the Mustache

Born in Piacenza, Italy, in 1897, Ettore Boiardi demonstrated a talent for cooking at an early age. He emigrated to the United States at age 16, took the name Hector, and proved his culinary prowess by working in the kitchens of several well-known restaurants, including the Plaza and Ritz-Carlton hotels in New York City and the Greenbriar in West Virginia, where he is said to have catered Woodrow Wilson's second wedding. He was often the kitchen's youngest chef, so he grew his trademark mustache to make himself appear older.

From the Kitchen to the Grocery Store

At 24, Boiardi moved with his family to Cleveland, where he worked at the Hotel Winton. A few years later, he opened his own restaurant, Il Giardino d'Italia, to rave reviews. Customers adored his spaghetti, and by popular demand, Boiardi started selling his homemade sauce in take-out containers.

The sale of his pasta sauces eventually exceeded his restaurant's business, so in 1936, Boiardi—who had grown tired of explaining the pronunciation of his last name—started selling prepackaged foods under the "Chef Boyardee" label. In 1946, he sold the operation to American Home Foods for $6 million. Today, the Chef Boyardee brand is owned by ConAgra, which continues to put Boiardi's face on its pasta products, including spaghetti and meatballs, ravioli, and lasagna. Boiardi died on June 21, 1985, but his legacy lives on in the millions of cans of heat-and-serve pasta sold each year in stores all over the world.

I Scream, You Scream, We All Scream for... Asian Ice Cream?

Ever have a taste for red bean ice cream? How about black sesame? These flavors—and many others that Westerners might find odd—are popular in Asian countries. Here are some of the more popular, and not-so-popular, Asian ice cream flavors.

✳ ✳ ✳ ✳

✳ **Red Bean:** It shouldn't come as a surprise to see beans used to flavor ice cream. After all, vanilla is a bean, right? Asian red bean ice cream is a bit different, though, mainly because this ice cream has real red beans inside it. Red beans are a common ingredient in Asian sweets. Ever have sweet red bean soup for dessert in a Chinese restaurant? Consider red bean ice cream part of the same family of Asian sweets.

✳ **Taro:** The taro plant, in its raw form, is considered toxic. Once it's cooked, though, the plant is perfectly safe to eat. That's a good thing, considering that taro-flavored ice cream is a popular treat in Asian countries. Taro ice cream actually uses the plant's root for its flavor. This root is lavender in color and has the texture of a potato. Taro ice cream itself, which is usually purple, boasts a nutty flavor.

✳ **Durian:** In Southeast Asia, the durian fruit is known as the "king of fruit" because it's so large and tasty. It does, however, have a drawback: It stinks. The fruit's odor is so overpowering, in fact, that some hotels in Asia have strict policies forbidding durian fruit on their premises. The taste is another story, though: Durian fruit has a rich custardlike flavor.

✳ **Horse Meat:** Even the most adventurous of Westerners might shy away from basashi ice cream. You can find this strange treat at the Ice Cream City shop in Tokyo. It's vanilla ice cream made with shreds of raw horse flesh. Yummy.

15 Fast Food Facts

For better or for worse, Americans love fast food. Here are some bite-size facts on the supersize industry.

✳ ✳ ✳ ✳

1. French fries weren't introduced at McDonald's until 1949. Until then, potato chips were the chain's side dish of choice.

2. Burger King claims that there are 221,184 possible ways its Whopper hamburger can be ordered.

3. "Taco Bell" comes from a family name: Glen Bell started working on the chain in the late 1940s in San Bernardino, California (the same city where McDonald's was born). His first venture was a hot dog stand called Bell's Drive-In. In the early '50s, he started adding Mexican items to the menu, eventually opening a secondary restaurant called Taco Tia.

4. The first actual Taco Bell opened in 1962. Glen Bell stepped down as chairman in 1975, and PepsiCo, Inc., bought the chain three years later.

5. Perhaps the most recognizable fast-food icon, Ronald McDonald first appeared in 1966 when Mickey D's aired its first television commercial. The Hamburglar, Grimace, and Mayor McCheese joined him five years later.

6. Wendy's joined the fast-food mix in 1969 when founder Dave Thomas opened his first restaurant in Columbus, Ohio. By 1976, there were 500 Wendy's locations sprinkled across the country. The number doubled to 1,000 by 1978.

7. Wendy's was named for Dave Thomas's daughter, Melinda Lou "Wendy" Thomas.

8. Arby's entered the restaurant biz in 1964. The first location opened in Boardman, Ohio, and featured the roast beef sandwiches that are still the chain's signature item.

9. The name "Arby's" actually represents the initials "R" and "B." The letters stand for "Raffel Brothers," in homage to founders Leroy and Forrest Raffel, although the company says that many suspect it also stands for "roast beef."

10. McDonald's has its own university (of sorts). Hamburger University opened in 1961. Graduates receive "Bachelor of Hamburgerology" degrees, which we're sure are incredibly impressive on any résumé. More than 5,000 employees attend the school each year.

11. Feeling the need to adapt to the variety of items offered by competitors, McDonald's altered its menu for the first time in 1963, adding the Filet-O-Fish, followed by such famous additions as the Big Mac and apple pie.

12. KFC's iconic founder, Colonel Sanders, was never in the military. The company claims that Kentucky Governor Ruby Laffoon named Harland Sanders an honorary colonel "in recognition of his contributions to the state's cuisine."

13. The Double Six Dollar Burger at Carl's Jr. is the unhealthiest hamburger option in America, according to an analysis by *Men's Health* magazine. It packs a whopping 1,520 calories and 111 grams of fat.

14. Chipotle's Mexican Grilled Chicken Burrito takes the prize for unhealthiest Mexican entrée, *Men's Health* says. One of these will put 1,179 calories into your body.

15. According to the book *Eat This, Not That*, Chick-fil-A is the healthiest overall fast-food chain. Subway, Jamba Juice, and Au Bon Pain also rank high on the list.

Living the Freegan Lifestyle

Move over vegetarians and vegans—there's a new eco-conscious lifestyle in town.

✳ ✳ ✳ ✳

The Freegan Basics

WHETHER YOU LIKE it or not, first-world countries in the West are monsters of consumption. We eat a lot, take up a lot of space, build a lot of stuff, and generally spend an incalculable amount of money on things we don't *really* need.

All that consumption can be annoying to anyone, but a new subculture of people who call themselves "freegans" are beyond fed up, and they're determined to opt out of the game altogether.

Using a combination of the words *free* and *vegan,* the freegan lifestyle was first applied to food. Freegans feel that the amount of perfectly good food thrown away by restaurants, grocery stores, and everyday folks is despicable when so many people in the world are starving. They grocery-shop in garbage cans and swear that they eat like kings simply by salvaging food from the trash. They never spend a penny, and they cut down on the waste produced by most everyone.

Granted, there aren't a lot of people interested in searching through trash bags for dinner; nevertheless, the freegan lifestyle spread across the United States and into Europe as well. Soon, "urban foragers" were discussing their philosophy online.

Look, Ma: No Money!

Most freegans aren't content with reducing the world's waste just by noshing on semi-wilted veggies. Many of these anti-consumerists find other ways to live off the fat of the land.

Clothes and appliances are easily scavenged. Dumpster divers boast of acquiring cast-off sporting equipment, artwork, furniture, musical instruments, and computers—much of it in near-perfect or perfect condition. Those handy with a needle and thread transform the apparel they find into designerlike duds. And many freegans participate in "freemeets," where anti-capitalists meet to trade goods they've found; no money is exchanged, so it is, quite literally, a free-for-all.

All of this swapping and foraging means these folks have little use for money, which means that the majority of freegans don't have jobs. One of the tenets of the freegan philosophy is that working as a slave to a corporation or unscrupulous business is both demoralizing and to blame for the disastrous results of mass consumerism—i.e., if you work for these people, you're implicated, too. Freegans don't let anyone off the hook.

Many hardcore freegans are squatters, living in abandoned, condemned, or otherwise vacant houses or buildings. Others are content to live in homes that are communal structures. Many freegans don't take their lifestyles quite this far, however: Quite a few live in apartments and houses they've furnished with items they've rescued from the curb.

The Freegan Flack

Not everyone thinks the scavenger way of life is a good thing. Some critics can get past the whole "I-found-this-tasty-sandwich-in-the-dump" part but take issue with the heart of the freegan philosophy. Some feel that living off the waste of others isn't what the freegans like to call "symbiosis," but more like "freeloading." Others point out the inherent hypocrisy in the freegan lifestyle: Freegans may effectively reduce a small amount of waste, but if there wasn't any waste, there wouldn't be any freegans. Still others claim that if freegans would buy goods from stores in the first place, there wouldn't be as much to throw away.

Although the world continues to get "greener" as the public's eco-consciousness grows, we're still a long way from being a waste-free society. As long as there are trash bins, there will be people to dig through them—whether you call them Freegans, saints, thieves, or nuts is really a matter of opinion.

Food Fight!

Sure, parties are fun, but add a lot of food and a reason to throw it around, and you've got yourself an international festival!

✳   ✳   ✳   ✳

Hunterville Huntaway Festival, Hunterville, New Zealand: The highlight of this annual October event is the Shepherd's Shemozzle, a grueling endurance and obstacle race in which shepherds and their dogs run through a course that includes a variety of culinary challenges, such as swallowing raw eggs and downing a delicious bowl of sheep's eyes mixed with cream. The winnings include a monetary prize and a supply of dog food.

La Tomatina, Buñol, Spain: Each year on the last Wednesday in August, thousands of people descend upon this sleepy little village for the world's most impressive tomato fight. Nearly 140 tons of juicy red tomatoes are trucked in from the countryside. The trucks dump off the tomatoes, the people fasten their goggles, and the fun begins. A word of advice: Don't wear your best clothes for this one.

Ivrea Carnival and Orange Battle, Ivrea, Italy: This is just like Spain's La Tomatina, only with oranges instead of tomatoes. It takes place 40 days before Lent, and it typically involves thousands of orange-hurling celebrants. Don't forget to duck— oranges can hurt!

Olive Oil Wrestling Competition, Edirne, Turkey: For centuries, hundreds of burly Turks have donned trousers made of water buffalo hide, slathered themselves with slippery olive oil, and wrestled each other. These contests occur throughout the country, but the most famous tournament takes place in the town of Edirne. The winner receives a cash prize and the right to call himself "Champion of Turkey."

West Virginia RoadKill Cook-off, Marlinton, West Virginia: This festival, which typically occurs in September, is exactly what its name suggests. One can try culinary delights such as Thumper Meets Bumper, Asleep at the Wheel Squeal, and Tire Tread Tortillas. Bottom line: If a critter's been smacked by a vehicle, it's as good as covered with gravy. Bon appétit!

Festival Gastronomico del Gato, La Quebrada, Peru: In this small Peruvian village, every September 21 cats go from favorite pets to delicious main dishes as part of this bizarre celebration (the name translates to "Gastronomic Festival of the Cat"). Participants sauté Mr. Whiskers to honor the village's original settlers: slaves who, at one time, were forced to live on cat meat.

Night of the Radishes, Oaxaca, Mexico: Held each year on December 23, this event is more of an art show than anything else as participants carve huge, gnarly radishes into elaborate figures. Ever wonder what the Virgin Mary would look like if she was carved into a radish? Here's your chance to find out.

Lopburi Monkey Festival, Lopburi, Thailand: This place is literally overrun with macaques, which in turn attract a lot of tourists. To show their appreciation, the locals host a party for the monkeys in November at the Prang Sam Yot temple. Thousands of pounds of food are presented to the pampered primates, including fruit, boiled eggs, cucumbers, and even soft drinks to wash it all down. Entrance to the festival costs 30 baht (approximately 90 cents). The ticket price includes a bamboo stick to ward off the more aggressive simians.

Demon-Chasing Ritual, Japan: Some Japanese are a bit superstitious, so every February 3, as winter gives way to spring, they chase away the evil from the previous year with a ritual known as *setsubun*. The ceremony involves shouting "Out with the demons, in with good luck!" while tossing roasted soybeans out the front door or at a family member wearing an Oni demon mask. Participants are also supposed to eat a soybean for every year they've been alive, plus an extra one for the coming year.

Religion and Food

Eating food is a universal human experience, so it's not surprising that people apply religious taboos to their diets. To uncover the connection between food and faith, we consulted someone in the know.

Q: What does *kosher* mean?

A: Kosher is a series of rules for food selection (no pork or shellfish, no dairy and meat at the same meal), harvesting (specific methods of slaughter and blood removal), and preparation (separate utensils for meat and dairy dishes). Kosher Jews (typically Conservative, Orthodox, and Chasidic, in ascending order of strictness) don't willingly or knowingly eat food that is traif—or nonkosher. The basic rules come from the Torah, with lots of arcane rabbinical clarification hashed out over the millennia.

Q: Why can't Hindus eat cheeseburgers?

A: Because cows are revered in Hinduism for their gentle nature and life-giving milk, the last thing a Hindu would do is kill one, much less eat one.

Q: Why do so many Catholics eat fish on Fridays?

A: It isn't so much eating fish as abstaining from meat. Devout Catholics make this sacrifice as an act of contrition for the sins of the past week, and Friday is chosen because it was the day of the Crucifixion— Jesus' sacrifice. Many Catholics abstain from eating meat on Fridays only during Lent.

Q: Who has the strictest dietary code?

A: Probably the Jainists. They eat according to their scripture: "Do not injure, abuse, oppress, enslave, insult, torment, torture, or kill any creature or living being." Jainist monks won't step on a bug, and they wear masks to keep from swallowing even a gnat. Some Jainists won't eat vegetables if harvesting kills the plant.

Beer Time

In 1933, when newly elected president Franklin D. Roosevelt lifted the country's ban on alcohol, he famously said, "I think this would be a good time for a beer." For some states, however, Prohibition continued long after that proclamation.

✻ ✻ ✻ ✻

THE VOLSTEAD ACT, also known as the Prohibition Act of 1919, gave U.S. authorities power to enforce the 18th Amendment to the Constitution, which banned the sale, manufacture, and transportation of intoxicating beverages. These were defined as any drink containing more than 0.5 percent alcohol by volume. The 18th Amendment took effect in 1920 and heralded an American era forever associated with gangsters, bootleggers, and speakeasies.

The Prohibition movement evolved from the religion-based Temperance movement of the late 19th century, in part as a response to the explosion in the number of saloons in the United States around the time of World War I. A number of counties and states, particularly in the South, adopted local prohibition laws prior to the national ban on alcohol in 1920. Prohibition quickly became unpopular, though, and created a nightmare for law-enforcement officials. In March 1933, shortly after he was elected president, Franklin Roosevelt amended the Volstead Act with the Cullen–Harrison Act, which allowed the manufacture and sale of "light" wines and 3.2 percent beer. In December 1933, he ratified the 21st Amendment to the Constitution, which ended national Prohibition.

In addition to being the first and only amendment to repeal a previous amendment to the Constitution, the 21st Amendment enabled individual states to use their own discretion in deciding when to repeal Prohibition. So, depending on where you lived, Prohibition may not have ended in 1933. Mississippi, the last dry state, did not repeal it until 1966.

Food Name Origins

Most food names are pretty straightforward: No one has to ask what's in a ham and cheese sandwich, and there's no confusion when it comes to pasta marinara. But some names might cause the cautious to get some backstory before opening wide.

✳ ✳ ✳ ✳

Deviled Eggs

EGGS MAY BE smelly sometimes, but what is it that makes deviled eggs so evil? This dish—in which halved egg whites are filled with mashed-up egg yolk, and sometimes topped with garnishes—was devilishly spicy in its original permutation. The mashed yolk is supposed to be blended and topped with an especially spicy ingredient, usually cayenne pepper.

Shepherd's Pie

The story behind the name of this popular dish seems straightforward at first: Shepherds herd lambs, and shepherd's pie is made of minced lamb. The peculiar thing is that it's not really a pie. The minced meat is covered with potatoes instead of crust, a practice that derives from the recipe's likely origin in peasant families of Northern England, who made this "pie" out of cheaper ingredients.

Sloppy Joes

This dish, which consists of ground meats mixed with sauce and spices and served on a bun, is definitely sloppy, but no one's quite sure who Joe was. Sloppy Joes first became popular as a cheap sandwich in rural areas, and from there became a diner staple. There was a popular restaurant called "Sloppy Joe's," which may be where the name came from—or it might just be an evocation of good old average Joe.

Baked Alaska

In 1804, physicist Benjamin Thompson Rumford discovered that stiffly beaten egg whites are resistant to heat. Thus began the creation of desserts in which ice cream was baked while covered in meringue, which is made from egg whites. The dessert would emerge from the oven steaming hot, but with rock-hard ice cream still frozen inside. Rumford deemed his invention "omelet surprise," but subsequent chefs of a more poetic inclination called it Baked Alaska.

Thousand Island Dressing

It turns out that this dressing owes its name to the Thousand Islands region in upstate New York, which actually boasts 1,800 small islands that stretch up into Canada. Back in the early 1900s, famous actress May Irwin was served this scrumptious salad dressing while at a dinner party in the Thousand Islands region. She named the dressing and spread the word.

Grasshopper Pie

There are no insects in this pie, which is filled with cream and créme de menthe, a mint-flavored alcohol. It was created in diners in the 1950s, a time when mint milk shakes and liquors were also in vogue. A popular alcoholic beverage known as a grasshopper was also made with créme de menthe. Some speculate that companies that wanted to promote their mint cream products invented the dessert.

---

* Incan women prepared beer by chewing grains and fruits; their saliva served to break down the sugars and begin the fermentation process.

* Ben and Jerry's Ice Cream company gives its ice cream waste to local hog farmers to use as feed. Apparently the hogs like every flavor except Mint Oreo.

* In 18th-century Europe, beer was a vital part of a meal for young and old. Pupils at London boarding schools were even served bread and beer for breakfast.

10 Weird and Wacky Food Festivals

Every year across North America, people gather to celebrate local food specialties. If you're tired of the same old rib and apple fests, you might want to check out some of these more, ahem, unusual food festivals.

✳   ✳   ✳   ✳

1. **Rocky Mountain Oyster Festival, Throckmorton, Texas:** Yep, this is what you think it is. If you find yourself in the mood for some bull testicles next May, the good people of Throckmorton, Texas, gather each year to celebrate these tasty treats. Enjoy!

2. **Turkey Testicle Festival, Huntley, Illinois:** For those of you who didn't get enough animal testicles in Texas, there's no need to worry—stop by the Parkside Pub in Huntley, Illinois, in November, and scoop up a handful.

3. **WAIKIKI SPAM JAM, Oahu, Hawaii:** Like meat in a can? You're in luck! Every April, Hawaiians celebrate SPAM with a street fair sponsored, not surprisingly, by Hormel.

4. **BugFest, Raleigh, North Carolina:** Each September, this celebration of all things arthropod is hosted by Raleigh's Museum of Natural Sciences. If you go, make sure to stop by Café Insecta for some stir-fried grasshoppers.

5. **RC Cola and Moon Pie Festival, Bell Buckle, Tennessee:** To Southerners, RC Cola and Moon Pies go hand in hand—they're a match made in heaven. To honor the long marriage of these treats, Bell Buckle, Tennessee, hosts the RC Cola and Moon Pie Festival each June. Events include a RC Cola/Moon Pie recipe contest and the cutting of the world's largest moon pie. Somehow, we don't think that too many people who have filled up on RC Cola and Moon Pies will be participating in the festival's opening event: a 10-mile run.

6. **Dandelion Festival, Dover, Ohio:** Most lawn owners consider the lowly dandelion to be a troublesome weed. Not so in Dover, Ohio, where the plant is actually celebrated each May during the Dandelion Festival. Dandelions reign supreme during this two-day event, which features dandelion wine-tastings and the chance to sample dandelion-inspired cuisine, including dandelion pizza, lasagna, jelly, and even dandelion ice cream.

7. **Rattlesnake Hunt, Waurika, Oklahoma:** Since 1961, celebrants have gathered in Waurika, Oklahoma, in April to hunt rattlesnakes and partake of a local delicacy: deep-fried rattlesnake meat. Organizers say this festival is "fangtastic!"

8. **Chitlin' Strut, Salley, South Carolina:** To many, the weekend after Thanksgiving officially kicks off the Christmas shopping season. But to folks in Salley, South Carolina, it's time to celebrate all things chitlin'. Events on tap for the Chitlin' Strut often include a beauty pageant, a Chitlin' Strut Idol contest, and an antique tractor show. Oh, and plenty of pig intestines.

9. **National Baby Food Festival, Fremont, Michigan:** If you're an adult who loves to eat baby food, then Fremont, Michigan, is the town for you. Every summer, baby-food aficionados convene in this Midwestern town to participate in baby-food cook-offs, bingo, and its main event: the Adult Baby Food Eating Contest.

10. **What the Fluff?, Somerville, Massachusetts:** Marshmallow Fluff is an integral ingredient in Rice Krispies treats and many other desserts. And nobody loves Fluff more than the residents of Somerville, Massachusetts. Each September, they celebrate the sticky stuff with the What the Fluff? festival. Attractions include a Marshmallow Fluff cooking contest and a tug-of-war battle in which the losing team falls into a pit of the gooey confectionary treat.

Freaky Facts: Taste Buds

❋ Babies are born with taste buds on the insides of their cheeks. They also have more taste buds than adults, but they lose some as they grow older.

❋ Adults have, on average, around 10,000 taste buds, although an elderly person might have only 5,000.

❋ One in four people is a "supertaster" and has more taste buds than the average person—more than 1,000 per square centimeter.

❋ Twenty-five percent of humans are "nontasters" and have fewer taste buds than other people their age—only about 40 per square centimeter.

❋ A taste bud is 30 to 60 microns (slightly more than $^1/_{1000}$ inch) in diameter.

❋ Taste buds are not just for tongues—they also cover the back of the throat and the roof of the mouth.

❋ Cats' taste buds cannot detect sweetness.

❋ The "suction cups" on an octopus' tentacles are covered in taste buds.

❋ A butterfly's taste buds are located on its feet and tongue.

❋ Attached to each taste bud are microscopic hairs called microvilli.

❋ Taste buds are regrown every two weeks.

❋ About 75 percent of what we think we taste actually comes from our sense of smell.

❋ Along with sweet, salty, sour, and bitter, there is a fifth taste called *umami*, which describes the savory taste of foods such as meat, cheese, and soy sauce.

Chapter 22

Religion

Mary and Her Message

Did the Virgin Mary appear to children in France and Portugal? Catholics seem to think so.

❋   ❋   ❋   ❋

PIOUS CATHOLICS HOLD a special place in their hearts for Mary, the mother of Jesus, praying to her daily for favors and blessings. Some say that the Virgin has appeared to them right here on Earth. Many of these claims are not verified— they are dismissed as products of overactive imaginations or as outright hoaxes. However, two incidents that defy scientific explanation have stood the test of time and remain highly cherished by Catholics around the world.

Our Lady of Lourdes

In February 1858, a poor, sickly 14-year-old peasant girl named Bernadette Soubirous was gathering firewood near a stream when she suddenly had a vision of a beautiful lady dressed in white. Overcome with fear, she rushed home to tell her mother, who told her to keep away from the place. However, Soubirous returned and repeatedly saw the Virgin—18 times in all. One message stood out: A chapel must be built on the site where Soubirous had first seen Mary.

Soubirous's parish priest was highly skeptical of the visions and dismissed the little girl's story. In spite of ridicule, Soubirous stuck to her account. It was only after people began reporting that their ailments had been cured after washing in the stream where the visions occurred that the Catholic Church decided to endorse the apparition. A shrine was built on the site; to this day, it is visited annually by millions of pilgrims who flock to Lourdes hoping to cure their physical ailments. Hundreds of miraculous healings have been reported, all of them verified by church and medical experts.

Our Lady of Fátima

In May of 1917, Lucia Santos and her cousins Jacinta and Francisco were tending their sheep in the town of Vila Nova de Ourém in the parish of Fátima, Portugal. Suddenly, they saw a tremendous flash of light. Thinking it was lightning, the children rushed for cover, only to see the same flash again. The children described seeing "a lady more brilliant than the sun." This was the first in a series of Marian apparitions reported by the children. Mary impressed upon the children the importance of daily prayer (especially reciting the rosary) and penance. She also told them there would be a second world war that would be much worse than the first. When World War II began almost two decades later, many saw it as the fulfillment of the prophecy. The Virgin also gave the children a brief glimpse of hell and revealed a mysterious secret, which Church authorities kept under wraps until 2000. (It was revealed to be a vision of the deaths of the pope and other individuals.)

Word of the apparitions soon spread, and on October 13, 1917, a crowd of 70,000—believers and skeptics alike— flocked to Fátima on the hot rumor that a miracle was about to occur. They were not disappointed. Newspaper reports of the day document that onlookers saw the sun burst through rain clouds and then begin dancing and spinning across the sky in a zigzag pattern, trailed by a brilliant array of colors. The so-called "Miracle of the Sun" solidified belief in the apparition, and a shrine was built on the site. Each year, scores of pilgrims visit the location, hoping to receive the graces of Mary.

One Bad Apple

If there ever was a piece of fruit with an undeserved reputation, it is the apple. In the biblical story, it's wrongly cited as the forbidden emblem of temptation.

✳ ✳ ✳ ✳

EVE PICKS A piece of fruit, eats it, and then—well, all hell breaks loose. But did you know that the fruit she bit into was probably not an apple? Apples aren't mentioned anywhere in this popular Bible story—only the word "fruit" is given in the text. You won't even find the phrase "forbidden" fruit in the book of Genesis.

So if it wasn't an apple, what kind of fruit *was* growing on that infamous tree? Pears? Cherries? Did Eve reach up and pick a lemon? Genesis describes rivers that include the Tigris and Euphrates in a region that approximates modern-day Iraq. Fruits that would have grown in this area during biblical times include figs, apricots, and pomegranates, but not apples.

The idea that the fruit Eve ate was an apple most likely stemmed from a bit of fifth-century wordplay. The monks who translated the Bible from Hebrew to Latin probably got a kick out of the fact that the Latin word for apple is *malum* and the Latin word for evil is *malus*. Using malum to mean "fruit," they could emphasize the evil of Eve's actions.

European painters are probably the most responsible for placing an apple in Eve's hand. In the Middle Ages, painters tried to relate biblical stories to the masses. The apple was a common fruit—in fact, the word "apple" was used to refer to any fruit. Artists often depicted biblical stories in medieval villages and biblical figures in period clothing. In the same way, apple trees in the Garden of Eden were representations that people would recognize and understand.

Murder in the Vatican

As head of the Catholic Church, popes serve as the Vicars of Christ and are among the world's most respected leaders. Yet dozens have met untimely fates at the hands of pagan oppressors, rivals to the papal throne, plotting aristocrats—and even outraged husbands.

❊　❊　❊　❊

MANY OF THE first 25 popes are believed to have been martyred by the Romans. Pontian (230–235) is the first pope recorded in history as having been murdered for his Christian beliefs. Arrested under the orders of Emperor Maximinus Thrax, Pontian was exiled to Sardinia, where he is believed to have died of starvation and exposure. Sixtus II (257–258) was another early martyr, killed in the persecutions of Emperor Valerian. Sixtus was arrested by Roman soldiers while giving a sermon and may have been beheaded on the spot. Martin I (649–653) began his papacy on bad terms with Emperor Constans II, who refused to recognize the pontiff's election. Martin made matters worse by condemning the doctrines of the Monothelite heretics, whose tenets were followed by many powerful Roman officials, including Constans II. Constans sentenced Martin to death and exiled him to the Crimea, where he died of starvation.

The martyrdom of popes passed with the fall of the Roman Empire, but the ascendancy of the Catholic Church was accompanied by endless papal intrigues. From the 9th to the 20th centuries—when popes served not only as head of the church but as rulers of the Papal State, a substantial kingdom in central Italy—rumors abounded that many pontiffs had been murdered.

Most documented papal murders occurred during the Middle Ages, particularly between 867 and 964, the so-called Iron Age of the Papacy, when the politically powerful families of Rome had pontiffs elected, deposed, and killed to advance their own ambitions. Seven popes were murdered during this period.

The first was Pope John VIII (872–882), who was so paranoid that he had several powerful bishops and cardinals excommunicated. Unknown conspirators convinced a relative to poison his drink. When the poison failed to kill him, he was clubbed to death by his own aides. According to some accounts, however, Pope John was actually Pope Joan—a female pope who was erased from the historical record when her true identity was uncovered. Though some historians believe Pope Joan is a myth, others point to an obscure Church ritual that began in the late 9th century, in which a papal candidate sat in an elevated chair with his genitals exposed, prompting passing cardinals to exclaim in Latin, "He has testicles, and they hang well!" Lacking a similar endowment, Pope Joan may have paid with her life.

Stephen VI (896–897) was the "mad pope" who placed his rival, Pope Formosus, on trial nine months after his death. Formosus had enraged Stephen by crowning one of the illegitimate heirs of Charlemagne as emperor (after having performed the same rite for Stephen's favorite candidate). Once Stephen was made pope, he directed Formosus's corpse to be disinterred, placed on a throne in the council chambers, and provided with legal counsel. When Stephen finished hurling abuse at the corpse, it was thrown from a balcony to a waiting mob, then dumped into the Tiber River. All ordinations performed by Formosus were annulled. In the tumult that followed among the Roman aristocracy, Stephen was imprisoned and ultimately strangled to death, making way for a saner pontiff.

The papal carnage continued over the next few hundred years. Adrian III (884–885) was allegedly poisoned. Leo V (903) was allegedly strangled. John X (914–928) may have been smothered with a pillow. Both Stephen VII (928–931) and Stephen VIII (939–942) met untimely ends due to palace intrigues.

John XII (955–964) was only 18 years old when he was elected pope. A notorious womanizer, he turned the papal palace into something resembling a brothel. He either suffered a heart attack while with a mistress or was murdered by a cuckolded husband. Pope Benedict V (964–966) raped a young girl and fled to Constantinople with the papal treasury, only to return to Rome when his coffers were empty. He was killed by a jealous husband, his corpse bearing 100 dagger wounds as it was dragged through the streets.

Benedict VI (973–974) and John XIV (983–984) had strangely parallel fates. Intriguers rebelled against Benedict VI after the death of his protector, Emperor Otto the Great. Benedict was strangled by a priest on the orders of Crescentius, brother of the late Pope John XIII. Boniface Franco, a deacon who supported Crescentius, became Pope Boniface VII but fled Rome due to the people's outrage over Benedict's murder, becoming Antipope Boniface. John XIV was chosen as a replacement by Emperor Otto II without consulting the Church. When Otto suddenly died, another new pope was left without allies. Antipope Boniface returned and had John thrown in prison, where he starved to death.

Popes Gregory V (996–999), Sergius IV (1009–1012), Clement II (1046–1047), and Damasus II (1048) were all allegedly poisoned or met otherwise suspicious ends. Boniface VIII (1294–1303) died from beatings by his French captors while he was held prisoner in Anagni. Benedict XI (1304–1305) may have also been poisoned.

Officially, no pope has been murdered in the modern age, though rumors suggest that Pope Clement XIV was poisoned in 1771 following his disbandment of the Jesuits. Such allegations arose again in 1978 with the sudden death of Pope John Paul I, who had planned reforms such as ordaining women as priests and welcoming gays into the church. In both cases, coroners and investigators found no evidence of foul play. John Paul's successor, John Paul II, was nearly murdered in St. Peter's Square in 1981 by Mehmet Ali Agca, a Turkish gunman who was part of a conspiracy involving the KGB and Bulgarian secret police.

Cross Country

Preacher Arthur Blessitt made history when he set out on a four-decade-long pilgrimage.

✳ ✳ ✳ ✳

ARTHUR BLESSITT WAS born in 1940, and at age 11 he converted to Christianity. By the time he was a young man in the '60s, he had moved to California and was preaching on Hollywood's Sunset Strip. On Christmas Day in 1969, Blessitt claims he was called by Jesus to make a pilgrimage on foot to Capitol Hill in Washington, D.C. He would take with him a 12-foot-tall, 45-pound wooden cross. Blessitt set out with the cross on his back and a mission to spread the word of Christ.

Not long after he began, Blessitt realized his cross, which dragged on the ground, was losing about an inch of wood every day thanks to rough roads. He added a tricycle wheel to the base of his cross, which helped facilitate the trek.

Blessitt didn't stop his walk when he got to Capitol Hill. He didn't stop at the Atlantic Ocean, either. No, the preacher boarded a plane—he had to break the cross down to get it into a ski bag because it wouldn't fit on the aircraft—and flew overseas.

Over the next decade, Blessitt continued his journey. By 1984, he had walked across 60 countries on six continents. Of course, his quest wasn't easy, and sometimes it was downright dangerous. Still, he garnered quite a following and became an evangelical celebrity of sorts, meeting with Yasser Arafat, Muammar al-Gaddafi, former UN Secretary Boutros Boutros-Ghali, and even Pope John Paul II.

Blessitt is credited with witnessing President George W. Bush's conversion to born-again Christianity, and he proudly admits that over time, his cross-bearing shoulder built up an extra inch of bone. By the end of the pilgrimage in 2008, Blessitt had covered more than 38,000 miles on all seven continents.

The Relished Relic

Relics are important elements in several of the world's major religions. These ancient holy artifacts—thought to be pieces of a saint's or significant leader's body, or one of their personal belongings—are said to be imbued with spiritual power and are highly protected. But are they authentic or not? As some believers would say, you just have to have faith.

✳ ✳ ✳ ✳

The Holy Prepuce

ACCORDING TO NEW TESTAMENT apocrypha (writings by early Christians about Jesus and his teachings that were not accepted into the holy canon), after baby Jesus was circumcised, an old Jewish woman saved his foreskin. But by the Middle Ages, several different foreskins were touted as the original and were worshipped as holy relics by various churches. Stories abound of prepuces gifted to monks, stolen by thieves, dismissed by popes, and marched in parades, all adding to the mystery of this particular (and particularly weird) relic.

The Tooth of Buddha

After the Buddha died (in approximately 500 B.C.), it's said that his body was cremated. As the story goes, after the cremation, a follower retrieved the Buddha's left canine tooth from the funeral pyre. The tooth was given to the king and quickly became legendary: Whoever claimed the tooth would rule the land. Wars were fought over possession of the tooth for centuries, and now it— or what's left of it 2,500 years later—is in a temple in Sri Lanka.

The Sacred Relics

From the 16th to 19th centuries, sultans of the Ottoman Empire collected religious items of the Islamic faith. Most were said to be relics of various Islamic prophets, though many of the pieces are of questionable origin. Included in the collection, now held in Istanbul, are Moses' staff, a pot belonging to Abraham, and a piece of the prophet Muhammad's tooth. Perhaps the most important of the relics is the Blessed Mantle, the black wool shawl that is said to have been placed on a poet's shoulders by Muhammad himself.

Relics of Sainte-Chapelle

If you find yourself in Paris, visit Notre Dame to behold the collection of Sainte-Chapelle relics, including shards of the True Cross (which are believed to be actual wood from Christ's cross), relics of the Virgin Mary, the Mandylion (a piece of fabric similar to the Shroud of Turin on which Christ's face is said to appear), and something called the Holy Sponge—a blood-stained sponge said to have been offered to Christ to drink from when he was languishing on the cross. The authenticity of these objects is as contested as any on this list, but the items are impressive for simply surviving the French Revolution, when many relics were destroyed or lost.

Veronica's Veil

According to tradition, a woman named Veronica (she's not mentioned by name in the Bible) wiped the face of Jesus when he was on his way to Calvary. The fabric she used was said to have taken the imprint of Jesus' face. The veil can now be found in St. Peter's Basilica in Rome. Or maybe it's in a friary outside Rome—there's another version of the veil there. Regardless, plenty of people claim to have seen the bloodstained face of Jesus in the fabric of Veronica's veil and continue to make pilgrimages to worship it.

The Shroud of Turin

Of all the relics on this list, the Shroud of Turin is the one whose authenticity remains the most hotly debated. Carbon dating originally proved that the material, purportedly the shroud laid over Christ at the time of his burial, was produced in the Middle Ages, but this has since been proven incorrect—the garment is, in fact, older. What's perhaps most fascinating about the Shroud is that the image itself is a negative; photographic methods were hardly known at the time of Christ's death, so how could anyone have faked such an image?

The House of David— God's Hairy Messengers

The only thing more startling than seeing men with waist-length hair and long beards playing baseball in the early 1900s might have been knowing that every player on this early barnstorming team was a member of a highly controversial religious sect known as the House of David.

✳ ✳ ✳ ✳

From Kentucky to the Second Coming

BASED IN BENTON HARBOR, Michigan, this religious sect centered around its charismatic leader, Benjamin Franklin Purnell, and his wife, Mary. The couple believed that they were God's appointed messengers for the Second Coming of Christ and that the human body could have eternal life on Earth. They also believed that both men and women should imitate Jesus by never cutting their hair. Purnell based his teachings on those of an 18th-century English group called the Philadelphians. Their teachings were developed from the prophecies of a woman named Joanna Southcott, who claimed she was the first of seven messengers to proclaim the Second Coming. Purnell deduced that he was also one of those seven.

Growing Hair, Religion, and Crowds

Born in Kentucky in March 1861, Purnell and Mary traveled around the country for several years while polishing their doctrine. After being booted out of a small town in Ohio— possibly because Purnell was accused of adultery with a local farmer's wife—they landed in Benton Harbor in March 1903. Members of a sect related to the Philadelphians called the Jezreelites lived in nearby Grand Rapids, and Purnell had been in touch with the Bauschke brothers of Benton Harbor, who were sympathetic to his cause.

With the backing of the Bauschkes and other prominent local citizens, Purnell soon attracted a crowd of believers and called his group the Israelite House of David. The 700 or so members lived chaste, commune-style lives on a cluster of farms and land, served vegetarian meals, and started successful cottage industries, such as a toy factory, a greenhouse, and a canning facility called House of David Jellies and Jams.

As word of the long-haired, oddly dressed sect members and their colony spread, curious folks began making Sunday trips to observe them. Purnell turned this into a moneymaking opportunity by opening an aviary, a small zoo, a vegetarian restaurant, an ice cream parlor, and, ironically, a barber shop. The crowds grew, and, in 1908, Purnell started work on his own amusement park, which included an expanded zoo and a miniature steam-powered railway.

Entertainment Evangelism

In the meantime, some members of the group formed a baseball team that also drew crowds, so Purnell added a large stadium next to the amusement park. The team traveled and added to their popularity with comical routines, such as hiding the ball under their beards. Building on the sports theme, the colony also featured exhibition basketball and, later, miniature car racing.

The colony also boasted a popular brass band, whose members capitalized on their long showy tresses by starting each concert facing away from the audience, hair covering half of their snazzy uniforms. They often played jazzy, crowd-pleasing numbers rather than the expected somber religious tunes.

Religious activities continued, too. Adopting the title "The Prince of Peace," Purnell often held teaching sessions, including one in which he was photographed allegedly changing water into wine.

Problems in Paradise

As happens with many large social enterprises, some members of the House of David became disgruntled and left; Purnell referred to them as "scorpions." Rumors grew concerning improper relations between Purnell and young females in the group, especially when the colony purchased an island in northern Michigan where they ran a prosperous lumber business. Newspaper reports alleged that rebellious group members were killed and buried there and that Purnell kept a group of young girls there as sex slaves. The public was also suspicious of the mass weddings he conducted. Lawsuits against Purnell had begun in Ohio, and they continued to mount even as the Michigan colony progressed.

In 1926, Purnell was finally arrested on charges that included religious fraud and statutory rape. He endured a lengthy trial, but he was ill for most of it, and much of his testimony was deemed incoherent. Most of the charges were eventually dismissed.

Purnell died in Benton Harbor on December 16, 1927, at age 66. Shortly before passing, he told his followers that, like Jesus, he would be back in three days. As far as anyone knows, he wasn't. His preserved remains were kept in a glass-covered coffin on the colony grounds for decades, although Mary's brother reportedly insisted that the body was not Purnell's but that of another colony member.

Remains of the Day

After Purnell's death, some of the believers declared their allegiance to his widow, Mary, who started a new colony called Mary's City of David. The colony still has a baseball league and runs a museum in Benton Harbor. The zoo closed in 1945 (the animals were given to Chicago's Lincoln Park Zoo), and the amusement park closed in the early 1970s. Today, many credit Benjamin Purnell as being the forerunner of later, high-style evangelical leaders such as Jim Bakker and Oral Roberts.

We Three-ish Kings

Tradition tells us that three kings came from the East to honor Jesus at his birth. But they weren't kings, and there were probably more than three of them.

※　※　※　※

ACCORDING TO THE gospel of Matthew, "After Jesus was born in Bethlehem in Judea, during the time of King Herod, magi from the east came to Jerusalem." The word *after* is important: Countless Christmas pageants notwithstanding, the magi ("wise men") were not present at the manger scene. Although it's not known exactly what country the wise men came from—it could have been Persia or Babylon—the journey must have taken weeks or even months. By the time the men arrived, Joseph and Mary had been able to find more comfortable lodging. (Matthew indicates this when he says, "On coming to the house, they saw the child with his mother, Mary.")

So who were the magi? They appear to have been religious or governmental officials—maybe both—who were well versed in astrology. They may have been followers of the Persian prophet Zoroaster. The wise men interpreted the mysterious star of Bethlehem as the birth announcement of an extremely important king, and they headed off to pay their respects. They took along valuable gifts of gold, frankincense, and myrrh, just as diplomats today might bear gifts representing the best of their countries' offerings.

Because Matthew mentions three gifts, early readers assumed that there were just three magi. They also began to think of them as kings, probably seeing the visit as a fulfillment of this verse in the Bible (Psalm 72:11): "All kings will bow down to him and all nations will serve him." But Matthew is unclear about their job titles and doesn't say anything about their numbers. So the song "We Three Kings," written by John H. Hopkins Jr. in 1857, is more poetic than accurate.

Drunken Reformers

If the squeaky-clean Puritans were known for their disdain of alcohol, why was there more sour mash than mashed potatoes at the first Thanksgiving?

✳   ✳   ✳   ✳

MANY PEOPLE ASSOCIATE the Puritans with the temperance movement that worked so hard to ban alcohol during the 19th and early 20th centuries. After all, wasn't it this fire-and-brimstone group that did everything it could to squeeze all the fun out of life? The truth is, the Puritans liked their hooch every bit as much as the average drunken sailor—and they loaded down their ships with the stuff to prove it.

"A Puritan is a person who pours righteous indignation into the wrong things," said writer G. K. Chesterton. Funny he should use the word *pour*. When the *Mayflower* sailed to America, its holds actually contained more beer than water. Although it was seemingly at odds with their religious precepts, the Puritans maintained a love affair with alcohol that would span the ages.

After setting up shop in the New World, colonial Puritans got busy producing their favorite beverage. Soon, the manufacture of rum would become colonial New England's largest industry. Even the hallowed halls of Harvard University included a brewery. Was nothing sacred?

Apparently not. John Hancock, the first signer of the Declaration of Independence, worked as an alcohol dealer. Thomas Jefferson, the document's author, penned its first draft at a Philadelphia tavern. Even superpatriot Patrick Henry ("Give me liberty or give me death") got in on the act as a bartender.

✳   **In the 1830s, during the Second Great Awakening of Puritanism, the average American consumed more than seven gallons of alcohol annually. Today, that figure is lower by two-thirds.**

Breaking the Bonds: Divorce Around the World

Scholars and experts have long studied the ways that people in different cultures around the world say, "I do." But how do they say, "I don't anymore"? Divorce laws date back as early as 2000 B.C., which means that for almost as long as couples have been binding together, they've been looking to cut the ties.

❋ ❋ ❋ ❋

I'd Like to Return This Faulty Wife, Please

WHO KNEW BRIDES had a warranty? In ancient Greece, a man simply had to send his wife back to her father—dowry in tow—to end their happily-ever-after. In ancient China, divorce was seen as a big no-no. One old proverb passed down to newlyweds was, "You're married until your hair turns white." Of course, that didn't mean a man couldn't divorce his wife—all he had to do was send an emissary to his wife's father, declaring that he "cannot worship at the ancestral shrine with your daughter any longer."

Hit the Road, Jack

Other cultures like to keep the separation simple. If an Eskimo couple wishes to separate, they just begin living apart from one another. Among certain Native American tribes, a woman could initiate divorce by placing her husband's moccasins outside the dwelling, indicating he should walk away and not come back. An Australian Aborigine woman has a tougher time dissolving her marriage: If she can't convince her husband to leave her, she must elope with someone else.

High-Speed Separation

If a quick separation is what you're after, check out the laws in Japan, Russia, and Sweden. In Japan, if both husband and wife agree that the marriage is kaput, they need only sign a document saying so and drop it off at city hall. In Russia, a couple has to go to court to separate only if children are involved or if one person doesn't want the divorce. It's a similar story in Sweden: Divorces are granted instantly upon application to couples who are on the same page about ending their marriage. If kids are involved or one party is against the idea, a "reconsideration period" of six months is required.

Under Islamic law, a man can divorce his wife by simply gathering two witnesses and announcing "I divorce thee" once. But recent technology has made the process even easier: Men can now deliver the decree via text message, leaving women to find out they're single via cell phone. (Worth noting: Arab women can, and do, divorce their husbands in court, but the process tends to be longer and more complicated.)

Technically Divorced

Technology hasn't just sped up the divorce process—in some cases, it's driving couples to split. In 2008, a UK couple divorced after the wife found her husband on the couch with another woman—well, a "virtual" woman on a "virtual" couch; after all, he was playing the online simulation game *Second Life*. In another case, two Jordanians trapped in an unhappy marriage found their soul mates in an online chatroom . . . only to meet in person and discover their soul mates were actually each other. The couple soon divorced, and each accused the other of cheating.

Look Before You Leave

Dissolving a marriage isn't quite so easy in Canada and Italy. Canadian couples can't divorce without going through a one-year trial separation (extreme situations notwithstanding), and divorce attorneys are required to encourage reconciliation or couples' counseling.

In Italy, couples must undergo a three-year trial separation, after which would-be divorcees face a lengthy legal process and low child-support benefits. No wonder so many Italian couples reconcile!

Pay As You Go

For some couples, however, the cost of separating is no match for the price of freedom.

In 1999, media mogul Rupert Murdoch paid his wife of 31 years $1.7 billion in assets to end the marriage, making it the most expensive divorce ever.

In 2006, Paul McCartney paid a rumored $38.5–50 million to walk away from his four-year marriage to Heather Mills. The settlement was topped by director James Cameron's $50 million payout to his wife of 17 months, actress Linda Hamilton, in 1999.

Their cheatin' ways cost Ted Danson, Michael Douglas, Kevin Costner, and Neil Diamond big money. Danson's dalliances cost him $30 million in a settlement with his first wife in 1992; Diandra Douglas left her husband with $45 million in her pocket in 1997; and Costner shelled out $80 million to his first wife in 1994. Also that year, Diamond coughed up $150 million in cash to his wife of 25 years. In 2007, basketball legend Michael Jordan paid his ex-wife $168 million. Slam dunk!

When Madonna and her ex, filmmaker Guy Ritchie, divorced in 2008, she was ordered to pay him $76 million.

Divorce Rates Worldwide

Ironically, cultures in which marriage is based on personal choice (that is, love and happiness) tend to have the highest divorce rates, whereas countries where the Catholic Church and arranged marriages dominate tend to have the lowest. In a 2008 study, 44 states and D.C. reported 2,162,000 marriages. The marriage rate in the study was 7.1 per 1,000 people (total population), and the divorce rate was 3.5 per 1,000 people.

Index

✳ ✳ ✳ ✳

Good Humor, 178
Goodman, Benny, 606
Google, 176, 471
Gorman, Margaret, 330
Gosling, James, 479
Graff, Frederick Sr., 477
Graham, Jack, 540–41
Grandma Moses, 114–15
Grant, Cary, 202–3
Gray, Pete, 100–101
Great Chicago Fire, 277
Great Race of Mercy, 145
Griffin, Howard, 206
Griffith, Clark, 91
Guillotin, Joseph-Ignace, 374–75
Guinness World Records, 186, 237, 633
Gunness, Peter, 542

H

Hachiko, 130, 131
Hackett, Joan, 486
Hagiwara, Makoto, 646
Haines, Mahlon, 430
Hale, Sarah Joseph, 304
Haley, Jack Jr., 11
Hall, Ed, 108–9
Hallaren, Mary A., 258
Hamilton, Alexander, 506
Hamilton, Scott, 211–12
Hamlet (play), 125
Hampton, Lionel, 606
Handler, Ruth, 454
Hansen, Betty, 552–53
Hansen, Henry O. "Hank," 412
Hanson, Chris, 82
Hanson, David, 118–19

Hanson, John, 528–30
Hanson Robotics, 118
Harding, Warren G., 505
Harlow, Jean, 322
Harry, Deborah, 211, 610
Hart, Gary, 506
Hartman, Mortimer, 202–3
Haskell, Carlos, 620, 621
Haskell, Martha Stewart, 620, 621
Haskell Free Library and Opera House, 620–21
Hastings, Lansford, 386
Haugen, Anders, 98
Haught, Deron, 64
Haydon, S. E., 358
Hayes, Woody, 108
H. D. Lee Company, 336
Helm, Benjamin Hardin, 253
Hemings, Sally, 506
Henderson, Oliver "Pappy," 368
Hendrix, Jimi, 494–95
Henie, Sonja, 75
Hennessy, Jill, 35
Henreid, Paul, 10
Henri II (French king), 323
Herman, Babe, 68
Hershey, 320
Hewlett, Jamie, 619
Heywood, John, 239
Hill, Barney, 355
Hill, Betty, 355
Hill, Faith, 212
Hill, Fred, 643
Hillerich & Bradsby Company, 92
Hilley, Carol, 543
Hilley, Frank, 543
Hilley, Marie, 543